Systems Analytics and Integration of Big Omics Data

Systems Analytics and Integration of Big Omics Data

Special Issue Editor

Gary Hardiman

MDPI • Basel • Beijing • Wuhan • Barcelona • Belgrade • Manchester • Tokyo • Cluj • Tianjin

Special Issue Editor
Gary Hardiman
Queen's University Belfast
UK

Editorial Office
MDPI
St. Alban-Anlage 66
4052 Basel, Switzerland

This is a reprint of articles from the Special Issue published online in the open access journal *Genes* (ISSN 2073-4425) (available at: https://www.mdpi.com/journal/genes/special_issues/Systems_Applications_Big_Omics_Data).

For citation purposes, cite each article independently as indicated on the article page online and as indicated below:

LastName, A.A.; LastName, B.B.; LastName, C.C. Article Title. *Journal Name* **Year**, *Article Number*, Page Range.

ISBN 978-3-03928-744-4 (Pbk)
ISBN 978-3-03928-745-1 (PDF)

Cover image courtesy of Ludivine Renaud.

Contents

About the Special Issue Editor

Gary Hardiman A native of Galway, Ireland, Gary Hardiman received B.Sc. Hons. and Ph.D. degrees in Microbiology/Molecular Biology from the National University of Ireland Galway (NUIG) in 1989 and 1993, respectively. He completed two post-doctoral research fellowships at DNAX Research Institute (MERCK), Palo Alto, CA (1993–1998) in the area of genomics and bioinformatics. He held faculty appointments at San Diego State University and the University of California, San Diego, USA. He was the Founding Director of the University of California, San Diego Biomedical Genomics Facility from 2000 to 2013. Most recently, Gary served as Scientific Director of the Center for Genomics Medicine Bioinformatics where he established a university-wide bioinformatics program. He was a Full Professor of Medicine and Public Health Sciences at the Medical University of South Carolina, Charleston, USA; Head of the Laboratory for Marine Systems Biology, Hollings Marine Laboratory; and a Visiting Scholar at Grice Marine Laboratories, College of Charleston, SC, USA. Professor Hardiman joined Queen's University Belfast in September 2018 as Professor of Systems Biology, School of Biological Sciences and Chair in Food Systems Biology at the Institute for Global Food Security (IGFS). As PI or co-Investigator on several active NIH-funded grants, his systems biology research program encompasses the following areas: 1) Studying the effects of man-made contaminants (e.g., microplastics, persistent organic pollutants, xenobiotics) on marine and human health; 2) Studying prostate cancer in the context of racial differences and nutritional deficiency; 3) Examining the impacts of long-term space travel—specifically the effects of nutrition, torpor, space radiation, and microgravity on hepatic and intestinal biology; 4) Developing a rat model of opioid abuse to better understand the biological basis for substance abuse disorders and advance development of preventive strategies and more efficacious treatments; and 5) Developing a robust toolkit for the better integration of Omics data sets into genotype–phenotype predictions. He has published more than 170 scientific articles, has filed more than 20 patents, and has served on scientific review panels in the USA, France, Italy, Singapore, Poland, Ireland, and Austria, as well as on five corporate scientific advisory boards. He has held consulting positions for private research institutes, investment banks, and pharmaceutical companies. He is a Fellow of the Royal Society of Biology. He is an Associate Editor of the "Psychopharmacology" section of Frontiers in Psychiatry, which focuses on Big Data) and serves on the Editorial Board of the journals *Pharmacogenomics, Expert Review of Molecular Diagnostics, Genes, High-Throughput, and Science of the Total Environment*. He is the editor of three books on genomics technologies: "'Microarray Methods and Applications", published by DNA Press, Inc. (2003); "Biochips as Pathways to Drug Discovery" with Dr. Andrew Carmen, published by CRC Press (2007); and "Microarray Innovations: Technology and Experimentation" (2009). He teaches on genomics technologies and bioinformatics and nutrition. He has mentored numerous graduate students, post-doctoral fellows, and junior faculty.

Preface to "Systems Analytics and Integration of Big Omics Data"

The emergence and global utilization of high-throughput (HT) technologies, including deep sequencing technologies (genomics) and mass spectrometry (proteomics, metabolomics, lipidomics), has allowed geneticists, biologists, and biostatisticians to bridge the gap between genotype and phenotype on a scale that was not previously possible.

Big data encompasses the collection of data sets derived from technologies. They are so large and complex that their processing is impractical using traditional data processing applications. Instead, challenges arise in collection: analysis, mining, sharing, transfer, visualization, archival and integration of big data.

As observed with DNA microarray analysis pipelines over a decade ago, and more recently with HT sequencing, better analytical tools are emerging primarily from open-source efforts, permitting additional analyses and enhanced information mining from raw data sets compared to the tool kits provided with the instruments themselves.

Administration and development strategies must take into account the ever-growing size of data, the public accessibility of analyzed data, software deprecations, software upgrades, user interface improvements, user account management, data archive, and security.

Against this backdrop, in this Special Issue we focused on the systems-level analysis of omics data, recent developments in pathway and network biology algorithm development, and the integration of omics data with clinical and biomedical data using machine learning.

As the Editor of this compilation, it is a privilege to have been associated with this publication. I am very grateful to those who have generously contributed material to this edition. I am appreciative of my colleagues at the Medical University of South Carolina and Queen's University Belfast over the past six years for the many discussions that helped shape this compendium.

I thank my parents Maureen and Joe. Finally I thank my wife Patricia and daughter Elena for their love and affection, continuing support and inspiration, and of course, their patience.

Gary Hardiman
Special Issue Editor

Editorial

An Introduction to Systems Analytics and Integration of Big Omics Data

Gary Hardiman [1,2]

[1]   School of Biological Sciences, Institute for Global Food Security (IGFS), Queen's University Belfast, BT7 1NN Belfast Northern Ireland, UK; g.hardiman@qub.ac.uk
[2]   Department of Medicine, Medical University of South Carolina, Charleston, SC 29425, USA

Received: 17 February 2020; Accepted: 20 February 2020; Published: 26 February 2020

A major technological shift in the research community in the past decade has been the adoption of high throughput (HT) technologies to interrogate the genome, epigenome, transcriptome, and proteome in a massively parallel fashion [1,2]. This has provided both unique discovery opportunities and challenges for computational and quantitative scientists in predicting phenotypic outcomes. 'Big Data' encompasses the collection of data sets derived from technologies and so large and complex that their processing is impractical using traditional data processing applications. Challenges arise in collection, analysis, mining, sharing, transfer, visualization, archival and integration of Big Data.

Genotype is one of three key factors that determine the phenotype, including inherited factors (DNA code), epigenetic factors (DNA methylation, histone modifications RNA-associated silencing) and non-inherited environmental factors [3]. In this special issue, there is a focus on systems level analysis of omics data, recent developments in pathway and network biology algorithm development, and integration of omics data with clinical and biomedical data using machine learning. The role of chromatin in genotype-phenotype is explored. Improvements to the Gene Ontology Resource to Facilitate More Informative Analysis and Interpretation of Alzheimer's Disease Data is covered.

One of the pressing challenges for integrative computational biology and statistical genetics is predicting genotype-to-phenotype maps of organisms in the context of environmental influences. As noted in the collection perspective by Lewis Frey, genotypes and phenotypes realized in Omics data collections are linked through the various nuclear and cellular processes that convert encoded genotype information into a macroscale manifestation of the organism phenotype [4]. The ability to identify the key drivers of genotype to phenotype is challenging among the multitude of interacting molecules. Frey makes a compelling argument for the application of artificial intelligence (AI) that can automate computable phenotypes and integrate them with genotypes. Challenges need to be overcome namely the rapid growth of data, the inaccessibility of data through issues with incompleteness, inaccuracies, and heterogeneity and data silos.

A review article in this collection by Núria Malats and colleagues explores the challenges that exist with the integration of Omics and Non-Omics (OnO) Data [5]. At present few omics-based algorithms that possess enough predictive ability are implemented in the clinic. Clinical/epidemiological data describe most of the variation in health-related traits. Effective modeling of this with omics data is urgently needed to increase the predictive ability of algorithms. Obstacles in OnO data integration are the nature and heterogeneity of non-omics data, the relationship between OnO data termed ascertainment bias, the presence of interactions, the fairness of the computational models, and the presence of sub-phenotypes. Most data to date is focused on RNA expression data and studies have incorporated non-omics data in a low-dimensionality manner. Integrative strategies typically adopt one of three modeling methods: Independent, conditional, or joint modeling. Joint modeling, where omics and non-omics data are modelled together in a supervised or unsupervised manner, are preferred for integrating large-scale OnO data, as they account for the correlation structure between the two data types. Additionally, they provide greater complexity than conditional or independent modeling [5].

Data from different sources (e.g., genome, epigenome, transcriptome, proteome, metabolome) tends to be analyzed in isolation using statistical and machine learning (ML) methods. Effective data integration poses new computational challenges [6]. State-of-the-art ML-based approaches for tackling five specific computational challenges associated with integrative analysis: namely the curse of dimensionality, data heterogeneity, missing data, class imbalance and scalability issues are reviewed by Peipei Ping and colleagues. Anagha Joshi and colleagues review Genotype to Phenotype via Chromatin [7]. They note that mapping mutations to causal genes and therapeutic targets to date has been quite limited. The majority of disease-associated mutations lie in inter-genic regions. An emerging trend is thus to focus on the epigenetic control of the disease to generate more complete functional genomic maps. Recent studies unravelling the mechanistic understanding of epigenetic processes in disease development and progression are reviewed [7].

This special issue presented new methodologies in the context of gene-environment, tissue-specific gene expression and how external factors or host genetics impact the microbiome [8–10]. Wolf and colleagues developed an analytical approach for identifying the main effects and interactions between genetic and environmental factors linked to a disease outcome [8]. The method involves selection of candidate genetic and/or environmental factors, utilization of a machine learning algorithm Logic Forest to identify the salient effects and interactions in the disease, followed by confirmation of the association between interactions identified by the algorithm using logistic regression. A case study examining the association between SNPs and cigarette smoke exposure with risk of developing systemic lupus erythematosus (SLE) is presented. This identified genetic and environmental risk factors, and potential interactions between exposure to secondhand smoke as a child and genetic variation in the Integrin alpha M (*ITGAM*) gene associated with increased risk of SLE [8].

Cai and colleagues exploited transcriptomic data from multiple tissues generated by the Genotype-Tissue Expression (GTEx) project [10,11] and developed a new methodology that integrates machine learning algorithms to identify genes widely expressed in human body tissues with different expression signatures that can distinguish different tissue types. The approach allows tissue classification via a 432 gene signature of quantitatively tissue-specific expression, suggesting that these genes could also play important roles in tissue development and function [10].

Three notable dynamic interactions play a role in phenotypic outcome. The first, is the association between the environment and the host; the second is that between the microbiome and host health or disease state; and the third is the linkage between the environment and the microbiome. Owing to this complexity the majority of observational and experimental study designs fail to fully assess the direct causal roles of the microbiome. To address this Big Omics challenge, Alekseyenko and colleagues developed a framework for multivariate omnibus distance mediation analysis (MODIMA). They exploited the power of energy statistics, to facilitate analysis of multivariate exposure-mediator-response triples [9].

An important resource for Big Omics data analysis is the Gene Ontology (GO, geneontology.org) which is used when performing gene enrichment analysis. Ruth Lovering and colleagues at University College London (UCL) describe improvements to the GO Resource to improve analysis and interpretation of Alzheimer's Disease data [12]. This project, funded by the Alzheimer's Research United Kingdom foundation and led by the UCL biocuration team, enhanced the GO resource by developing new neurological GO terms, and annotating gene products associated with dementia. Of the total 2055 annotations contributed for the prioritized gene products, 526 had associated proteins and complexes with neurological GO terms. To ensure that these descriptive annotations could be provided for Alzheimer's-relevant gene products, over 70 new GO terms were created. This important novel resource will benefit the scientific community and enhance the interpretation of dementia data [12].

Functional enrichment analyses often result in long lists of biological terms associated to proteins that can be difficult to digest and interpret. Fiero and colleagues addressed this Big Omics data analysis challenge via the development of Network-based Visualization for Omics (NeVOmics).This tool provides a hypergeometric distribution test to compute significantly enriched biological terms. It enables analysis of cluster distribution and relationship of proteins to biological processes and pathways [13]. Even though databases such as the Cancer Cell Line Encyclopedia (CCLE), the Cancer Therapeutics Response Portal (CTRP), and The Cancer Genome Atlas (TCGA) are available it remains challenging for researchers to explore the relationship between drug response and the underlying genomic features due data heterogeneity. Sung Min Ahn and colleagues address this via the development of the Integrated Pharmacogenomic Database of Cancer Cell Lines and Tissues (IPCT) [14]. The IPCT allows users to identify new linkages between drug responses and genomic features. It also allows comparison of the genomic features of sensitive cell lines or small molecules with the genomic features of tumor tissues.

30% of all genes in mammalian cells are predicted to be regulated by microRNA (miRNAs) miRNAs. Da Silveira and Renaud and colleagues describe a new tool, "miRmapper", which identifies the most dominant miRNAs in a miRNA–mRNA network and recognizes similarities between miRNAs based on commonly regulated mRNAs. The most relevant miRNAs are not necessarily those with the greatest change in expression levels between healthy and diseased tissue. Differentially expressed (DE) miRNAs that modulate a large number of messenger RNA (mRNA) transcripts ultimately have a greater influence in determining phenotypic outcomes and are more important in a global biological context than miRNAs that modulate just a few mRNA transcripts. Da Silveira and Renaud exploit this concept to analyze data from a nonmetastatic and metastatic bladder cancer cell lines and demonstrated that the most relevant miRNAs in a cellular context are not necessarily those with the greatest fold change [15].

In summary, the emergence and global utilization of high throughput (HT) technologies, including deep sequencing technologies (genomics) and mass spectrometry (proteomics, metabolomics, lipids), has allowed geneticists, biologists, and biostatisticians to bridge the gap between genotype and phenotype on a scale that was not possible previously. In this special issue integration strategies for systems level analysis of Omics data, recent developments in gene ontology pathway and network algorithm development are explored as is the integration of Omics data with clinical and biomedical data.

Funding: G.H. acknowledges support from NIH/NIDA 1U01DA045300-01A1, NIH/NIMHD 5U54MD010706-02 and start-up funding from Queens University Belfast.

Conflicts of Interest: The author declares no conflict of interest.

References

1. Bhasker, C.R.; Hardiman, G. Advances in pharmacogenomics technologies. *Pharmacogenomics* **2010**, *11*, 481–485. [CrossRef] [PubMed]
2. Hardiman, G. Applications of microarrays and biochips in pharmacogenomics. *Methods Mol. Biol. (Clifton N.J.)* **2008**, *448*, 21–30. [CrossRef]
3. Benfey, P.N.; Mitchell-Olds, T. From genotype to phenotype: Systems biology meets natural variation. *Science* **2008**, *320*, 495–497. [CrossRef] [PubMed]
4. Frey, L.J. Artificial intelligence and integrated genotype–phenotype identification. *Genes* **2018**, *10*, 18. [CrossRef] [PubMed]
5. Lopez de Maturana, E.; Alonso, L.; Alarcon, P.; Martin-Antoniano, I.A.; Pineda, S.; Piorno, L.; Calle, M.L.; Malats, N. Challenges in the integration of omics and non-omics data. *Genes* **2019**, *10*, 238. [CrossRef] [PubMed]
6. Mirza, B.; Wang, W.; Wang, J.; Choi, H.; Chung, N.C.; Ping, P. Machine learning and integrative analysis of biomedical big data. *Genes* **2019**, *10*, 87. [CrossRef] [PubMed]
7. Romanowska, J.; Joshi, A. From genotype to phenotype: Through chromatin. *Genes* **2019**, *10*, 76. [CrossRef] [PubMed]
8. Wolf, B.J.; Ramos, P.S.; Hyer, J.M.; Ramakrishnan, V.; Gilkeson, G.S.; Hardiman, G.; Nietert, P.J.; Kamen, D.L. An analytic approach using candidate gene selection and logic forest to identify gene by environment

interactions (G x E) for systemic Lupus Erythematosus in African Americans. *Genes* **2018**, *9*, 496. [CrossRef] [PubMed]

9. Hamidi, B.; Wallace, K.; Alekseyenko, A.V. MODIMA, a method for multivariate omnibus distance mediation analysis, allows for integration of multivariate exposure-mediator-response relationships. *Genes* **2019**, *10*, 524. [CrossRef] [PubMed]

10. Li, J.; Chen, L.; Zhang, Y.H.; Kong, X.; Huang, T.; Cai, Y.D. A Computational method for classifying different human tissues with quantitatively tissue-specific expressed genes. *Genes* **2018**, *9*, 449. [CrossRef] [PubMed]

11. Human genomics. The Genotype-Tissue Expression (GTEx) pilot analysis: multitissue gene regulation in humans. *Science* **2015**, *348*, 648–660. [CrossRef] [PubMed]

12. Kramarz, B.; Roncaglia, P.; Meldal, B.H.M.; Huntley, R.P.; Martin, M.J.; Orchard, S.; Parkinson, H.; Brough, D.; Bandopadhyay, R.; Hooper, N.M.; et al. Improving the gene ontology resource to facilitate more informative analysis and interpretation of Alzheimer's disease data. *Genes* **2018**, *9*, 593. [CrossRef] [PubMed]

13. Zuniga-Leon, E.; Carrasco-Navarro, U.; Fierro, F. NeVOmics: An enrichment tool for gene ontology and functional network analysis and visualization of data from OMICs technologies. *Genes* **2018**, *9*, 569. [CrossRef] [PubMed]

14. Shoaib, M.; Ansari, A.A.; Haq, F.; Ahn, S.M. IPCT: Integrated pharmacogenomic platform of human cancer cell lines and tissues. *Genes* **2019**, *10*, 171. [CrossRef] [PubMed]

15. da Silveira, W.A.; Renaud, L.; Simpson, J.; Glen, W.B., Jr.; Hazard, E.S.; Chung, D.; Hardiman, G. miRmapper: A tool for interpretation of miRNA–mRNA interaction networks. *Genes* **2018**, *9*, 458. [CrossRef] [PubMed]

Perspective

Artificial Intelligence and Integrated Genotype–Phenotype Identification

Lewis J. Frey [1,2]

[1]   Department of Public Health Sciences, Biomedical Informatics Center, Medical University of South Carolina, Charleston, SC 29425, USA; frey@musc.edu or lewis.frey@va.gov; Tel.: +1-843-792-4216

[2]   Health Equity and Rural Outreach Innovation Center (HEROIC), Ralph H. Johnson Veteran Affairs Medical Center, Charleston, SC 29401, USA

Received: 9 December 2018; Accepted: 21 December 2018; Published: 28 December 2018

Abstract: The integration of phenotypes and genotypes is at an unprecedented level and offers new opportunities to establish deep phenotypes. There are a number of challenges to overcome, specifically, accelerated growth of data, data silos, incompleteness, inaccuracies, and heterogeneity within and across data sources. This perspective report discusses artificial intelligence (AI) approaches that hold promise in addressing these challenges by automating computable phenotypes and integrating them with genotypes. Collaborations between biomedical and AI researchers will be highlighted in order to describe initial successes with an eye toward the future.

Keywords: artificial intelligence; genotype; phenotype; deep phenotype; data integration; genomics; phenomics; precision medicine informatics

---

1. Introduction

Genotypes and phenotypes expressed in genomic and phenomic data are related through the processes that converts molecular-scale genotype information into a macroscale manifestation of a particular phenotype of an organism. Integrated multi-omic processes drive this metamorphosis of genomic information stored in the nucleus of the cell. The ability to identify the drivers of this transformation is elusive among the plethora of interacting components that obfuscate our view. Through integrating data into knowledge networks and reasoning over them with artificial intelligence (AI), we can more vividly clarify this transformation.

Alan Turing, in his seminal 1950 paper in the journal *Mind* [1], laid the foundation for the field of AI through framing the task of building and testing machine intelligence using an imitation game, where the machine imitates the interactions of an individual communicating with two players: an adversary and an interrogator. Moreover, he conjectured that discussions about intelligent machines would become commonplace by the end of the millennium through improved computational speed, memory and algorithms.

Viewing Figure 1, the cost of computing in gigaflops has gone from tens of billions of dollars in the 1960s to pennies today, with a similar pattern for a gigabyte of random access memory. Combined with such improvements in computing and memory, Turing suggested two avenues of research in algorithms to advance intelligent machines: abstract activity modeling, such as the game of chess; and sensory perception approaches. The timeline in the lower half of Figure 1 is based on algorithmic advances in AI as described in Buchanan's brief history of AI and is extended with deep learning [2,3]. The labeled events above the timeline in Figure 1 show milestones in abstract activity modeling using predicate logic and knowledge representation approaches resulting in machines being able to imitate and exceed human performance in the game of chess before the year 2000 [2,4,5]. Medical publications related to AI, shown in Figure 1 (thin red dash-dot line) as cumulative counts of PubMed references, have increased following the development of biomedical expert systems such as Mycin [6]. The labeled events below the timeline are those associated with perception approaches such as the perceptron, back propagation, neural networks and deep learning [3,7,8]. Machines imitating and exceeding human performance in the game of Go occurred in 2016 through a combination of both types of approaches: deep learning and tree-based knowledge representation trained through reinforcement learning [9].

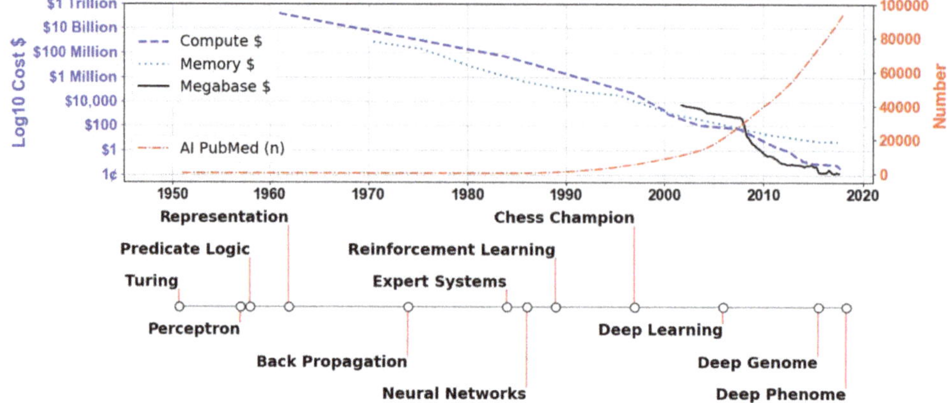

Figure 1. The cost of technology in 2017 US dollars on a log10 scale is plotted in relation to the left axis, and the cumulative number (n) of artificial intelligence (AI) publications in PubMed is plotted in relation to the right axis across time up to and including 2017. The costs of three technologies are compared: Compute, Memory and Megabase. Compute corresponds to the computing costs in gigaflops (one billion floating point operations per second), memory corresponds to the cost of one gigabyte of random access memory, and Megabase corresponds to the cost per megabase sequenced. The cumulative number of PubMed AI-related publications was calculated from identical scripts run for each year starting in 1950. The bottom timeline represents events in the history of AI beginning with Turing's 1950 publication of "Computing Machinery and Intelligence" and ending with deep learning applied to biomedical phenomic data in 2018 (Supplementary File).

The concept of a learning machine was articulated by Turing in terms of evolution, with hereditary material, mutations and natural selection being analogous to the structure of the machine, changes to the machine and performance evaluation providing feedback to the machine at a faster rate than natural selection [1]. In such a framework, machine learning constitutes performance improvement on an evaluation task through the use of labeled data for supervised learning and unlabeled data in unsupervised learning. The speed of reinforcing feedback using deep learning methods over large memory representations of data has made the difference for improving performance on difficult tasks such as object recognition, speech recognition, and playing the game of Go [3,9]. Learning algorithms can miss the mark of the true target, due to the "curse of dimensionality" when too many features in the data space result in overfitting of random variance in the data [10].

Deep learning approaches have managed to navigate the complex space of overfitting and underfitting the data through the use of large amounts of sample and powerful learning algorithms [3]. Deep learning algorithms achieve this high performance through learning multiple layers of non-linear features in the data. Different features are learned depending on initial conditions, making consistent interpretation difficult [11]. The "black box" nature of deep learning approaches highlights differences between the abstraction and perception approaches proposed by Turing to solve the imitation game [1]. The former has well formulated feature representations that are more easily interpreted, but can underfit the data [10]. The latter is not biased to underfit the data, but lacks interpretability due to highly complex features [3,11]. Biomedical research, with the need for biological interpretation, will likely benefit from a hybrid of the two approaches in much the same way as AlphaGo identifies solutions to the game of Go [9]. As can be seen in Figure 1, the solid black line represents the cost per megabase sequenced of genomic information, which went from thousands of dollars to pennies at a faster rate than a gigabyte of random access memory dropped to a few dollars [12–14]. Having inexpensive genomic data for building large repositories, that can be integrated and harnessed by these data-hungry AI methods, will have a far-reaching impact on discovery in biomedical domains [15–17]. The expertise to integrate and analyze these data resides in both biomedical and AI researchers who have an opportunity to drive a new wave of discovery through high-dimensional analysis of deep genomes and deep phenomes [18–21].

An important emerging area is the field of Precision Medicine Informatics which takes on the challenges of big data by integrating, in a knowledge network (i.e., a general high-level conceptualization of knowledge represented as facts connected by relationships between facts), multi-omic data on individuals to increase access and discover new knowledge based on a new taxonomy of disease [22–24]. Biomedical layers of data at different scales are positioned to be integrated in knowledge networks that can be computationally reasoned over to accelerate discoveries [22,24]. Ontologies can be used to formalize knowledge networks through the description of facts, concepts and properties over which logical reasoning engines can be run to generate new facts or inconsistencies implicit in the ontology [25]. Reasoning over knowledge networks can include AI approaches (e.g., Never Ending Learning) that go beyond function approximation methods to reason over a network of documents through a set of AI modules and measure the consistency of the knowledge learned [26,27].

The gene ontology community has developed a knowledge network of molecular functions, cellular components and biological processes [28]. Human phenotype ontology represents a knowledge network of human disease phenotypes that provide a mechanism for connecting genomic and phenomic medical data [29]. The community managing these ontologies is challenged by scalability issues related to the manual curation of data given the exponential growth in genomic and medical phenomic data. To mitigate these issues, AI methods are needed to automate deep phenotyping in the electronic health record (EHR) by incorporating longitudinal data to improve predictive modeling and integrate phenomic and genomic data [30]. The following will discuss specific examples of AI applied to genomic and phenomic data and how they are making headway against the challenge of exponentially expanding data sets and the goal of advancing scientific knowledge [20,21,31].

2. Genomics

The discovery of new treatments will be advanced through understanding the mechanisms by which genomics drives the expression of disease. As indicated, the gene ontology community is iteratively building a knowledge network that can inform biomedical research on the mechanisms and processes that impact the expression of phenotypes [28,32]. The ontology is evaluated on how it performs over time as the ontology is incrementally improved. This is achieved by adding sequences that are annotated with protein function information. These sequences are being added to the system at an exponential rate. However, the validation through experimental findings that confirm or invalidate the protein function information in gene ontology is being added at a linear rate [32]. Thus, there is a vast gap between the number of experimentally validated protein functions and the number of sequences in gene ontology. To address this, there are community-wide evaluation approaches to assess protein function predictions through competitions using a variety of approaches to predict protein function, and to generate candidate predictions at a rate that matches sequence accumulation [28]. Findings on some of the challenges have shown that the use of AI methods, such as the multiple data source k-nearest neighbor algorithm, combined with biological knowledge can give superior results [32].

AI techniques have been used to extract information down to the level of binding properties of genomic sequences that influence the transcriptional networks of cells. Specifically, deep learning methodologies have been used to predict sequence specificity of DNA and RNA binding proteins [20]. The AI approach scanned for motifs in DNA and RNA and identified binding protein promoter sites that would change their binding properties based on variance in single nucleotide polymorphisms, deletions or insertions. The approach identified gain of function mutations or loss of function mutations based on the changing binding affinity of the DNA sequence that the mutations impacted. The deep learning approach is powerful in its ability to discover new knowledge around regulatory processes and biological systems and identifying causal disease variants (i.e., those disease variants that, when changed, change the binding affinities of key regulatory genes in disease processes). The approach also worked on RNA-binding and integration of transcriptomic and genomic analysis [20]. The scale and complexity of data and the techniques now available position AI to be integral to the process of accelerating scientific discovery. The next step is to integrate genomic and phenomic data at different scales.

3. Phenomics

AI combined with phenomics can improve disease state detection when the right criteria are used to recognize the drivers. For example, deep learning has been used to identify histological markers of metastatic breast cancer in lymph nodes that pathologists have difficulty identifying, particularly under standard time constraints [31]. For this study, the deep learning algorithm was assessed via an immunochemistry test that verifies whether metastatic cancer was in the tissue or not. The deep learning algorithm performs at an area under the receiver operating characteristic (ROC) curve of 99% compared with 81% for the pathologists who were given about a minute per slide. A strength of deep learning is the capacity to visually identify histopathology phenomics (i.e., phenotypes in histology images) to improve classification of clinically relevant data. Since knowing that breast cancer has metastasized to the lymph node is critical for treatment decisions, the adoption of such AI technologies for decision support will enhance early detection and improve clinical decision making. It also will improve automated phenotype identification from images that will make new genotype and phenotype identification feasible.

Phenomics of EHR data can be developed from formal ontologies of the phenotype criteria. They can also be developed using deep learning to automate phenotype construction for predictive models that distinguish important categories of individuals [29,30]. Deep phenotyping with AI approaches have demonstrated empirically that incorporating temporal information (e.g., lab values over 24 h after hospital admission) into phenomic models improves accuracy of predicting mortality, length of hospital stay and diagnosis at discharge [21]. In this example, deep learning technology was applied to adult EHR data including both laboratory values and clinical notes on a timeline of events prior to hospitalization and for at least 24 h after admission in order to predict their in hospital mortality. The algorithm incorporated time stamped tokens of events in the EHR and used them to improve predictions of mortality. The algorithms predicted in hospital mortality at 93 to 94% area under the ROC curve compared with 91% for existing clinical predictive models [21]. Notably, the analysis was undertaken without the need to harmonize data across multiple hospital centers. The major strength of the AI deep learning approach is the incorporation of temporal information while eliminating the need to curate the phenotype collection manually and harmonize the data manually [21]. The work demonstrates the benefit of incorporate temporal information in patients' phenotypes, through automated and efficient strategies that show utility in predicting the outcomes of interest, and moves towards an individual focused knowledge network of precision medicine.

4. Conclusions

Over 60 years ago, Turing postulated that we would experience a change in perspective on how learning machines are perceived. Breakthrough AI approaches have brought this to pass and have expanded our ability to recognize drivers of phenotypes resulting from single nucleotide variations, valid protein function mechanisms in biological systems, cancer disease states and deep phenotypes automatically constructed from the EHR. Through combining and expanding on these approaches in a collaborative effort, the biomedical community will accelerate discovery and improve our understanding of mechanisms in the genomic and phenomic expression of disease.

Supplementary Materials: The following are available online at http://www.mdpi.com/2073-4425/10/1/18/s1.

Funding: The work of was supported in part by National Institutes of Health (NIH) grant U54-MD010706 and Health Equity and Rural Outreach Innovation Center grant CIN 13-418.

Conflicts of Interest: The authors declare no conflict of interest.

References

1. Turing, A.M. Computing machinery and intelligence. *Mind* **1950**, *59*, 433–460. [CrossRef]
2. Buchanan, B.G. A (very) brief history of artificial intelligence. *AI Magazine* **2005**, *26*, 53.
3. LeCun, Y.; Bengio, Y.; Hinton, G. Deep learning. *Nature* **2015**, *521*, 436. [CrossRef] [PubMed]
4. McCarthy, J. *Programs with Common Sense*; RLE and MIT Computation Center: Cambridge, MA, USA, 1960.
5. Minsky, M. Steps toward Artificial Intelligence. *Proc. IRE* **1961**, *49*, 8–30. [CrossRef]
6. Buchanan, B.G.; Shortliffe, E.H. *Rule-Based Expert Systems: The MYCIN Experiments of the Stanford Heuristic Programming Project*; Addison-Wesley: Reading, MA, USA, 1984; ISBN 9780201101720.
7. Rosenblatt, F. *The Perceptron, a Perceiving and Recognizing Automaton Project Para*; Cornell Aeronautical Laboratory: Buffalo, NY, USA, 1957.
8. Rumelhart, D.E.; McClelland, J.L.; PDP Research Group. *Parallel Distributed Processing*; MIT Press: Cambridge, MA, USA, 1987; Volume 1.
9. Silver, D.; Huang, A.; Maddison, C.J.; Guez, A.; Sifre, L.; van den Driessche, G.; Schrittwieser, J.; Antonoglou, I.; Panneershelvam, V.; Lanctot, M.; et al. Mastering the game of Go with deep neural networks and tree search. *Nature* **2016**, *529*, 484. [CrossRef] [PubMed]
10. Domingos, P. A Few Useful Things to Know about Machine Learning. *Commun. ACM* **2012**, *55*, 78–87. [CrossRef]
11. Hinton, G. Deep Learning—A Technology with the Potential to Transform Health Care. *JAMA* **2018**, *320*, 1101–1102. [CrossRef] [PubMed]
12. FLOPS. Available online: https://en.wikipedia.org/wiki/FLOPS (accessed on 7 December 2018).
13. Mearian, L. CW@50: Data Storage Goes from \$1M to 2 Cents per Gigabyte. Available online: https://www.computerworld.com/article/3182207/data-storage/cw50-data-storage-goes-from-1m-to-2-cents-per-gigabyte.html (accessed on 7 December 2018).
14. Wetterstrand, K.A. DNA Sequencing Costs: Data from the NHGRI Genome Sequencing Program (GSP). Available online: https://www.genome.gov/sequencingcostsdata/ (accessed on 7 December 2018).
15. Camacho, D.M.; Collins, K.M.; Powers, R.K.; Costello, J.C.; Collins, J.J. Next-Generation Machine Learning for Biological Networks. *Cell* **2018**, *173*, 1581–1592. [CrossRef] [PubMed]
16. Pirih, N.; Kunej, T. Toward a Taxonomy for Multi-Omics Science? Terminology Development for Whole Genome Study Approaches by Omics Technology and Hierarchy. *OMICS* **2017**, *21*, 1–16. [CrossRef]
17. Stephens, Z.D.; Lee, S.Y.; Faghri, F.; Campbell, R.H.; Zhai, C.; Efron, M.J.; Iyer, R.; Schatz, M.C.; Sinha, S.; Robinson, G.E. Big Data: Astronomical or Genomical? *PLoS Biol.* **2015**, *13*, e1002195. [CrossRef]
18. Frey, L.J. Data Integration Strategies for Predictive Analytics in Precision Medicine. *Per. Med.* **2018**, *15*, 543–551. [CrossRef] [PubMed]
19. Kitano, H. Artificial intelligence to win the Nobel Prize and beyond: Creating the engine for scientific discovery. *AI magazine* **2016**, *37*, 39–49. [CrossRef]
20. Alipanahi, B.; Delong, A.; Weirauch, M.T.; Frey, B.J. Predicting the sequence specificities of DNA- and RNA-binding proteins by deep learning. *Nat. Biotechnol.* **2015**, *33*, 831–838. [CrossRef] [PubMed]
21. Rajkomar, A.; Oren, E.; Chen, K.; Dai, A.M.; Hajaj, N.; Hardt, M.; Liu, P.J.; Liu, X.; Marcus, J.; Sun, M. Scalable and accurate deep learning with electronic health records. *NPJ Digital Medicine* **2018**, *1*, 18. [CrossRef]
22. Frey, L.J.; Bernstam, E.V.; Denny, J.C. Precision medicine informatics. *J. Am. Med. Inform. Assoc.* **2016**, *23*, 668–670. [CrossRef] [PubMed]
23. Collins, F.S.; Varmus, H. A New Initiative on Precision Medicine. *N. Engl. J. Med.* **2015**, *372*, 793–795. [CrossRef] [PubMed]
24. National Research Council; Division on Earth and Life Studies; Board on Life Sciences; Committee on a Framework for Developing a New Taxonomy of Disease. *Toward Precision Medicine: Building a Knowledge Network for Biomedical Research and a New Taxonomy of Disease*; National Academies Press: Washington, DC, USA, 2011; ISBN 9780309222259.

25. Baader, F.; Calvanese, D.; McGuinness, D.; Patel-Schneider, P.; Nardi, D. *The Description Logic Handbook: Theory, Implementation and Applications*; Cambridge University Press: Cambridge, UK, 2003; ISBN 9780521781763.
26. Mitchell, T.; Cohen, W.; Hruschka, E.; Talukdar, P.; Yang, B.; Betteridge, J.; Carlson, A.; Dalvi, B.; Gardner, M.; Kisiel, B. Never-ending Learning. *Commun. ACM* **2018**, *61*, 103–115. [CrossRef]
27. Nickel, M.; Murphy, K.; Tresp, V.; Gabrilovich, E. A Review of Relational Machine Learning for Knowledge Graphs. *Proc. IEEE* **2016**, *104*, 11–33. [CrossRef]
28. *The Gene Ontology Handbook*; Methods in Molecular Biology; Dessimoz, C.; Škunca, N. (Eds.) Humana Press: New York, NY, USA, 2017; ISBN 9781493937417.
29. Köhler, S.; Vasilevsky, N.A.; Engelstad, M.; Foster, E.; McMurry, J.; Aymé, S.; Baynam, G.; Bello, S.M.; Boerkoel, C.F.; Boycott, K.M.; et al. The human phenotype ontology in 2017. *Nucleic Acids Res.* **2016**, *45*, D865–D876. [CrossRef]
30. Frey, L.J.; Lenert, L.; Lopez-Campos, G. EHR Big Data Deep Phenotyping: Contribution of the IMIA Genomic Medicine Working Group. *Yearb. Med. Inform.* **2014**, *9*, 206–211.
31. Bejnordi, B.E.; Veta, M.; van Diest, P.J.; van Ginneken, B.; Karssemeijer, N.; Litjens, G.; van der Laak, J.A.W.M.; Hermsen, M.; Manson, Q.F.; Balkenhol, M.; et al. Diagnostic Assessment of Deep Learning Algorithms for Detection of Lymph Node Metastases in Women with Breast Cancer. *JAMA* **2017**, *318*, 2199–2210. [CrossRef] [PubMed]
32. Jiang, Y.; Oron, T.R.; Clark, W.T.; Bankapur, A.R.; D'Andrea, D.; Lepore, R.; Funk, C.S.; Kahanda, I.; Verspoor, K.M.; Ben-Hur, A. An expanded evaluation of protein function prediction methods shows an improvement in accuracy. *Genome Biol.* **2016**, *17*, 184. [CrossRef] [PubMed]

Review

Challenges in the Integration of Omics and Non-Omics Data

Evangelina López de Maturana [1], **Lola Alonso** [1], **Pablo Alarcón** [1],
Isabel Adoración Martín-Antoniano [1], **Silvia Pineda** [1], **Lucas Piorno** [1], **M. Luz Calle** [2,*]
and **Núria Malats** [1,*]

[1]   Genetic and Molecular Epidemiology Group, Spanish National Cancer Research Centre (CNIO),
       and CIBERONC, Melchor Fernández Almagro 3, 28029 Madrid, Spain; melopezdm@cnio.es (E.L.d.M.);
       lalonso@cnio.es (L.A.); pabloalarconmoreno@gmail.com (P.A.); iamartin@ceu.es (I.A.M.-A.);
       spineda@ext.cnio.es (S.P.); lucasp01012002@gmail.com (L.P.)
[2]   Biosciences Department, University of Vic—Central University of Catalonia, Carrer de la Laura 13,
       08570 Vic, Spain
*   Correspondence: malu.calle@uvic.cat (M.L.C.); nmalats@cnio.es (N.M.)

Received: 2 January 2019; Accepted: 14 March 2019; Published: 20 March 2019

Abstract: Omics data integration is already a reality. However, few omics-based algorithms show enough predictive ability to be implemented into clinics or public health domains. Clinical/epidemiological data tend to explain most of the variation of health-related traits, and its joint modeling with omics data is crucial to increase the algorithm's predictive ability. Only a small number of published studies performed a "real" integration of omics and non-omics (OnO) data, mainly to predict cancer outcomes. Challenges in OnO data integration regard the nature and heterogeneity of non-omics data, the possibility of integrating large-scale non-omics data with high-throughput omics data, the relationship between OnO data (i.e., ascertainment bias), the presence of interactions, the fairness of the models, and the presence of subphenotypes. These challenges demand the development and application of new analysis strategies to integrate OnO data. In this contribution we discuss different attempts of OnO data integration in clinical and epidemiological studies. Most of the reviewed papers considered only one type of omics data set, mainly RNA expression data. All selected papers incorporated non-omics data in a low-dimensionality fashion. The integrative strategies used in the identified papers adopted three modeling methods: Independent, conditional, and joint modeling. This review presents, discusses, and proposes integrative analytical strategies towards OnO data integration.

Keywords: data integration; omics data; genomics; RNA expression; non-omics data; clinical data; epidemiological data; challenges; integrative analytics; joint modeling

---

1. Introduction

Most health-related traits are complex in nature. They result from the interaction of multiple internal features/alterations with multiple external conditions over a lifespan [1]. Understanding these complex systems requires modeling exhaustive and appropriate data that characterizes in detail such features and conditions.

Big data in the biomedical field may refer to different scenarios encompassing large numbers of clinical (e-medical/e-health records, EMR/EHR) and epidemiological registries (hereinafter, non-omics data), as well as large biomarker datasets characterizing biological features, such as genomics, transcriptomics, proteomics, metabolomics, and metagenomics, among others. The latter type of data are commonly named omics data. While non-omics data are usually obtained through a pre-elaborated process done either by the subject when s/he reports on her/his life-style habits or symptoms, or by the physician/pathologist when s/he evaluates the characteristics of the disease or the tumor, omics data are generated by high-throughput biotechnological platforms delivering hundreds of thousands of raw (non-elaborated) variables. Recently, imaging-based high-throughput data is also generated and named radiomics.

Omics data integration has been addressed in recent years by several important reviews [2–4], and integrative efforts have been successfully conducted with already available examples of studies that integrated ≥ 2 different omics sets [5–7]. However, only a few of them resulted in omics-based algorithms with enough, though still controversial, predictive ability to be implemented into clinics or public health domains [8,9]. The relatively poor predictive ability of genomic data may partly be explained by the large variation of health-related traits explained by non-omics data, such as clinical and epidemiological variables [10]. Therefore, it is crucial to integrate omics and non-omics (OnO) data in the same models. This provides the opportunity to get insights into biological systems of health and disease. Unquestionably, this endeavor poses several challenges regarding data generation, capture, curation, sharing, analysis, visualization, as well as information privacy and storage.

What does OnO data integration mean in the biomedical arena? While it certainly refers to the inclusion and analysis of these two types of data in the same model/algorithm, several scenarios can be contemplated according to the number of each considered data type. There is no doubt that modeling > 1 omics data sets with > 1 non-omics variables falls under this integrative concept. However, should we consider integration when one omics data set (i.e., genome) is jointly modelled with only one non-omics variable (i.e., age or tumor stage)? In this scenario, the boundaries of the integrative picture become blurred and the definition depends on the purpose of the analysis and whether the inclusion of the non-omics variables aims only to control for a potential confounding effect or whether its prediction ability is being assessed in combination with the omics data. As a onsequence of this confusion, the benefit of models including OnO data, is still unclear. This supports the need for a thorough dissection of the field to diagnose the challenges of the OnO data integrative endeavor and to identify the analytical strategies to reduce the variability of the study results.

In this review, we focus on the integration of OnO data to investigate complex traits, including disease risk and prognosis, according to the definition provided above. We first outline and examine the challenges of integrating the two types of data, we then present the integrative analytical strategies available, we describe the integrative attempts published in the literature, and we further propose statistical methods to be used in the analysis of OnO integrative models before concluding.

2. Challenges in Integrating Omics and Non-Omics Data

In this section, we focus mainly on the challenges of OnO data integration which are primarily related to the nature of both types of data and to the relationship between them, since much attention has already been paid to the integration challenges of only-omics data in previous reviews [3,4].

2.1. Challenges Due to the Nature of Non-Omics Data

2.1.1. Non-Omics Data Are Complex and Heterogeneously and Subjectively Defined

There is an increasing awareness of the need for standards for non-omics data to integrate them in both predictive and inference models. Epidemiological data are subject to a survey mode, survey question standardization, and also context, which may influence data quality and comparability, and ultimately, the contribution of these variables in the outcome prediction. Standards are yet to be adopted in epidemiological data generated by different scientists or organizations through different procedures (i.e., questionnaires) to provide uniformity and consistency in this type of data, which may help scientists and data analysts to better use, share, and integrate them.

Clinical variables may also be affected by the complexity of their definition. A tumor stage, for instance, results from a combination of pathology and imaging information. Regarding clinical standardization, there are some initiatives as CDISC (Clinical Data Interchange Standards Consortium, http://www.cdisc.org) that harmonizes definitions and develops standards across the clinical space (i.e., the Study Data Tabulation Model (SDTM) and the Analysis Data Model (ADaM)) to enable information systems' interoperability to improve medical research and related areas of healthcare.

Another important challenge relates to the nature of the aforementioned types of non-omics data, because they are subjective assessments that result from a complex elaboration process based on skills and previous knowledge of the evaluator which may lead to reporting biases (i.e., grading/staging, clinical decisions, or reporting past occupational exposures). In this regard, non-omics assessments totally differ from omics variables that are completely homogeneous and standardized data within the same data set. Integrating these different types of data poses challenges in the analytics strategy since the transformation or weighting of data may be required.

2.1.2. Heterogeneity Across Non-Omics Data

The lack of uniformity of non-omics data, including qualitative and quantitative variables measured with different scales even to characterize a unique trait/exposure, also limits their integration in an OnO model to predict the outcome of interest and imposes both a conceptual challenge and a hurdle in practical data analysis. Moreover, data transformation (i.e., integrating variables with zero values) and data normalization procedures may be necessary prior to integration analysis, to avoid getting biased parameter estimates when the normality assumption required by some methods is violated.

2.1.3. Large Scale Non-Omics Data

To date, the inclusion of non-omics data into integrative prediction models has been at a low dimension. However, the hype generated by so-called Big Data has also affected the healthcare industry. The advent of Big Data in the clinical setting has increased by the availability of EHRs (e-health records), unstructured medical text, and image data. These "large in scale, high in dimension" non-omics data, along with the design of well-characterized large and longitudinal epidemiological studies at an unprecedented scale, has led to the need for the integration of high dimension non-omics data in models. The use of other digital data sources coming from different wearable devices, such as smart watches, wristbands or wearable health equipment, are also expected to revolutionize epidemiology. The availability of longitudinal data concerning vital signs or environmental variables is expected to shed light on the knowledge of disease dynamics [11]. In addition to the high volume of data, other challenges of using digital epidemiology data are related to the collection, mining, access (i.e., limited and costly access), and data sharing (i.e., variability in definition/standardization of variables and subjective filters applied to the raw data which are needed to analyze those data).

The high dimensionality in non-omics data also implies the presence of (1) correlation structure between these variables, (2) large scale longitudinal data, (3) data sparseness (i.e., medications, laboratory or diagnosis tests), and (4) data missingness, which in contrast to omics data, are not independent on the participating individuals. In this regard, multi-dimensional approaches need samples with all the OnO data measured in the same individuals. All of these aspects must be taken into account in integration models.

Moreover, the advent of using EHRs will also be challenging in processing both objective and subjective traits, as well as structured and unstructured data. Subjective traits were defined by Jette as phenotypes that the "physician cannot assess directly with confidence and have to rely on patient (i.e., pain, physical, social, and emotional function) [12]. On the contrary, objective outcomes are those which the "physician can assess directly with confidence" [13]. Unstructured data, as the physician notes, which are in many cases embedded within semi-structured EHR data, are the most frequent data in the medical records. Although they have been mostly ignored, they are needed to understand the whole of a patient, and it will be needed to process and utilize them.

2.2. Challenges Due to the Relationship between Non-Omics and Omics Data

2.2.1. Ascertainment Bias

In a case-control design, the integration of OnO data may be affected by the presence of ascertainment bias. In this type of epidemiological design, individuals are enriched for the risk factors of the study. If omics data are generated on the basis of the subject's exposure, ascertainment could induce additional correlation between all OnO data [14]. It is known that omics profiles are not independent of demographic factors [15]. For example, age and gender may be associated with DNA methylation values [16,17]. In the clinical setting, genomic variables may be correlated with clinical variables due to population stratification [18]. Furthermore, when survival is the outcome, an insufficient clinical follow-up and the larger incompleteness affecting the clinical variables, in contrast to the completeness of high-throughput molecular data, may bias the effect estimates of the remaining clinical variables in a greater manner than their counterparts of omics variables. On the contrary, traits identified from an observational resource, such as medical records, may also be subject to the presence of ascertainment bias, since the probability that a particular phenotype is recorded is not uniform across patients or diseases.

2.2.2. Interactions between Omics and Non-Omics Data

In order to understand the underlying mechanisms of the disease of interest, it is important to consider the combined interactions between the factors included in the model, irrespectively of their nature (omics vs. non-omics). The interaction between data types can be complex as well: gene expression changes may imply phenotypic abnormalities, and this results in a more complex relationship between molecular and clinical data.

2.3. Other Challenges

2.3.1. Fairness

According to Van de Geer, a fair model is a model where all variable blocks, each block representing a set of variables sharing similar characteristics, contribute equally, in contrast to a model dominated by only a few of the different sets [19]. In OnO integrative modeling, should each variable or block contribute equally to the outcome? How can we prevent clinical variables from being penalized when combined with a high-throughput dataset?

2.3.2. Presence of Subphenotypes

The consideration of heterogeneous phenotypes in the model may also add complexity to the OnO model definition. However, ignoring the presence of subphenotypes may affect the performance of the OnO model [20].

3. Integrative Analytical Strategies

The strategies for building hybrid models that contain both omics and non-omics data can be classified as: Independent modelling, conditional modelling, and joint modelling (Figure 1). While the joint modeling strategy is the most proper integrative approach, independent and conditional modeling are also commonly used approaches to jointly model OnO data.

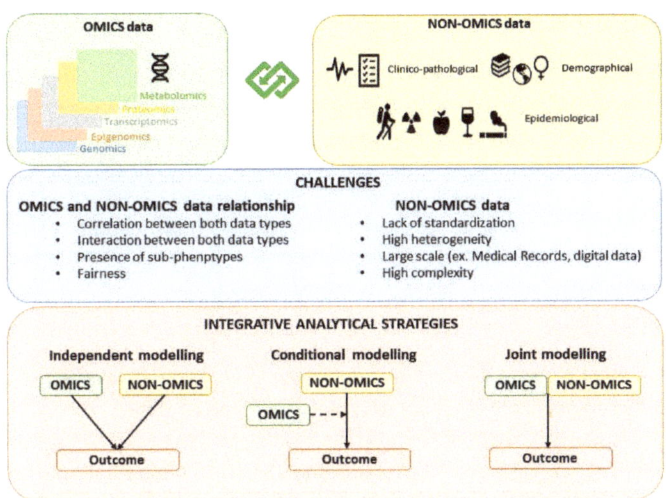

Figure 1. Classification of the strategies for building OnO models.

3.1. Independent Modeling Approach

This strategy, also known as late integration, implies that both the omics and the non-omics data models are built independently [21]. The non-omics data model is built independently of the omics variables by fitting a model that only includes clinical/epidemiological variables or already well-established risk or prognostic score/factors identified and reported in previous efforts. In parallel, the omics variables are selected by considering a model only including omics variables. Both modelling processes typically require variable selection or dimension reduction. The independently selected omics and non-omics variables are then combined in a final model. The predictive accuracy of the combined model is compared with that of the non-omics data model.

Although independent modeling is the simplest integrative approach and, probably, the most common strategy for combining OnO data, this approach cannot capture the correlation/interaction structure of the datasets of different natures. To overcome this limitation, Nevins et al. [22] and Pittman et al. [23] proposed tree-based approaches to combine clinical and molecular scores in such a way possible interactions among OnO data are considered. Whether this approach is also applicable to omics data should be elucidated. Another caveat of the independent modeling strategy is that the predictive power of omics data tends to be overestimated since the trait is also used in the feature selection process.

3.2. Conditional Modeling Approach

This strategy consists in first defining a clinical model with non-omics variables and second, adding omics variables to the already built non-omics model. In other words, in the conditional modelling approach, the selection of omics variables is performed by considering a model that contains or adjusts for the previously selected clinical/epidemiological covariates. The key point of this conditional modeling approach is to decide which omics variables should be added to the clinical model. There are different ways to implement this strategy, the simplest one, though not recommended, is univariate selection, where each omics variable is tested individually and added to the clinical model if there is an increase in the prediction accuracy. As discussed in Bovelstad et al. [24], univariate selection performs poorly, usually yielding worse predictions than the clinical model approach. A more powerful strategy is to perform partial dimension reduction, which consists in considering the joint model with all omics and clinical variables and applies a dimension reduction process only to omics variables. One of such dimension reduction approach is least squares-partial least squares (LS-PLS) [25]. The major caveat of this method is that it suffers from convergence problems, and its performance also depends on the level of collinearity between the two types of data. Alternative approaches to LS-PLS, when the outcome is binary, are partial least square regression [26], ridge regression [27], and LASSO [28] performing dimensionality reduction only on omics data [24]. Other approaches for time dependent variables are described in [29,30]. Binder et al. proposed the algorithm CoxBoost which implements Cox penalized regression that allows some covariates (clinical variables) to be unpenalized [29] and Li et al. applied partial dimension reduction of the supergenes identified after estimating principal components with the omics variables, meanwhile considering the clinical covariates [30]. A common drawback of all the above methods is that they are computationally intensive.

3.3. Joint Modeling Approach

Under this strategy, omics and non-omics data are jointly modelled in a supervised or unsupervised manner. While there is a growing body of articles on multi-marker and multi-omics data integration [2,4,7,20,31,32], the literature that explicitly addresses how to integrate omics and non-omics data in a joint modeling approach is scarce. Following Ritchie's suggestion [3], we can further classify the joint modeling approaches of OnO data into multi-staged (i.e., separate analysis of the associations between the different data types and subsequently with the outcome of interest) and meta-dimensional analyses (i.e., simultaneous analysis of the different data types). One of the first examples of a meta-dimensional approach is the study by Sun et al. that performs concatenation-based integration and joint variable selection of both OnO data using the i-relief algorithm [33]. Those classified as meta-dimensional analyses were further classified into three groups as concatenation-based integration, transformation-based integration, and model-based integration.

4. Attempts of OnO Data Integration in Clinical and Epidemiological Studies

We searched the PubMed electronic database using keywords to identify studies integrating OnO data towards their association with or prediction of the trait of interest, as well as to evaluate their joint classification performance. The search strategy included a combination of keywords related to omics, non-omics, and data integration, for the period 1 December 2009 to 1 October 2018. The logic terms used were: ((integration AND (risk OR score OR prediction OR prognosis) AND (epidemiological OR clinical OR environmental OR exposure) AND (genomic OR GWAS OR genetic OR transcriptomics OR proteomics OR metabolomics OR gene expression OR epigenomics OR epigenetic OR microbiome OR metagenomics))).

The search strategy generated 1,634 records. In this review, we only considered those articles integrating non-omics and high-throughput generated omics data sets in the modeling of the disease/trait as defined in the Introduction. The search resulted in a total of 16 studies almost all of them belonging to the cancer research area (see Table 1). We were first surprised by the small number of published studies at present that performed a "true" integration of OnO data. Although this contribution does not intend to be a systematic review, we consider that the identified papers constitute a representative sample of the attempts done in the field up to date. Hereinafter, we describe the objectives of the OnO integration, the outcomes and the OnO data types considered in the models, as well as the integrative analytical strategies applied in the selected papers (Table 1).

Table 1. Main features of the identified studies conducting omics and non-omics data integration.

Reference	Title	Outcome	Big Data: Omics and Image Data	Non-Omics	Objective	Model Performance	Approach
[25]	Classification based on extensions of LS-PLS using logistic regression: application to clinical and multiple genomic data	Dataset1: Response of childhood malignant embryonal tumors of the CNS to therapy Dataset2: ER status in breast cancer	Dataset1: gene expression data Dataset2: somatic CNA	Dataset1: sex, age, chemo CX, chemo VP. Dataset2: grade, tumor stage, HER2 status, tumor size, progesterone receptor status	Dataset1: Prediction performance: AUC Misclassification rates Dataset2: predict ER stratification of a novel breast tumor to select appropriate treatment: AUC Misclassification rates	Dataset1: $AUC_{non\text{-}omics} = 0.60$ $AUC_{omics} = 0.92$ $AUC_{OnO} = 0.82-0.90$ Dataset2: $AUC_{non\text{-}omics} = 0.87$ $AUC_{omics} = 0.84$ $AUC_{OnO} = 0.93$	Conditional modeling
[34]	Whole-genome multi-omic study of survival in patients with glioblastoma multiform	Survival time in glioblastoma	TCGA: SNP + methylation + CNV + gene expression	TCGA: Sex + use of temozolomide	Predictive ability: AUC	$AUC_{non\text{-}omics} = 0.71$ $AUC_{omics} = $ non-provided $AUC_{OnO} = 0.72$	Joint modeling
[24]	Survival prediction from clinico-genomic models—a comparative study	BrCa dataset: Survival time DLBCL dataset: Survival time Neuroblastoma dataset: Survival time	BrCa dataset: gene expression DLBCL dataset: gene expression Neuroblastoma dataset: microarray gene expression	BrCa dataset: tumor diameter, lymph node status, grade and NPI DLBCL dataset: International Prognostic Index (IPI) Neuroblastoma dataset: NB2004 stratification index	Prediction performance: deviance	Breast cancer dataset: $DiD_{non\text{-}omics} = -12.5$ $DiD_{omics} = -14$ to -2 $DiD_{OnO} = -14$ to -10 DLBCL dataset: $DiD_{non\text{-}omics} = -12.5$ $DiD_{omics} = 0$ to -9 $DiD_{OnO} = -8$ to -19 Neuroblastoma dataset: $DiD_{non\text{-}omics} = -42$ $DiD_{omics} = -28$ to -45 $DiD_{OnO} = -40$ to -50	Conditional and joint modeling
[35]	IPF-LASSO: Integrative L_1-penalized regression with penalty factors for prediction based on multi-omics data	AML dataset: OS BrCa dataset: Distant relapse free survival time Pathological response (binary)	AML dataset (TCGA): microarray gene expression somatic CNA BrCa dataset: microarray gene expression	AML dataset (TCGA): age, % blast cells in bone marrow, white blood cell count per mm^3, and sex BrCa dataset: age, nodal status, tumor size, grade, estrogen receptor, and progesterone receptor	Predictive ability: Prediction error curves, Brier score, integrated Brier score	AML dataset: $IBS_{non\text{-}omics} = $ non-provided $IBS_{omics} = $ non-provided $IBS_{OnO} = 0.211-0.196$ Breast cancer dataset: $IBS_{non\text{-}omics} = $ non-provided $IBS_{omics} = $ non-provided $IBS_{OnO} = 0.134-0.127$	Joint modeling

Table 1. *Cont.*

Reference	Title	Outcome	Big Data: Omics and Image Data	Non-Omics	Objective	Model Performance	Approach
[36]	Deep learning based multi-omics integration robustly predicts survival in liver cancer	TCGA dataset: Survival LIRI-JP cohort: Survival NCI cohort: Survival Chinese cohort: Survival E-TABM-36 cohort: Survival Hawaiian cohort: Survival	TCGA dataset: RNA-seq, miRNA-seq, DNA methylation LIRI-JP cohort: RNA-seq NCI cohort: microarray gene expression Chinese cohort: miRNA E-TABM-36 cohort: gene expression Hawaiian cohort: DNA methylation	TCGA dataset: Stage, grade, race, gender, age, and risk factor	Predictive ability: C-index, Brier score Long-rank p-value	LIRI-JP[a]: $C\text{-index}_{\text{non-omics}} = 0.55$ $C\text{-index}_{\text{omics}} = 0.75$ $C\text{-index}_{\text{OnO}} = 0.74$ NCI [a]: $C\text{-index}_{\text{non-omics}} = 0.45$ $C\text{-index}_{\text{omics}} = 0.67$ $C\text{-index}_{\text{OnO}} = 0.65$	Independent modeling
[37]	A strategy for multimodal data integration: application to biomarkers identification in spinocerebellar ataxia	SCA dataset: SCA subtypes and controls	SCA dataset: 754 metabolites	SCA dataset: MRS of the cerebellum, calorimetry information, volume of the pons	Graphical Reliability of parameter estimates % of times a variable has a non-null weight	Non-provided	Joint modeling
[38]	Prediction of years of life after diagnosis of breast cancer using omics and omic-by-treatment interactions	METABRIC dataset: Log (survival time)	METABRIC dataset: CNV, gene expression (GE) Also interactions with treatment: GExCT, GExRT, GExHT	METABRIC dataset: Age, cancer subtype, histological type, Nottingham Prognostic Index (tumor size, grade and nodal involvement), treatment	Prediction accuracy: AUC % of variance explained Definition of two groups: high and low risk	$AUC_{\text{non-omics}} = 0.72\text{-}0.77$ $AUC_{\text{omics}} = \text{non-provided}$ $AUC_{\text{OnO}} = 0.74\text{-}0.81$ [b]	Joint modeling
[39]	Determination of prognosis in metastatic melanoma through integration of clinic-pathologic, mutation, mRNA, microRNA, and protein information	MIA dataset: Survival grouped into good prognosis and poor prognosis	MIA dataset: mRNA, somatic mutation, microRNA expression, protein expression	MIA dataset: Nineteen clinical/pathological data: ulceration and thickness of primary tumor, number of lymph nodes with metastases, and size of nodal metastasis at the time of staging and others	Prediction error rate (ER) Kaplan-Meier curves	$ER_{\text{non-omics}} = 30\%$ $ER_{\text{mRNA}} = 25\%$ $ER_{\text{protein}} = 35\%$ $ER_{\text{microRNA}} = 37\%$ $ER_{\text{OnO}} = 29\text{--}33\%$	Independent modeling
[40]	Whole Genome Prediction of Bladder Cancer Risk With the Bayesian LASSO	SBC/EPICURO dataset: BC risk	EPICURO/SBC dataset: SNP	SBC/EPICURO dataset: age + gender + region + smoking	Prediction: AUC	$AUC_{\text{non-omics}} = 0.65$ $AUC_{\text{omics}} = 0.53$ $AUC_{\text{OnO}} = 0.65$	Joint modeling
[20]	Prediction of non-muscle invasive bladder cancer outcomes assessed by innovative multimarker prognostic models	SBC/EPICURO dataset: Time to first recurrence (TFR) Time to progression (TP)	EPICURO/SBC dataset: SNP	SBC/EPICURO dataset: TFR: Area + gender + # of tumors + TSG + tumor size + treatment TP: Area + age + # of tumors + TSG + # of recurrences + treatment	Prediction: AUC, R^2	TFR [c]: $AUC_{\text{non-omics}} = 0.62$ $AUC_{\text{omics}} = 0.55$ $AUC_{\text{OnO}} = 0.61$	Joint modeling

Table 1. *Cont.*

Reference	Title	Outcome	Big Data: Omics and Image Data	Non-Omics	Objective	Model Performance	Approach
[41]	A pathway based data integration framework for prediction of disease progression	METABRIC dataset: Survival vs. not survival at 2000 days	METABRIC dataset: Gene expression, CNV	METABRIC dataset: ER status only, disease &treatment group, grade of disease, stage, histological type, HER2 status, age, tumor size, NPI (tumor size, lymph node, grade), tumor cellularity, PAM50-based subtype	Accuracy	$Acc_{non\text{-}omics}$ = non-provided Acc_{omics} = 0.64–0.71 Acc_{OnO} = 0.66–0.80	Joint modeling
[42]	A methylation-to-expression feature model for generating accurate prognostic risk scores and identifying disease targets in clear cell kidney cancer	TCGA dataset: OS in clear cell renal cell carcinoma	TCGA dataset: Gene expression (RNA seq), DNA methylation profile	TCGA dataset: Age, sex, tumor stage	Classification performance: C-index	$C\text{-index}_{non\text{-}omics}$ = 0.776 $C\text{-index}_{omics}$ = 0.702 $C\text{-index}_{OnO}$ = 0.792	Independent modeling approach
[43]	Methylation-to-expression feature models of breast cancer accurately predict overall survival, distant-recurrence free survival, and pathologic complete response in multiple cohorts	TCGA dataset: OS (time) Terunuma dataset: OS (time) Kao dataset: OS (time) Hatzis1 dataset: Distant recurrence-free survival (time) Hatzis2 dataset: Distant recurrence-free survival (time) Pathologic complete response in BC (binary)	TCGA dataset: gene expression and methylation profiles Terunuma dataset: gene expression Kao dataset: gene expression Hatzis1 dataset: gene expression Hatzis2 dataset: gene expression	All datasets: AJCC stage, Age, ER status, PR status, HER2 status, PAM50-based subtype	Classification performance: C-index, AUC	OS TCGA [d]: $C\text{-index}_{non\text{-}omics}$ = 0.75 $C\text{-index}_{omics}$ [e] = 0.69 $C\text{-index}_{OnO}$ = 0.79	Independent modeling
[44]	Integration of Clinical and Gene Expression Data Has a Synergetic Effect on Predicting Breast Cancer Outcome	Vijver dataset: Poor/good outcome Other datasets: Poor/good outcome	Vijver dataset: Expression data Other datasets: Expression data	Vijver dataset: 45 clinical variables Otherdatasets: age, tumor size, grade, ER stats, lymph node, NPI	Classification performance: error rate, AUC	Vijver dataset: $AUC_{non\text{-}omics}$ = 0.75 AUC_{omics} = 0.74 AUC_{OnO} = 0.74–0.78	Independent and Joint modeling
[45]	A Comprehensive Genetic Approach for Improving Prediction of Skin Cancer Risk in Humans	Kreger's dataset: Skin cancer risk	Kreger's dataset: SNPs	Kreger's dataset: Age, ethnicity	Classification performance: AUC	$AUC_{non\text{-}omics}$ = 0.53–0.54 AUC_{omics} = 0.63-0.64 AUC_{OnO} = non-provided	Joint modeling

Table 1. *Cont.*

Reference	Title	Outcome	Big Data: Omics and Image Data	Non-Omics	Objective	Model Performance	Approach
[46]	Integrating Clinical and Multiple Omics Data for Prognostic Assessment across Human Cancers	TCGA datasets: Survival prediction in multiple cancer types	TCGA datasets: Significant SNPs (PRS, as fixed effect), somatic mutation, mRNA, miRNA, methylation, copy number, immune and metagenes signatures	TCGA datasets: Age, stage, Lauren classification (STAD) Also depends on the cancer type MammaPrint and PAM50 gene signatures	Classification performance: C-index	Ovarian and HNSC [f]: $\text{C-index}_{non\text{-}omics} = 0.60; 0.61$ $\text{C-index}_{omics} = 0.61; 0.61$ $\text{C-index}_{OnO} = 0.64; 0.64$	Independent modeling

CNS: Central Nervous System; ER: Estrogen receptor; NPI: Nottingham Prognostic Index; NB2004: German neuroblastoma trial; OS: Overall Survival; AML: Acute myeloid leukaemia; CT: Chemotherapy; RT: Radiotherapy; CT: Chemotherapy; TSG: Tumor stage and grade; PRS: Polygenic risk score; HNSC: Head and neck squamous cell carcinoma; DLBCL: Diffuse large B-cell lymphoma; IBS: Integrated Brier score. [a] Models performance in the largest datasets. [b] It corresponds to the AUC of COV+GE+GExHT model. [c] No improvement in classification performance was also obtained in TP. [d] We provide only the results for OS, when no external validation was considered. Similar performances were obtained when the external validation was performed. [e] Performance of M2EFM Meth+Exp model. [f] We report the C-index results for the cancers where the largest prognostic power was achieved.

4.1. Study Objective

All selected papers aimed to evaluate the prediction performance of the OnO integrative models.

4.2. Study Outcome

Garali et al. identified OnO variables discriminating cases with spinocerebral ataxia from controls [37]. Only two out the 16 selected studies integrated OnO data to predict the risk of skin [45], and bladder [40] cancers. The rest of papers integrating OnO data analyzed cancer outcomes. Among the cancers analyzed were breast [24,35,38,41,43,44], central nervous system [24,25,34], liver [36], hematological [24], melanoma [39], bladder [20], kidney [42], and several cancers [46]. Six studies integrated both data types to evaluate the ability to predict the survival time [24,35,36,42,43,46]. Four studies transformed the survival time into a binary outcome (i.e., survival at a given time) [35,39,41,44]. López de Maturana et al. [20] transformed each time to event into several binary outcomes by accounting for censoring and time. Two studies analyzed the logarithm of survival time also accounting for censoring [34,38] . And two studies assessed the treatment prediction response as a categorical variable: Responders vs. non-responders [25,43].

4.3. Omics Data

Most of the papers only integrated one type of omics data [20,24,25,36,40,43–45]; five papers integrated two omics data types [35,38,41–43]; and four papers integrated > 2 omics data [34,36,39,46]. Gene expression data was the most commonly used high-dimensional omics data [24,25,34–36,38,39,42–44,46] followed by copy number alterations (CNA) [25,34,35,38,41,46], and SNPs [20,34,40,45,46]. Methylation data was considered by five selected papers [34,36,42,43,46] and three studies integrated microRNA (miRNA) data [36,39,46]. Only Jayawardana et al. [39] integrated protein expression data and Garali et al. [37] integrated 754 metabolite biomarkers in a predictive model. In those studies that integrated > 1 omics data set, gene expression was the most informative type in terms of prognostic utility [39,46], followed by microRNAs, and DNA methylation profiles [46].

4.4. Non-Omics Data

All the selected papers incorporated the non-omics data in a low-dimensionality fashion, meaning that only a few variables were integrated in the models. Non-omics information was a quite heterogeneous group of data formed by both categorical and continuous variables. The majority of non-omics data were clinico-pathological variables, including treatment, tumor stage, tumor size, lymph status, histological type, estrogen receptor status, progesterone receptor status or human epidermal growth factor receptor (see Table 1 for further details). Specific tumor scales, such as Breslow thickness and Clark's level in melanoma [39] or classifications as Lauren classification in stomach adenocarcinoma [46] or the international prognostic index in lymphoma [24] were also used as non-omics clinico-pathological variables. Moreover, cancer subtype definition based on gene-expression signatures as PAM50 signature for breast cancer [42,43,46] as well as Mammaprint [46] were also considered. Jayawardana et al. were the only ones integrating metabolic imaging obtained by magnetic resonance spectroscopy, along with pons volume [39]. In addition, demographical data as age, gender, ethnicity, or region were also considered. Smoking status was the only epidemiological/life-style variable included in the risk models [40]. None of the papers considered large scale clinical or epidemiological data in their models.

4.5. Integrative Analytical Strategies

The integrative strategies used in the identified papers adopted the three different modeling methods described before (see Table 1): (1) Independent modeling, (2) conditional modeling, and (3) joint modeling, which were implemented using one-step or two-step designs. The published studies applied these methods assuming low-dimensional non-omics data, while omics data were high dimensional and required some variable selection, dimension reduction, or regularization process.

Examples applying the independent modeling approach are found in [36,39,42–44,46]. Thompson and Marsit [42], combined both multistage and meta-dimensional elements in a Methylation-to-Expression Feature Model (M2EFM) by first defining a molecular score that combined DNA methylation and gene expression and then performing a second regression to integrate clinical variables in a prognostic model for clear cell kidney cancer. van Vliet et al. [44] determined the optimal sets of features from each data type separately by using different classifiers such as the nearest mean, the simple Bayes, the 3-Nearest-Neighbor, the support vector machine, and the Tree Classifier. They then used these sets of features and all training samples in the final integrative model. Jayawardana et al. [39] used multiple types of omics data (i.e., microRNA, mRNA and protein expression) to integrate them with clinico-pathological variables also using an independent modeling approach. Briefly, they selected an optimized set of omics features integrated as a molecular signature of each data type (known as pre-validated vector) and then modelled them in combination with the clinico-pathological data, creating a combined prognostic signature. Chaudhary et al. [36] applied a transformation-based integration of multi-omics data independently from the clinical variables by using a deep-learning approach to integrate RNA-seq, miRNA-seq, and DNA methylation data to identify subgroups of hepatocellular carcinoma. Zhu et al. [46] did similarly, which led to substantially improved prognostic performance over the use of clinical variables alone in half of the cancer types examined. Particularly, they used the kernel-fusion Cox model as the multi-omics kernel learning method for prognostic prediction. Their approach consisted of three steps: (1) They built a kernel reflecting the similarity of the individuals based on each omics data including mRNA, miRNA, CNA, methylation and mutational status; (2) they applied a kernel alignment approach to evaluate whether the similarity matrix built using an omics data set aligned well with its counterpart defined by another omics data type; and (3) they evaluated the prognostic performance of the molecular profile of each individual, which was assumed to follow a multivariate normal distribution with mean zero and (co)variance matrix K corresponding to a fused kernel. This resulted from the linear combination or fusion of each omics similarity matrices (somatic mutation, mRNA, miRNA, methylation, and copy number profiles), along with the clinical prognostic score and the polygenic risk score based on odds ratios reported in the literature. Through this way, prognosis-relevant signals from multiple pathways and involving a large number of omics biomarkers became visible only when aggregated. Zhu et al. applied, by far, the most comprehensive integrative approach [46].

Two studies applied the conditional modeling approach. Bazzoli and Lambert-Lacroix [25] adopted it using a one-step approach. They adapted the Least Squares—Partial Least Squares (LS-PLS) procedure to accommodate logistic regression hybrid models resulting into three different approaches: LS-PLS-IRLS (where IRLS denotes Iteratively Reweighted Least Squares algorithm), R-LS-PLS, and IR-LS-PLS differing in the way PLS is used in the classification context. The three approaches involved the incorporation of PLS scores resulting from the application of PLS regression on omics data into the OLS equations in an iterative way to obtain a one-step hybrid model accommodating OnO data. Bovelstad et al. [24] proposed a Cox regression model including OnO variables and applying different methods for dimensionality reduction only to omics data and found that the improvement of the OnO model varied among diseases: Whereas large improvements were obtained when OnO model was applied to diffuse large B-cell lymphoma (DLBCL) and neuroblastoma datasets, similar performance was obtained using gene expression data only vs. the integrative model.

Joint modeling integration was the most commonly used approach by the identified studies. Particularly, the majority of the studies applied the transformation-based meta-dimensional analysis, which combined multiple data sets after transforming each data type into an intermediate form, such as a graph or a kernel matrix. Three studies applied Bayesian Reproducing Kernel Hilbert spaces regressions as a modeling framework able to incorporate clinical risk factors and high-dimensional omics profiles [34,38,45], González-Reymúndez et al. also assessed the interactions between OnO factors [38]. Seoane et al. proposed a multiple kernel learning strategy implementing feature selection separately for each data type and by pathway membership [41]. Examples of the concatenation-based integration meta-dimensional analyses are found in [20,24,35,40,44]. Boulesteix et al. [35] applied the IPF-LASSO, a penalized regression method that allows different penalty terms to the different layers of information, whereas López de Maturana et al. [20,40] implemented a Bayesian LASSO coupled threshold modeling with different priors imposed for OnO data. Bovelstad et al. [24] proposed a Cox regression model including OnO variables and applying different methods for dimensionality reduction only to omics data. Van Vliet et al. [44] applied five classifiers (nearest mean, the simple Bayes, the 3-nearest-neighbor, the support vector machine, and the tree classifier) concatenating the omics and clinical features.

In addition, Garali et al. implemented a regularized generalized canonical correlation analysis (RGCCA) and a sparse generalized canonical correlation analysis (SGCCA) model-based integration approaches, in which each data type is analyzed separately and then combined in a final integrative model [37]. Rather than operating sequentially on parts of the measurements, this integrative approach aims at summarizing the relevant information between and within blocks of variables. Particularly, RGCCA incorporates a variable selection procedure and SGCCA allows both the extraction of biomarkers and the reduction of the multiblock datasets into a few meaningful components.

4.6. OnO Data Integrative Models Performance

The performance of the models considered in the selected papers was retrieved, whenever provided, and is displayed in Table 1. In general, the selected papers showed that the OnO data integrative models perform better in terms of classification performance than the only-clinical/ epidemiological or only-omics model [24,25,34,38,39,41–44,46]. However, there were studies reporting no/slight improvement in terms of classification performance of OnO data integrative models [20,24,36,40,46]. The variability in terms of predictive improvement observed when applying OnO modeling could depend on different factors, such as the outcome, the omics and clinical/epidemiological variables, and the integrative method implemented. For example, Bovelstad et al. [24] found that the improvement of OnO model varied among diseases: While large model performance improvements were obtained when OnO data integration was applied to DLBCL and neuroblastoma datasets, no gain in performance was observed when gene expression data was integrated with clinic-pathological variables in the breast cancer dataset. Furthermore, the SNPs performed poorly in the outcome prediction across cancer types [20,40,46].

5. Recommended Integration Strategies

As previously discussed, we distinguished three different strategies for building hybrid models containing both omics and non-omics data: Independent modeling, conditional modeling, and joint modeling. Selected papers have applied these approaches to the integration of low-dimensional non-omics data and high dimensional omics data, which requires some variable selection, dimensionality reduction or regularization process before or during their modeling. However, these integrative modeling strategies also apply to high dimensional non-omics data, a scenario that is becoming more frequent because of new technological advances that constantly increase our capacity for obtaining additional information from many different sources (e.g., EHRs or wearable sensors).

Joint modeling approaches, where omics and non-omics data are jointly modelled in a supervised or unsupervised manner, are those recommended to integrate both large-scale OnO data, because they account for the correlation structure between the two data types and capture a larger complexity than the conditional or independent modeling. The decision of which modeling strategy (multi-staged or meta-dimensional) to follow should be done in accordance with the main objective of the analysis: Association testing or risk prediction [47]. Multi-staged analysis that models the relationship between the different layers of information will probably be preferable when the interest is to increase our biological knowledge of the disease mechanisms. On the other hand, the meta-dimensional approach will be more suitable when the goal is to improve prediction or prognosis for personalized medicine and modeling the mechanisms is not so relevant, although they are not exclusive for these purposes. Concatenation-based integration combines the different data types into a joint data matrix and performs a variable selection or dimension reduction to the whole data set. The concatenation approach cannot ignore that the different data types are expected to have different relevance to the outcome and the joint analysis should take this into account.

In addition to the methods used by the identified studies described previously, the following modeling strategies could be considered in jointly modeling OnO data. The kernel-fusion Cox model used in Zhu et al. [46] initially designed as multi-omics kernel learning method could be extended to include also non-omics data in a kernel reflecting also the similarity between the profiles for each multimodal data. iCluster and iCluster2 are examples of a model-based integration strategy and could also accommodate non-omics variables [6,48]. Briefly, they perform a joint latent variable model-based clustering method, where the latent component connects the different data specific models, inducing dependencies across the different data types. Furthermore, deep-learning methods could also be used in a model-based integrative approach [49,50]. Another machine learning approach, the tensor factorization, allows the integration of multiple data modalities and supports dimensionality reduction and identification of latent groups [51]. A tensor factorization is a multidimensional array where each modality spans one axis and helps identifying group-wise interaction. Since it is an unsupervised method, it may be used to identify phenotypes, as it has been done in the Multi-Ethnic Study of Atherosclerosis (MESA) for discovering subgroups of heart failure patients. A drawback of this method is the interpretability of the results.

6. Concluding Remarks

Disentangling a complex trait requires not only the understanding of its "complex" biological system but also the combinatorial effects of other factors (i.e., host-related, environmental, socio-economics, etc.). The integration of OnO data can lead to finding new risk factors of a disease, propose better predictive models, distinguish patients with favorable response to treatment, and therefore help in the future of personalized medicine [52]. Unfortunately, OnO data integrative efforts are still scarce, although they are expected to become more frequent because of the advent of Big Data in the medical field.

In general, integrating both molecular and clinical data results in better prognostic models than either type alone as has been shown by several authors [39,43,44,46]. Possible explanations are that individual classifiers collect associations with the outcome of interest and their redundancy leads to a better prediction; that the clinical set of features adds some additional information which is not captured by the omics data; and that relevant signals may come from multiple pathways and involve a large number of omics biomarkers, the effect of which may be visible only when aggregated. However, model improvement has not always been observed when OnO data is integrated [36,40].

In any case, exploring OnO data integration becomes a must in the biomedical field. It requires method development, validation, and standardization. This review represents an endeavor towards these aims by identifying the challenges that OnO data integration presents, as well as discussing and proposing integrative analytical strategies. We hope it guides OnO data integrative efforts.

Author Contributions: Conceptualization, E.L.d.M., L.A., M.L.C., N.M.; methodology, E.L.d.M., L.A., S.P., I.A.M.-A., L.P., M.L.C., N.M.; identification and review of articles, E.L.d.M., L.A., P.A., I.A.M.-A., M.L.C.; writing—original Draft Preparation, E.L.d.M., P.A., M.L.C., N.M.; writing—review & editing, E.L.d.M., L.A., S.P., I.A.M.-A., L.P., M.L.C., N.M.; supervision, L.A., M.L.C., N.M.; funding acquisition, N.M.

Funding: This research was partly funded by Fondo de Investigaciones Sanitarias (FIS), Instituto de Salud Carlos III, Spain (#PI12-00815, #PI15/01573); European Cooperation in Science and Technology (COST Action #BM1204: EUPancreas), World Cancer Research (#15-0391); and Ministerio de Economía y Competitividad, Spain (MTM2015-64465-C2-1-R).

Conflicts of Interest: The authors declare no conflict of interest. The sponsors had no role in the design, execution, interpretation, or writing of the study.

References

1. Civelek, M.; Lusis, A.J. Systems genetics approaches to understand complex traits. *Nat. Rev. Genet.* **2014**, *15*, 34–48. [CrossRef]
2. Hasin, Y.; Seldin, M.; Lusis, A. Multi-omics approaches to disease. *Genome Biol.* **2017**, *18*, 83. [CrossRef] [PubMed]
3. Ritchie, M.D.; Holzinger, E.R.; Li, R.; Pendergrass, S.A.; Kim, D. Methods of integrating data to uncover genotype-phenotype interactions. *Nat. Rev. Genet.* **2015**, *16*, 85–97. [CrossRef] [PubMed]
4. Kristensen, V.N.; Lingjærde, O.C.; Russnes, H.G.; Vollan, H.K.M.; Frigessi, A.; Børresen-Dale, A.L. Principles and methods of integrative genomic analyses in cancer. *Nat. Rev. Cancer* **2014**, *14*, 299–313. [CrossRef] [PubMed]
5. Robinson, D.R.; Wu, Y.M.; Lonigro, R.J.; Vats, P.; Cobain, E.; Everett, J.; Cao, X.; Rabban, E.; Kumar-Sinha, C.; Raymond, V.; et al. Integrative clinical genomics of metastatic cancer. *Nature* **2017**, *548*, 297–303. [CrossRef]
6. Shen, R.; Olshen, A.B.; Ladanyi, M. Integrative clustering of multiple genomic data types using a joint latent variable model with application to breast and lung cancer subtype analysis. *Bioinformatics* **2009**, *25*, 2906–2912. [CrossRef]
7. Pineda, S.; Gomez-Rubio, P.; Picornell, A.; Bessonov, K.; Márquez, M.; Kogevinas, M.; Real, F.X.; Van Steen, K.; Malats, N. Framework for the Integration of Genomics, Epigenomics and Transcriptomics in Complex Diseases. *Hum. Hered.* **2015**, *79*, 124–136. [CrossRef] [PubMed]
8. Brandão, M.; Pondé, N.; Piccart-Gebhart, M. Mammaprint™: A comprehensive review. *Futur. Oncol.* **2019**, *15*, 207–224. [CrossRef]
9. Wallden, B.; Ferree, S.; Ravi, H.; Dowidar, N.; Hood, T.; Danaher, P.; Mashadi-Hossein, A.; Wright, G.; Schaper, C.; Justin, J. Development of the molecular diagnostic (MDx) DLBCL Lymphoma Subtyping Test (LST) on the nCounter Analysis System. *J. Clin. Oncol.* **2015**, *33*.
10. Lichtenstein, P.; Holm, N.V.; Verkasalo, P.K.; Iliadou, A.; Kaprio, J.; Koskenvuo, M.; Pukkala, E.; Skytthe, A.; Hemminki, K. Analyses of Cohorts of Twins from Sweden, Denmark, and Finland. *N. Engl. J. Med.* **2000**, *343*, 78–85. [CrossRef]
11. Salathé, M.; Bengtsson, L.; Bodnar, T.J.; Brewer, D.D.; Brownstein, J.S.; Buckee, C.; Campbell, E.M.; Cattuto, C.; Khandelwal, S.; Mabry, P.L.; et al. Digital epidemiology. *PLoS Comput. Biol.* **2012**, *8*, 1–5. [CrossRef]
12. Jette, A.M. Measuring subjective clinical outcomes. *Phys. Ther.* **1989**, *69*, 580–584. [CrossRef] [PubMed]
13. Tugwell, P.; Bombardier, C. A methodologic framework for developing and selecting endpoints in clinical trials. *J. Rheumatol.* **1982**, *9*, 758–762. [PubMed]
14. Balliu, B.; Tsonaka, R.; Boehringer, S.; Houwing-Duistermaat, J. A Retrospective Likelihood Approach for Efficient Integration of Multiple Omics Factors in Case-Control Association Studies. *Genet. Epidemiol.* **2015**, *39*, 156–165. [CrossRef] [PubMed]
15. Spiegl-Kreinecker, S.; Lötsch, D.; Ghanim, B.; Pirker, C.; Mohr, T.; Laaber, M.; Weis, S.; Olschowski, A.; Webersinke, G.; Pichler, J.; et al. Prognostic quality of activating TERT promoter mutations in glioblastoma: Interaction with the rs2853669 polymorphism and patient age at diagnosis. *Neuro. Oncol.* **2015**, *17*, 1231–1240. [CrossRef] [PubMed]
16. Boks, M.P.; Derks, E.M.; Weisenberger, D.J.; Strengman, E.; Janson, E.; Sommer, I.E.; Kahn, R.S.; Ophoff, R.A. The relationship of DNA methylation with age, gender and genotype in twins and healthy controls. *PLoS ONE* **2009**, *4*, e6767. [CrossRef] [PubMed]
17. Horvath, S.; Zhang, Y.; Langfelder, P.; Kahn, R.S.; Boks, M.P.M.; van Eijk, K.; van den Berg, L.H.; Ophoff, R.A. Aging effects on DNA methylation modules in human brain and blood tissue. *Genome Biol.* **2012**, *13*, R97. [CrossRef]
18. Marchini, J.; Cardon, L.R.; Phillips, M.S.; Donnelly, P. The effects of human population structure on large genetic association studies. *Nat. Genet.* **2004**, *36*, 512–517. [CrossRef] [PubMed]
19. Van de Geer, J.P. Linear relations among k sets of variables. *Psychometrika* **1984**, *49*, 79–94. [CrossRef]
20. López de Maturana, E.; Picornell, A.; Masson-Lecomte, A.; Kogevinas, M.; Márquez, M.; Carrato, A.; Tardón, A.; Lloreta, J.; García-Closas, M.; Silverman, D.; et al. Prediction of non-muscle invasive bladder cancer outcomes assessed by innovative multimarker prognostic models. *BMC Cancer* **2016**, *16*, 351. [CrossRef]

21. Gevaert, O.; De Smet, F.; Timmerman, D.; Moreau, Y.; De Moor, B. Predicting the prognosis of breast cancer by integrating clinical and microarray data with Bayesian networks. *Bioinformatics* **2006**, *22*, e184–e190. [CrossRef] [PubMed]

22. Nevins, J.R.; Huang, E.S.; Dressman, H.; Pittman, J.; Huang, A.T.; West, M. Towards integrated clinico-genomic models for personalized medicine: combining gene expression signatures and clinical factors in breast cancer outcomes prediction. *Hum. Mol. Genet.* **2003**, *12*, R153–R157. [CrossRef] [PubMed]

23. Pittman, J.; Huang, E.; Dressman, H.; Horng, C.-F.; Cheng, S.H.; Tsou, M.-H.; Chen, C.-M.; Bild, A.; Iversen, E.S.; Huang, A.T.; et al. Integrated modeling of clinical and gene expression information for personalized prediction of disease outcomes. *Proc. Natl. Acad. Sci. USA* **2004**, *101*, 8431–8436. [CrossRef]

24. Bøvelstad, H.M.; Nygård, S.; Borgan, Ø. Survival prediction from clinico-genomic models—A comparative study. *BMC Bioinform.* **2009**, *10*, 413. [CrossRef] [PubMed]

25. Bazzoli, C.; Lambert-Lacroix, S. Classification based on extensions of LS-PLS using logistic regression: Application to clinical and multiple genomic data. *BMC Bioinform.* **2018**, *19*, 314. [CrossRef]

26. Nygård, S.; Borgan, Ø.; Lingjærde, O.C.; Størvold, H.L. Partial least squares Cox regression for genome-wide data. *Lifetime Data Anal.* **2008**, *14*, 179–185. [CrossRef]

27. Hoerl, A.E.; Kennard, R.W. Ridge Regression: Biased Estimation for Nonorthogonal Problems. *Technometrics* **1970**, *12*, 55–67. [CrossRef]

28. Tibshirani, R. Regression Selection and Shrinkage via the Lasso. *J. R. Stat. Soc. B* **1996**, *58*, 267–288. [CrossRef]

29. Binder, H.; Schumacher, M. Allowing for mandatory covariates in boosting estimation of sparse high-dimensional survival models. *BMC Bioinform.* **2008**, *9*, 14. [CrossRef]

30. Li, L. Survival prediction of diffuse large-B-cell lymphoma based on both clinical and gene expression information. *Bioinformatics* **2006**, *22*, 466–471. [CrossRef] [PubMed]

31. Pineda, S.; Real, F.X.; Kogevinas, M.; Carrato, A.; Chanock, S.J.; Malats, N.; Van Steen, K. Integration Analysis of Three Omics Data Using Penalized Regression Methods: An Application to Bladder Cancer. *PLoS Genet.* **2015**, *11*, e1005689. [CrossRef] [PubMed]

32. Manzoni, C.; Kia, D.A.; Vandrovcova, J.; Hardy, J.; Wood, N.W.; Lewis, P.A.; Ferrari, R. Genome, transcriptome and proteome: The rise of omics data and their integration in biomedical sciences. *Brief. Bioinform.* **2018**, *19*, 286–302. [CrossRef] [PubMed]

33. Sun, Y.; Goodison, S.; Li, J.; Liu, L.; Farmerie, W. Improved breast cancer prognosis through the combination of clinical and genetic markers. *Bioinformatics* **2007**, *23*, 30–37. [CrossRef]

34. Bernal Rubio, Y.L.; González Reymúndez, A.; Wu, K.-H.H.; Griguer, C.E.; Steibel, J.P.; de Los Campos, G.; Doseff, A.; Gallo, K.; Vazquez, A.I. Whole-Genome Multi-omic Study of Survival in Patients with Glioblastoma Multiforme. *G3 (Bethesda)* **2018**, *8*, 3627–3636. [CrossRef] [PubMed]

35. Boulesteix, A.L.; De Bin, R.; Jiang, X.; Fuchs, M. IPF-LASSO: Integrative L1-Penalized Regression with Penalty Factors for Prediction Based on Multi-Omics Data. *Comput. Math. Methods Med.* **2017**, *2017*, 1–14. [CrossRef] [PubMed]

36. Chaudhary, K.; Poirion, O.B.; Lu, L.; Garmire, L.X. Deep learning–based multi-omics integration robustly predicts survival in liver cancer. *Clin. Cancer Res.* **2018**, *24*, 1248–1259. [CrossRef] [PubMed]

37. Garali, I.; Adanyeguh, I.M.; Ichou, F.; Perlbarg, V.; Seyer, A.; Colsch, B.; Moszer, I.; Guillemot, V.; Durr, A.; Mochel, F.; et al. A strategy for multimodal data integration: Application to biomarkers identification in spinocerebellar ataxia. *Brief. Bioinform.* **2018**, *19*, 1356–1369. [CrossRef] [PubMed]

38. González-Reymúndez, A.; De Los Campos, G.; Gutiérrez, L.; Lunt, S.Y.; Vazquez, A.I. Prediction of years of life after diagnosis of breast cancer using omics and omic-by-treatment interactions. *Eur. J. Hum. Genet.* **2017**, *25*, 538–544. [CrossRef]

39. Jayawardana, K.; Schramm, S.J.; Haydu, L.; Thompson, J.F.; Scolyer, R.A.; Mann, G.J.; Müller, S.; Yang, J.Y.H. Determination of prognosis in metastatic melanoma through integration of clinico-pathologic, mutation, mRNA, microRNA, and protein information. *Int. J. Cancer* **2015**, *136*, 863–874. [CrossRef]

40. De Maturana, E.L.; Chanok, S.J.; Picornell, A.C.; Rothman, N.; Herranz, J.; Calle, M.L.; García-Closas, M.; Marenne, G.; Brand, A.; Tardón, A.; et al. Whole Genome Prediction of Bladder Cancer Risk With the Bayesian LASSO. *Genet. Epidemiol.* **2014**, *38*, 467–476. [CrossRef]

41. Seoane, J.A.; Day, I.N.M.; Gaunt, T.R.; Campbell, C. A pathway-based data integration framework for prediction of disease progression. *Bioinformatics* **2014**, *30*, 838–845. [CrossRef]

42. Thompson, J.A.; Marsit, C.J. A Methylation-To-Expression Feature Model for Generating Accurate Prognostic Risk Scores and Identifying Disease Targets in Clear Cell Kidney Cancer. *Pacific Symp. Biocomput.* **2017**, *2017*, 509–520. [CrossRef]

43. Thompson, J.A.; Christensen, B.C.; Marsit, C.J. Methylation-to-Expression Feature Models of Breast Cancer Accurately Predict Overall Survival, Distant-Recurrence Free Survival, and Pathologic Complete Response in Multiple Cohorts. *Sci. Rep.* **2018**, *8*, 5190. [CrossRef] [PubMed]

44. Van Vliet, M.H.; Horlings, H.M.; van de Vijver, M.J.; Reinders, M.J.T.; Wessels, L.F.A. Integration of clinical and gene expression data has a synergetic effect on predicting breast cancer outcome. *PLoS ONE* **2012**, *7*, e40358. [CrossRef] [PubMed]

45. Vazquez, A.I.; de los Campos, G.; Klimentidis, Y.C.; Rosa, G.J.M.; Gianola, D.; Yi, N.; Allison, D.B. A comprehensive genetic approach for improving prediction of skin cancer risk in humans. *Genetics* **2012**, *192*, 1493–1502. [CrossRef] [PubMed]

46. Zhu, B.; Song, N.; Shen, R.; Arora, A.; Machiela, M.J.; Song, L.; Landi, M.T.; Ghosh, D.; Chatterjee, N.; Baladandayuthapani, V.; et al. Integrating Clinical and Multiple Omics Data for Prognostic Assessment across Human Cancers. *Sci. Rep.* **2017**, *7*, 16954. [CrossRef]

47. Abraham, G.; Inouye, M. Genomic risk prediction of complex human disease and its clinical application. *Curr. Opin. Genet. Dev.* **2015**, *30*, 10–16. [CrossRef]

48. Shen, R.; Mo, Q.; Schultz, N.; Seshan, V.E.; Olshen, A.B.; Huse, J.; Ladanyi, M.; Sander, C. Integrative subtype discovery in glioblastoma using iCluster. *PLoS ONE* **2012**. [CrossRef]

49. Angermueller, C.; Pärnamaa, T.; Parts, L.; Stegle, O. Deep learning for computational biology. *Mol. Syst. Biol.* **2016**, *12*, 878. [CrossRef] [PubMed]

50. Min, S.; Lee, B.; Yoon, S. Deep learning in bioinformatics. *Brief. Bioinform.* **2017**, *18*, 851–869. [CrossRef]

51. Luo, Y.; Wang, F.; Szolovits, P. Tensor factorization toward precision medicine. *Brief. Bioinform.* **2017**, *18*, 511–514. [CrossRef]

52. López de Maturana, E.; Pineda, S.; Brand, A.; Van Steen, K.; Malats, N. Toward the integration of *Omics* data in epidemiological studies: still a "long and winding road". *Genet. Epidemiol.* **2016**, *40*, 558–569. [CrossRef] [PubMed]

Review

Machine Learning and Integrative Analysis of Biomedical Big Data

Bilal Mirza [1,2,*], Wei Wang [1,3,4,5], Jie Wang [1,2], Howard Choi [1,2,5], Neo Christopher Chung [1,6] and Peipei Ping [1,2,4,5,7,*]

1   NIH BD2K Center of Excellence for Biomedical Computing, University of California Los Angeles, Los Angeles, CA 90095, USA; weiwang@cs.ucla.edu (W.W.); jw744@g.ucla.edu (J.W.); cjh9595@g.ucla.edu (H.C.); nchchung@gmail.com (N.C.C.)
2   Department of Physiology, University of California Los Angeles, Los Angeles, CA 90095, USA
3   Department of Computer Science, University of California Los Angeles, Los Angeles, CA 90095, USA
4   Scalable Analytics Institute (ScAi), University of California Los Angeles, Los Angeles, CA 90095, USA
5   Department of Bioinformatics, University of California Los Angeles, Los Angeles, CA 90095, USA
6   Institute of Informatics, Faculty of Mathematics, Informatics and Mechanics, University of Warsaw, Banacha 2, 02-097 Warsaw, Poland
7   Department of Medicine (Cardiology), University of California Los Angeles, Los Angeles, CA 90095, USA
*   Correspondence: bmirza@mednet.ucla.edu (B.M.); pping38@g.ucla.edu (P.P.); Tel.: +1-310-267-5624 (P.P.)

Received: 2 December 2018; Accepted: 21 January 2019; Published: 28 January 2019

Abstract: Recent developments in high-throughput technologies have accelerated the accumulation of massive amounts of omics data from multiple sources: genome, epigenome, transcriptome, proteome, metabolome, etc. Traditionally, data from each source (e.g., genome) is analyzed in isolation using statistical and machine learning (ML) methods. Integrative analysis of multi-omics and clinical data is key to new biomedical discoveries and advancements in precision medicine. However, data integration poses new computational challenges as well as exacerbates the ones associated with single-omics studies. Specialized computational approaches are required to effectively and efficiently perform integrative analysis of biomedical data acquired from diverse modalities. In this review, we discuss state-of-the-art ML-based approaches for tackling five specific computational challenges associated with integrative analysis: curse of dimensionality, data heterogeneity, missing data, class imbalance and scalability issues.

Keywords: machine learning; multi-omics; data integration; curse of dimensionality; heterogeneous data; missing data; class imbalance; scalability

---

1. Introduction

Technological advancements in high-throughput cell biology have enabled researchers to examine the landscape of biomolecules (i.e., DNA, RNA, proteins, metabolites, etc.) associated with a phenotype of interest. Next-generation sequencing technologies [1–3] have revolutionized the profiling of DNA and messenger RNA (mRNA), allowing genomes and transcriptomes to be sequenced quickly and economically. Mass spectrometry [4,5] allows us to efficiently identify and quantify proteins, metabolites and lipids in cells, capturing underlying cellular variations in response to physiological and pathological changes. Consequently, large-scale studies on the genome, the transcriptome, the proteome, the metabolome, the lipidome, etc. have created a plethora of data associated with these "-omes" also known as "omics" data. In this regard, machine learning (ML) algorithms [6–10] have been developed to elucidate complex cellular mechanisms, identify molecular signatures, and predict clinical outcomes from large biomedical datasets [11,12]. Traditionally, ML-based single-omics analyses provide assorted perspectives on cellular processes with respect to a particular -ome [13–16]. However, isolated omics studies frequently fall short when identifying the cause of multifaceted diseases such as cancer [17], cardiac diseases [18], diabetes [19], etc. This evidence suggests that an inclusive view of cellular processes, constructed by integrating information within and across -omes, is required to provide a comprehensive picture of the biological mechanisms [20].

ML-empowered integrative analysis has emerged as a key player in studies involving multiple omics data [21–25]. By analyzing different omics layers together, ML-based integrative methods provide a holistic view of biological processes, offer new mechanistic insights on the phenotype of interest, and facilitate the advancements in precision medicine [26]. For example, Hoadley et al. employed ML-based integrative clustering in a comprehensive study of twelve different types of cancer which resulted in a new molecular taxonomy of diverse tumor types [21]. They integrated genomics, epigenomics, transcriptomics, and proteomics data utilizing cluster-of-cluster-assignments (COCA) to obtain clinically relevant sub-types. In [22], canonical correlation analysis (CCA) with dimensionality reduction was employed for jointly analyzing microRNA (miRNA) and gene expression data. This analysis provided insight into the mechanisms of head and neck squamous cell cancer and its response to treatment via cetuximab. In another study, Arelaguet et al. [23] performed integrative analysis of somatic mutations, RNA expression, and DNA methylation data associated with chronic lymphocytic leukemia (CLL). This study identified new factors predictive of clinical outcome by employing a latent variable modeling approach. To identify markers of body fat mass changes in obesity [24], proteomics and metabolomics data were integrated to create a "transomic" dataset whose individual features went through z-score transformation prior to independent component analysis (ICA). It was noted that a combined transomics dataset better discriminates lean and obese subjects as compared to single-omics data. For improving drug sensitivity in breast cancer, genomics, epigenomics, and proteomics, data were integrated using a multiview multiple kernel learning (MKL) approach [25]. This study showed that the predictive performance achieved by multiview learning was found to be better than that obtained by any individual view, where a 'view' describes a particular representation of the input data.

Integrative analysis of biomedical data with ML can be performed in a variety of ways. For example, the simplest approach is to construct a large feature matrix by directly concatenating features from different datasets [27]. Each feature may go through z-transformation for standardization across all biological samples, followed by ML-based feature selection for molecular signature extraction and biomarker identification. Another common integrative analysis approach is to transform data from heterogenous sources into joint latent profiles. Latent (hidden) profiles are the transformations of data that can capture hidden sources of variation. ML-based clustering is then performed in common latent sub-space for the identification of clinically relevant patient sub-groups [28]. In addition, there are ML-based frameworks that fuse data as a step toward building a model, e.g., multiple kernel learning or network modelling approaches [25,29]. Notably, the accumulation of large biomedical data and the inevitable benefits of studying multiple omics together present new challenges and opportunities for developing novel computational approaches customized for integrative analysis. For example, heterogeneous data with mixed variable types, and missing values in one or more omics can substantially hinder the data integration and analysis. In addition, when integrating multiple omics data, the dimensions of the dataset can grow into hundreds or thousands of variables, while the number of observations or biological samples remains limited. This disparity is called the curse of dimensionality or the $p \gg n$ problem, where p is the number of variables and n the number of samples. Moreover, the rarity or class imbalance in the data can also lead to results that are biased or less accurate. A class imbalance problem arises when rare events are analyzed and compared against events that happen much more frequently, a common occurrence in omics datasets. Furthermore, standard integrative frameworks may not be suitable for large-scale multi-omics analysis due to computational and storage limitations.

Fortunately, advancements in the field of data science are constantly improving the precision of biomedical research, and machine learning is well poised to enable seamless integration of molecular and clinical data. In addition, deep learning architectures [30–32], which better recognize complex features through representation learning with multiple layers, can facilitate the integrative analysis by effectively addressing the challenges discussed above. In this article, we review some of the integrative computational approaches recently proposed for analyzing biomedical data from multiple sources. Specifically, we discuss state-of-the-art ML approaches that can address five important challenges in multi-omics integrative analysis: the curse of dimensionality, data heterogeneity, missing data, class imbalance, and scalability issues.

2. Curse of Dimensionality

In the integrative analysis of multi-omics, the number of variables or features to study is increased, but the number of samples is generally the same, since the measurements from multiple platforms essentially belong to the same biological sample. For example, in the stratification of ovarian cancer patients (samples) based on their DNA methylation, miRNA expression and gene expression measurements (variables), the number of variables can be substantially higher than the number of samples (thousands of variables measured on just few hundred patients) [33]. This is the so-called curse of dimensionality or the $p \gg n$ problem in machine learning [25,34]. The increased dimensionality in the number of variables, with the same sample size, makes most ML methods vulnerable to an overfitting problem, i.e., highly accurate on training data but poor generalization on unseen test data [33]. This is due to that fact that the same samples now cover a much smaller fraction of input feature space [7]. The addition of more features may carry new information; however, the benefit of new information can be outweighed by the curse of dimensionality. Dimensionality reduction (DR) is commonly employed in omics studies as datasets from genomics, proteomics, transcriptomics, medical imaging, and clinical trials are frequently faced with the $p \gg n$ problem. DR techniques are employed either as feature extraction (FE) or feature selection (FS) [35,36]. Feature extraction projects the data from high-dimensional space to lower dimensional space, while feature selection reduces the dimensionality by identifying only a relevant subset of original features [34–36].

Feature extraction facilitates data visualization, data exploration, latent (hidden) factor profiling, compression, etc. Principal component analysis (PCA), a popular FE method, reduces the dimensionality of the data by orthogonally transforming the high-dimensional features to linearly uncorrelated principal comments (PC). Given orthogonality constraints, the top PCs capture maximal variance in the dataset. PCA in combination with clustering is an intuitive way for exploratory data analysis (EDA), e.g., visualization of sub-groups in a molecular dataset which otherwise are uninterpretable due to high dimensionality. Non-negative matrix factorization (NMF) is another FE method that achieves dimensionality reduction by finding two non-negative matrices whose product approximate the original non-negative matrix. Unlike PCA in which decomposition matrices have both positive and negative values, the resulting matrices from NMF only have positive values; thus, original data is represented only by additive combinations of latent variables. t-distributed stochastic neighbor embedding (t-SNE) [37] is an FE algorithm increasingly applied for the visualization of high-dimensional data. t-SNE is a nonlinear method and hence performs better when the relationships in the data are not linear. The similarity between data points are used to construct joint probability distributions in such a way that the divergence between joint probabilities in low-dimension embedding and original high dimensions is minimal. Autoencoder, a building block of many deep learning networks, can also be employed for nonlinear FE by restricting the number of hidden layer nodes to less than the number of original input nodes [38,39].

Feature extraction approaches are typically used in unsupervised integrative analysis, i.e., when response or group labels are unknown. ML-based FE can facilitate the discovery of disease specific sub-groups in multi-omics studies. In recent years, many feature extraction methods have been proposed for integrative omics exploratory analysis, with many of them based on PCA [40]. For example, multi-omics factor analysis (MOFA) was proposed recently as a generalization of PCA to multi-omics data to identify biomarkers in CLL [23]. Specifically, somatic mutations, DNA methylation and RNA expression were profiled together with ex vivo drug responses and MOFA disentangled sources of systematic variation (latent factors) arising from disease heterogeneity based on the multi-omics data. The latent factors identified by MOFA were shown to be predictive of clinical outcomes. Joint and individual variation explained (JIVE) [41], another extension of PCA, was proposed to identify individual and combined variations between miRNA and gene expression data for the same set of 234 Glioblastoma Multiforme (GBM) tumor samples. JIVE is an integrative EDA method that decomposes a dataset into a sum of three terms: two low-rank approximation terms, one for capturing joint structure across data types and other for capturing structure individual to each data, and a term for residual noise. In order to integrate protein and gene expression datasets from National Cancer Institute (NCI)-60 cell-lines, the multiple co-inertia analysis (MCIA) [42] employed FE methods like PCA on each data set separately to project them to similar (lower) dimensional space for EDA. In MCIA, the diverse sets of variables were transformed to the same scale to easily combine genes and proteins features, providing better biological pathway interpretation. Joint NMF [43] and intNMF [44] performed integrated data exploration with gene expression, DNA methylation and miRNA expression data to facilitate the identification of clinically distinct patient sub-groups by utilizing the NMF concept. In addition, integrative-NMF (iNMF) [45] was able to identify the heterogenous and homogenous factors across different types of data. Non-linear FE techniques including *t*-SNE and autoencoders also play key roles in multi-omics studies. For example, *t*-SNE was employed to facilitate the visualization and clustering in an integrated multi-omics study of transcriptional and epigenetic states in the human adult brain [46], and the integration of single-cell transcriptomic data across different conditions, technologies, and species [47]. In a precision oncology study of cancer cell lines involving gene expression, copy number, mutation status and drug sensitivity data, the dimensionality of the integrated data was effectively reduced by a deep autoencoder [48]. The autoencoder was able to extract cellular state features that were highly predictive of drug sensitivity. Moreover, representation learning [49] or the automatic extraction of meaningful representation of raw data (embeddings), which makes predictive models much more accurate, was also considered for integrated analyses [50,51]. For example, representation learning was employed to generate node embeddings that consequently produced informative edges in biological knowledge graphs [50]. Many life sciences databases make their data available as Linked Data, i.e., data having biological entities and their connections standardized with unique identifiers for better interoperability across resources. In [50], Linked Data, biomedical ontologies and ontology-based annotations were integrated, facilitating functional prediction and the predictions of protein–protein interaction (PPI), drug target relations, candidate genes of diseases, etc. In another study [51], a Multi-view Factorization Autoencoder was proposed for integrating multi-omics data with domain knowledge. This deep representation learning method effectively tackled the $p \gg n$ problem in datasets, and learned feature embedding and patients embedding simultaneously.

In biomedicine, ML-based feature selection methods are frequently applied to identify small subsets of key molecules or molecular signatures [33,52–55]. FS methods are classified into three main types:

(1)   Filter methods,
(2)   Wrapper methods,
(3)   Embedded methods.

Filter methods are used to select a subset of relevant features independent of any model. Many of the filter methods are univariate and provide statistical test scores for each feature-outcome combination. Examples in this category include ANOVA, Pearson's correlation, information gain (IG), etc. In addition, maximal-relevance and minimal-redundancy (mRMR), correlation-based FS (CFS) and ReliefF [56,57] are some advanced filter methods which consider feature combinations. For example, mRMR identifies features which are most relevant to the outcome but are not highly correlated among themselves [56]. Wrapper methods try to search for the best feature combination by training a particular predictive model repeatedly for various feature subsets and keep aside the best or worst performing subsets. Therefore, wrapper methods provide the best performing feature combination on that predictive model. Recursive feature elimination (RFE) [58], boruta [59], and jackstraw [60] are popular wrapper methods that repeatedly construct a model (e.g., random forest) and remove features with low weights. Whereas Boruta selects features with critically large variable importance measures in Random Forest, the jackstraw methods identify statistically significant features with respect to latent variables. Wrapper methods can be computationally expensive on a large dataset. Embedded methods are in between filter and wrapper methods in terms of computational complexity. These are the algorithms with built-in feature selection methods, i.e., they perform feature selection as a step toward predictive model building. Least absolute shrinkage and selection operator (LASSO) is a popular embedded FS method due to its simplicity. It is essentially a linear regression method with an $L1$-penalty (regularization) which shrinks many of the coefficients to zero. The features with non-zero coefficients in LASSO are considered relevant variables. However, when the features are correlated, LASSO tends to randomly pick only one feature. Various modifications are proposed to circumvent this problem, including stability selection [61–63] and elastic net [64]. Stability selection performs random subsampling and constructs many models on these bootstrap samples. Elastic net strikes the balance between $L1$ and $L2$-regularized regression penalty terms, with $L1$-penalty preferring a parsimonious model and $L2$-penalty retaining some correlated features such as co-expressed molecules.

Feature selection is generally employed in supervised ML-based integrative analysis (response or group labels are known) including classification and regression applications. In multi-omics studies, FS are commonly employed on each omics dataset prior to integration as datasets are high-dimensional and all the variables in individual datasets may not be informative [65–67]. This reduction in the number of variables as a pre-processing step attenuates noise prior to integration [67–70]. In [71], supervised feature selection for multi-omics data was proposed for Cox regression analysis that identified more true signature genes in cancer prognosis. In [70], an mRMR-based feature selection method was developed to identify epigenetic markers from cancer datasets using gene expression and methylation data. The markers identified through this approach were most relevant and least redundant in prostate carcinoma and leukemia datasets. mRMR was also employed to identify key features in predicting ovarian cancer grade or patient survival using concatenation of genomic, imaging, and proteomic data [72]. In [73], various FS methods including CFS, IG, ReliefF, fast clustering-based feature selection algorithm (FAST) and support vector machine based on RFE (RFE-SVM) were employed to identify features with the highest classification accuracy, in the identification of breast cancer sub-types using protein, gene expression and methylation data. Wrapper and embedded FS methods are multivariate, i.e., they can extract relationships among different features and hence particularly suited to multi-omics studies. RFE is one of the commonly used wrapper FS algorithms in biomedicine [52,53,58,74] and has been recently applied to integrative analysis [33]. In [69], mixOmics R package incorporated $L1$-penalized embedded FS into various supervised omics-integration methods to enable molecular signature extraction. In addition, $L1$-penality based regularization was implemented in unsupervised integrated clustering [28,75], as well as in the integrated predictive modelling framework to allow for genetic feature selection [76].

Figure 1 shows the taxonomy of ML-based approaches for dimensionality reduction.

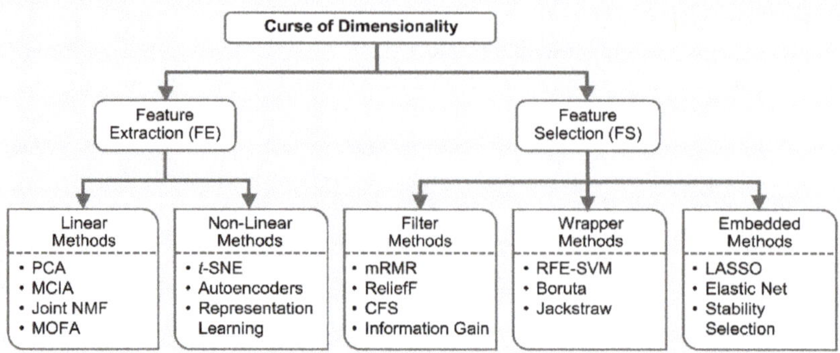

Figure 1. Machine learning (ML) with curse of dimensionality. ML-based dimensionality reduction (DR) approaches, for tackling the curse of dimensionality, can be classified into feature extraction (FE) and feature selection (FS). FE methods project data from a high-dimensional space to a lower dimensional space, while FS methods identify a small relevant subset of original features in order to reduce the dimensionality. Principal component analysis (PCA), multi-omics factor analysis (MOFA), multiple co-inertia analysis (MCIA), and joint non-negative matrix factorization (NMF) are some examples of FE methods applied in integrative analysis. These FE approaches assume linear relationships in the dataset. Nonlinear FE methods also exist including *t*-SNE, autoencoders, representation learning, etc. ML-based FS is broadly divided into filter, wrapper and embedded methods. Filter methods such as maximal-relevance and minimal-redundancy (mRMR), correlation-based FS (FCS), ReliefF and Information Gain are employed as a pre-processing step before training any model, while wrapper methods such as recursive feature elimination-support vector machine (RFE-SVM) and Boruta incorporate a predictive model to judge the importance of features. Embedded methods which include least absolute shrinkage and selection operator (LASSO), Elastic Net, stability selection, etc., perform feature selection as part of the model building process.

3. Heterogenous Data

One of the biggest challenges in multi-omics integrated analysis is the heterogeneity of data. Reasons for such heterogeneity include, but are not limited to, substantially different number of variables, mismatched distributions and scaling, diverse data modalities, i.e., continuous signals, discrete counts, intervals, ordered and unordered categorical, pathways, etc. For example, Glioblastoma Multiforme is a highly aggressive type of brain cancer whose prognostic prediction can be improved by considering multiple data types together [77], i.e., clinical data, gene expression, miRNA expression, DNA methylation, and copy number alterations (CNA). However, integration of these diverse data types in a single predictive model is challenging due to heterogeneities mentioned above. In the case of naive data integration, i.e., by concatenating features from different data sources, decision trees (DT) may work well with mixture of continuous and categorical variables. The decision rules in DT are well interpretable, unlike most nonlinear models which are generally considered black-box. In addition, DT has the inherent mechanism of ranking features based on their importance in decision making. However, decision trees are known to suffer from the overfitting problem; consequently, an ensemble of DTs or random forest (RF) [78] is preferred over DT.

Penalized linear models with $L1/L2$ regularization also minimize the risk of overfitting and perform feature selection. Therefore, they are also attractive for feature concatenation-based integrative analysis. For example, elastic net [64] was employed for multi-omics analysis in drug–response prediction from the collection cancer cell line encyclopedia (CCLE) [79] encompassing 36 tumor types with diverse variables including gene expression, copy number, mutation values, etc. All of these variables were assembled into a matrix and each feature went through z-score transformation across all cell lines. As discussed in the previous section, being a penalized linear regression model, elastic net can perform FS-based dimensionality reduction. However, the final list of key predictors obtained using this model (and tree-based approaches) can be dominated by the variables from a dataset with the largest number of variables. One way to overcome this problem is to perform block-scaling [80], i.e., scaling each variable by the inverse of the number of variables in the corresponding data block. Moreover, it was pointed out in [81] that the results obtained by elastic net with simultaneous analysis of various molecular data types in drug-response studies (containing both continuous and binary variables) are usually dominated by gene expression data (continuous variables). Consequently, the TANDEM method [81] employed a two-stage FS approach where the first stage uses all the binary variables, referred to as upstream data, and the second stage uses continuous gene expression variables or the downstream data. The model selected by TANDEM was more interpretable by preferentially focusing on upstream features while maintaining predictive power comparable to other integrative methods.

Simple feature concatenation-based integration is not feasible in many scenarios because different heterogeneities may be present in datasets and are not known *a priori*. Multiple kernel learning (MKL) [82] has become a popular approach to integrate data by calculating individual kernel matrices for each data type and fusing them into a global model. While kernel matrix encodes similarity between samples, different data sources may have different notions of similarity. Therefore, in MKL, data from each source has a separate kernel matrix. MKL [77] was successfully applied to GBM prognosis from different data types including, gene expression, CNA, DNA methylation, etc., employing the simpleMKL algorithm [83]. Similarly, Speicher et al. [84] integrated DNA methylation, gene and miRNA expression profiles using MKL, and later performed unsupervised clustering to discover cancer sub-types. Bayesian multitask MKL, the top performing algorithm, introduced as a result of a collaborative effort between the NCI and the dialogue on reverse engineering assessment and methods (DREAM) project [25], was applied to integrate data from different profiling sources including, CNA, DNA methylation, gene expression, reverse phase protein array (RPPA), etc., for predicting drug sensitivity in breast cancer cell lines. It employed a Gaussian kernel for real-valued data and the Jaccard similarity coefficient for categorical data. The Multitask MKL algorithm integrated different views from different data types by constructing a global similarity matrix as a weighted sum of the view-specific kernel matrices, where kernel weights reflect the relevance of each view.

Network-based approaches for integrative analysis can also leverage the concept of similarity fusion. Similarity network fusion (SNF) framework aggregated mRNA expression, DNA methylation and miRNA expression data for cancer patients, and used networks as a basis for integration [29]. SNF fused individual similarity networks obtained from different data sources to obtain single similarity network that captures complementary information. It employed scaled exponential similarity kernel in which Euclidean distance was used for continuous variables, chi-squared distance for discrete variables, and agreement-based measure for binary variables. Recently, GloNetDRP [85] was proposed, which built a heterogenous network using cell-line similarity networks from omics data of cell lines, and drug similarity network by exploiting chemical similarity between drugs. Probabilistic graphical models (PGMs) [86] are also a good candidate to integrate mixed data types [87]. For example, in a study of long-term body weight change in the general population [88], a multi-omics partial correlation network was constructed by first employing weighted correlation network analysis (WGCNA) [89] on metabolomics and transcriptomics data separately, and then integrating them using Gaussian graphical model (GGM) [90]. PAthway Recognition Algorithm using Data Integration on Genomic Models (PARADIGM) [91], a factor graph-based PGM approach that was proposed to integrate copy number and gene expression with curated pathway information from NCI, provides patient-specific inference of genetic pathway activities. PARADIGM inferred cellular activities helped classify patients into clinically relevant sub-groups. In [92], sparse graphical models were proposed for accurate group-wise expression quantitative trait loci (eQTL) mapping, by capturing the joint effect of a set of single-nucleotide polymorphisms (SNPs) on a set of genes. This approach used two types of hidden variables, one extracted set associations between SNPs and genes, and the other extracted confounders. Recently, a Network-based Integration of Multi-omics Data (NetICS) [93] method was proposed to prioritize cancer genes by integrating heterogenous multi-omics data into a directed functional interaction network. This interaction network expresses the directionality of the interactions, which is essential as it can explain how aberration events in one gene or miRNA can lead to expression changes of its interaction partners in the network. In addition, heterogenous information networks (HINs) [94,95] which capture multi-level interactions in heterogenous datasets can play important roles in integrative analysis of biomedical data. For example, HeteroMed [96], extracted latent low dimensional embeddings form EHR data (comprising raw text, numeric, categorical formats) for robust medical diagnosis. This method can potentially be extended to the integrative analysis of EHR with other data types.

Another prominent integrative analysis approach involves transforming data from heterogenous sources to latent sub-space, e.g., using PCA or NMF, then performing joint latent analysis or integrative clustering [44,45,97]. This approach allows joint modeling, with a combination of distributions, to include different variable types like continuous (Gaussian), binary (Bernoulli) and count (Poisson) [23]. An integrative clustering method iCluster [28], based on latent variable modelling, was proposed to identify clinically relevant disease sub-types in latent sub-space from two cancer datasets; breast cancer and lung cancer [28] as well from Glioblastoma dataset [75]. Instead of finding clusters of tumor sub-types for each dataset separately and later manually integrating the results, iCluster allowed automated integrated cluster assignment and performed dimensionality reduction simultaneously. This was achieved by leveraging the connection between PCA, latent variable modelling and LASSO-type penalty. Recently, iCluster was upgraded to iCluster+ to incorporate diverse data modalities including, binary, categorical and continuous values such that somatic mutation, CNA and gene expression were integrated and distinct tumor sub-groups were identified [75]. To achieve this iCluster+ assumed different distribution for different data types, e.g., Poisson, normal linear, logistic, multilogit, etc. Recently, the Scluster method had been shown to outperform iCluster and SNF methods in identifying cancer sub-types by jointly analyzing mRNA expression, miRNA expression, and DNA methylation data [97]. A latent factor-based clustering method referred to as mixed variable restricted Boltzmann machine (MV-RBM) [98] was proposed to aggregate data from highly heterogenous sources including demographics, diagnosis, pathologies and treatments in diabetes mellitus studies. With MV-RBM, the datasets were aggregated into latent profiles (homogenous representation), and these profiles facilitated the extraction of patient sub-groups by performing unsupervised affinity propagation (AP) clustering [99]. This approach has the potential to be extended to multi-omics integrative analysis.

Deep learning approaches have been getting attention from biomedical researchers to integrate heterogenous data. Specifically, in [100], omics data from multiple sources (gene expression, miRNA expression, and DNA methylation) were combined with clinical data to perform integrated clustering based on multimodal deep belief networks (DBN) [101]. Multimodal DBN is a network of stacked RBMs that seamlessly handles continuous and categorical data, and helps in discovering disease sub-types in cancer patients. In addition to integrative clustering, this method can identify signature genes and miRNAs that may play key roles in the pathogenesis of different cancer subtypes. In [32], a deep learning-based method was proposed to predict cancer prognosis using CNA, DNA methylation, gene expression, and somatic mutation data. This method is an extension of Clustering and PageRank (CPR) algorithm [102] to address the heterogeneity in multi-omics cancer datasets. In [103], three separate deep neural networks (DNN) were trained on gene expression, copy number and clinical data, respectively, for prognosis prediction of human breast cancer. Later, score level fusion was performed to get final multimodal deep network. Hepatocellular carcinoma (HCC) is the most prevalent type of liver cancer in the U.S. and to better understand HCC heterogeneity among patients using gene expression, miRNA expression, DNA methylation and clinical information, a deep learning framework was proposed [104]. This framework employed an autoencoder to perform nonlinear FE on the heterogenous data, which resulted in the aggregation of genes that share similar pathways. Autoencoder transformations led to the discovery of two liver cancer sub-types with significant differences in survival. Recently, a Deep Neural Network Synergy model with Autoencoders (AuDNNsynergy) was proposed that integrated multi-omics with chemical structure data to accurately predict drug combinations in cancer therapy [105]. This model utilized three autoencoders for gene expression, copy number and mutation data. A deep neural network combined the output of three autoencoders with physicochemical properties of drugs, predicting synergy value of given pair-wise drug combination against specific cancer cell lines. Figure 2 lists diverse ML-based approaches available for integrative analysis from heterogenous data.

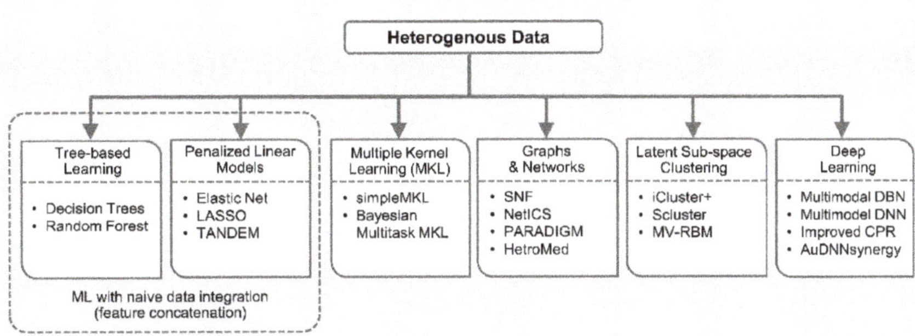

Figure 2. Machine learning with heterogenous data. ML algorithms can handle heterogenous data in different ways. For naive feature concatenation-based data integration, tree-based methods (e.g., decision trees and random forest), and penalized linear models (e.g., elastic net and LASSO) can be employed. A two-stage elastic net-based approach like TANDEM is useful if data sources with continuous features (e.g., gene expression) dominate the data sources with binary features (e.g., mutation). Multiple kernel learning (MKL), a robust integrative analysis approach with heterogenous data, employs different kernels or similarity functions for data from different sources and fuses them into a global matrix. Bayesian multitask MKL and simpleMKL are notable examples in this category. Network fusion methods such as similarity network fusion (SNF) employ similarity network for each data type and fuse heterogenous networks. PAthway Recognition Algorithm using Data Integration on Genomic Models (PARADIGM) can incorporate different heterogenous data including gene expression, copy number and curated pathways. Network-based Integration of Multi-omics Data (NetICS) integrates multi-omics data on a directed functional interaction network. Heterogenous information networks like HetroMed can handle raw text, numeric, and categorical data in electronic health records (EHRs) for medical diagnosis. Integrative methods including iCluster+, Scluster and mixed variable restricted Boltzmann machine (MV-RBM) first transform data from heterogenous sources into latent sub-space, and then perform clustering on the latent profiles. Deep learning models such as improved Clustering and PageRank (CPR), Deep Neural Network Synergy model with Autoencoders (AuDNNsynergy), multimodal deep belief networks (DBN) and deep neural networks (DNN) have been employed to perform integrative analysis of heterogenous data by learning complex features through data transformations at multiple layers.

4. Missing Data

Data acquired from high-throughput omics platforms are known to have missing observations due to various reasons, such as low coverage of next-generation sequencing, low sensitivity in protein and peptide detection, and faltered metabolite measurement by tandem mass spectrometry, etc. [106,107]. The problem of missing data is exacerbated in multi-omics studies as there can be more samples with missing values [108]. For example, a CLL study involving simultaneous analysis of DNA methylation, somatic mutation and gene expression measurements against drug response can have up to 40% of the biological samples with some but not all omics data, i.e., missing values in 40% of the samples [23]. Given that the biological samples are the same, it is statistically plausible to infer missing values in one omics from observed values and in other omics by exploiting any existing correlations found through complete cases. Complete case refers to the samples with measurements available on all variables under consideration [106,107,109,110]. Generally, most modern missing data methods focus on *item non-response* case, i.e., when data is missing on some variables for some biological samples [106,111,112]. Other cases include data missing on all variables for some biological samples, known as *unit non-response*, and data missing on a variable for all samples, known as *latent variable*. Missing data methods should be able to maximally utilize the available information, properly estimate the uncertainty in missing values and minimize bias [113].

Most statistical approaches rely on certain assumptions to tackle the missing data problem [111]. Suppose data is missing on variable Y while another variable X is always observed. The strongest assumption is that data is missing completely at random (MCAR), meaning that the probability of missingness on Y does not depend on X as well as on Y itself. For example, in a clinical study, it may be difficult to obtain a particular test result because the test itself is costly, hence it is only available for 30% of the samples. For the remaining 70%, the data is MCAR. Note that, if data is MCAR, the complete data subsample is just a random sample from the original target sample. The MCAR assumption is required by conventional methods, which is frequently violated in practical applications. However, most modern approaches work well with a weaker assumption of data missing at random (MAR). MAR assumes the probability of missingness on Y does not depend on Y, after controlling for the observed variable X, i.e., once dependence on X is adjusted, the probability of missingness on Y does not depend on Y itself. Again, consider the clinical study example in which cholesterol levels are missing for many subjects and the probability of missingness depends on subject's sex, i.e., females may be less likely to report cholesterol levels than males. However, within each gender type, subjects with higher cholesterol levels are neither more nor less likely to report than subjects with lower cholesterol levels. We can say that the cholesterol level variable has data missing at random because, after adjusting for subjects' gender, the missingness of the cholesterol level variable does not depend on whether the cholesterol level is high or not. MCAR is a special case of MAR, i.e., if data is missing completely at random then they are also missing at random. If the data is not missing at random (NMAR) then the missing data mechanism has to be modelled [113,114], i.e., simultaneous estimation of the scientific model and missing data mechanism is required.

The simplest approach to deal with missing data is a complete case analysis also known as listwise deletion. Listwise deletion means that the entire sample is excluded from analysis if data is missing on any variable for that sample. However, it may result in substantial information loss if the missing data percentage is high. In addition to complete case analysis, traditional single imputation methods are also very popular due to their ease of implementation. Any approach which estimates or guesses the missing values is called imputation. Missing values on a variable can be imputed by replacing it with a mean or median of the variable over all the available samples. Imputation based on regression or conditional mean imputation trains any type of regression model for the variable with missing data based on observed values. Subsequently, the model is used to generate predicted values for the cases with missing data. The k-nearest neighbors approach is also commonly employed for imputation of missing values.

In multi-omics studies, imputation based on k-nearest neighbors for profiles and genes expression [76], autocorrelation with cubic interpolation for spectral analysis of time series molecular data [115], fully conditional specification (FCS) for metabolite concentrations [88], etc., were employed for one or more data types separately, prior to integration [33]. In [107], stochastic gradient boosted trees (GBT) was employed to predict protein abundance for undetected proteins by exploiting the nonlinear correlations between available transcriptomics and proteomics data [107,116]. A multi-omics imputation method that considers correlations across microRNA, mRNA and DNA methylation data, and iteratively performs self-imputations (with features from same omics data) and cross-imputations (with features from different omics data) was implemented by employing an ensemble regression framework [110]. In general, it is recommended that any deterministic imputation should be done multiple times to account for the uncertainty in imputed values [113,117]. Consequently, various multiple imputation (MI) methods have been proposed [118–121]. In MI, instead of imputing single value for each missing data point, multiple values are imputed, resulting in multiple completed datasets rather than just one [122]. The observed values are the same in each dataset, but imputed values are slightly different. This difference is generally achieved by making random draws from error distribution of the regression model and adding those random draws to the values predicted by that regression model. Moreover, instead of explicitly assuming that regression parameters are true parameters and not estimates, these parameters can be randomly drawn from their posterior distribution for each dataset separately [113,118–120]. MI is an attractive approach for missing data because of its sound statistical properties and robustness established by extensive simulations.

Recently, a MI-based approach, referred to as MI for multiple factor analysis (MI-MFA), was proposed for multi-omics data integration [123]. MI-MFA used hot-deck imputation, which is a non-parametric method commonly used in big surveys due to its scalability to a large number of variables with missing values. To perform hot-deck imputation, the missing value on a variable is replaced with an observed value from a similar sample or donor. Some other popular iterative MI methods include Markov-chain Monte Carlo (MCMC) [118], fully conditional specification, also known as, sequential generalized regression or multivariate imputation by chained equation (MICE) [119] and AMELIA II [120]. MCMC is a general method used in Bayesian statistics for various applications. MCMC assumes a comprehensive joint distribution of all variables with missing data, generally applied under multivariate normal assumption. A key feature of MCMC is that imputed values are never used as the basis for predicting other missing values, i.e., imputations are only performed based on observed data. Given all assumptions are met and enough iterations are run, MCMC is guaranteed to converge to the correct posterior distribution for the imputed values. However, due to multivariate assumption and having one comprehensive model for all of the variables, MCMC may not be preferred for datasets with both quantitative and categorical variables. MICE, also an iterative algorithm like MCMC, is preferred in mixed-type datasets which builds a separate regression model for each variable depending on its type. MICE can also incorporate methods for imputing data that are not normally distributed [121]. Unlike MCMC, MICE does not have any theoretical proof of convergence and it can also be computationally much more expensive than MCMC. There is a risk of overfitting associated with any data imputation technique, but MI methods are generally less prone to this problem than single imputation methods [124]. However, most software packages available for MI methods assume data is MAR. When data is NMAR, extra care must be taken in date imputation to avoid overfitting and the introduction of bias in downstream analyses. Various plausible models should be tried, e.g., MI with pattern–mixture models [125]. This should be accompanied by sensitivity analysis to verify the consistency of the results across models.

In addition to MI methods for missing data, there are several ways to get maximum likelihood estimates with missing data based on multivariate assumption, including expectation–minimization (EM) and direct maximization of the likelihood or full information maximum likelihood [114,123]. Maximum likelihood is a general method commonly employed for parameter estimations in linear models. Compared to MI, maximum likelihood approaches generally have more rigorous mathematical proofs related to parameter estimation with missing data. Maximum likelihood chooses as parameter estimates those values which maximize the likelihood function, given the observation, i.e., maximize the probability of observing the data. The main disadvantage of maximum likelihood is that it is restricted to the type of model you want to estimate, e.g., linear or logistic regression. To obtain maximum likelihood with missing data, you need software that is specifically designed for the model you want to estimate, which is not always available, whereas MI methods are more general and can be employed in different types of analyses. There are many software packages that automatically generate multiple imputed datasets and combine results from multiple linear regression analyses in programming software R including, jomo, mice, Amelia II, etc. [120,126]. Notably, most missing data approaches exist only for linear analysis.

For nonlinear analysis with missing data, a two-stage MI and learning workflow based on Gaussian mixture model (GMM) and extreme learning machine (ELM) is available [127]. In order to include nonlinearity in MICE imputation, a random forest-based MICE algorithm was proposed for epidemiological study of angina patients [128]. This method can accommodate nonlinearities in the datasets and provide better parameter estimates, confidence intervals under MAR assumption. Deep learning techniques were recently applied to handle missing data in biomedical datasets [129–132]. The success of many of these data imputation methods can be contributed to autoencoder-based nonlinear FE. In [129], a multilayer autoencoder with dropout-based imputation on EHR datasets for amyotrophic lateral sclerosis (ALS) clinical trials was shown to outperform popular MI techniques including MICE. In addition, a denoising autoencoder (DAE)-based MI (MIDA) was also proposed very recently [130]. MIDA outperformed MICE algorithm on multiple datasets from various domains including bioinformatics.

AutoImpute [132], inspired by recommender systems (collaborative filtering) in information retrieval, is an autoencoder-based method for single cell RNA-seq (scRNA-seq) gene expression imputation. This method learns the distribution of scRNA-seq data and imputes the dropout (i.e., missing) gene expressions accordingly. In scRNA-seq analysis with missing data, matrix factorization-based imputation techniques are also popular in replacing the dropout with non-zeros values. For example, adaptively-thresholded low-rank approximation (ALRA) [133] computed a low-rank approximation of original matrix with missing data using singular value decomposition (SVD), followed by a thresholding to ensure that the biological zeros are preserved and technical zeros were imputed. SVD-based imputation techniques have traditionally been used in biomedical datasets due to their simplicity and superior performance rather than simple mean imputation [134]. Recently, the Sparse Recovery (SparRec) framework [135], also inspired by a low-rank matrix factorization model, was proposed for genetic data imputation for genome-wide association study (GWAS). It is a flexible imputation method that can be applied to large-scale meta-analysis, even without a reference panel. Sequencing To Imputation Through Constructing Haplotypes (STITCH) [136] is another notable imputation technique for quick and cost-effective genotyping from sequence data without reference panel. The imputation in STITCH is based on hidden Markov model (HMM) and EM algorithms.

In multi-omics and clinical big data analytics for precision medicine, missing data is a challenging problem [106,117] and conventional methods are prone to adding biases. Specialized integrative methods, such as ensemble regression imputation [110], can perform integrative imputation by combing the estimates from individual omics data itself as well as other omics. Similarly, MOFA [23] can leverage information from multiple omics layers to accurately impute missing data in integrative analysis. Specifically, it discovers latent factors by means of multi-omics FE and uses those factors to impute missing data. In addition, recently proposed Late Fusion Incomplete Multi-View Clustering (LF-IMVC) [137] is also attractive for multi-omics studies with missing data, where each data source with missing values can be treated as an incomplete view. LF-IMVC employs a kernel matrix for each view, and performs imputation and clustering simultaneously. To this end, modern statistical and machine learning methods such as MI, maximum likelihood, matrix factorization, autoencoders and integrative imputation methods can play key roles in facilitating integration of datasets with different missingness patterns. Figure 3 summarizes statistical and machine learning based solutions for handling missing data.

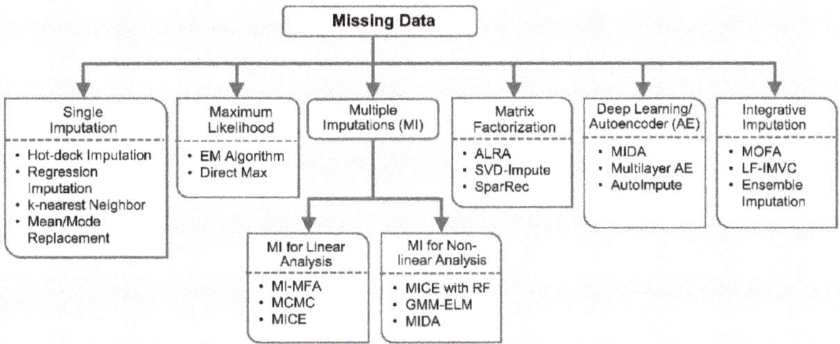

Figure 3. Machine learning with missing data. Conventional single imputation methods for handling missing data include replacement with mean or mode values, hot-deck imputation, regression imputation, *k*-nearest neighbor, etc. Maximum Likelihood approaches including those based on an expectation-minimization (EM) algorithm and Direct Maximization have attractive statistical properties compared to the conventional methods that often result in biased parameter estimates. Multiple imputation (MI) methods like Markov-chain Monte Carlo (MCMC) and multivariate imputation by chained equation (MICE) are also statistically robust, compared to conventional single imputation methods, as they take into account the uncertainty in the imputed values. MI for multiple factor analysis (MI-MFA) tackles the missing data problem in multi-omics analysis by performing MI based on hot-deck imputation. MI for nonlinear analysis can be performed using random forest (RF) and extreme learning machine (ELM). Adaptively-thresholded low-rank approximation (ALRA), singular value decomposition (SVD)-impute and SparRec methods employ matrix factorization for data imputation. In addition, imputation methods based on autoencoder and deep learning like denoising autoencoder-based MI (MIDA), AutoImpute and multilayer autoencoder (AE) have been proposed for high-dimensional datasets with missing data. Recently, integrative imputation methods such as ensemble regression imputation, multi-omics factor analysis (MOFA) and Late Fusion Incomplete Multi-View Clustering (LF-IMVC) are also available.

5. Rarity and Class Imbalance

In omics studies, ML-based models are often faced with the rarity in the target class or the class imbalance problem [12,138]. For example, a machine learning classifier trained to predict the location of enhancer in the genome suffers from the class imbalance problem, i.e., the dataset has many more negative samples (non-enhancer) compared to positive samples (enhancer) [12,139]. Similarly, ML-based contact map prediction in a protein structure dataset also suffers from the imbalance problem because of the sparseness of the contacts, i.e., of all possible amino acid pairs in a protein, only about 2% are in contact [140]. Prediction of post-translation modifications (PTM) sites in a protein sequence also encounters the same problem as occurrence of PTM is a sparse event [141], i.e., most of the amino acid residues are not modified. Other examples of the imbalanced problem in omics studies include prediction of protein-DNA binding residues from primary sequences [142], miRNAs identification [143], mutations incidence prediction [144], DNA methylation status/sites prediction [145,146], PPI sites prediction [147,148], identification of antimicrobial peptides (AMP) functional types [149], etc. In addition, the class imbalance problem in clinical datasets is prevalent due to the intrinsic imbalance in case-control pairing. Experimentally, it is often challenging and costly to generate data from a treatment group as compared to a control group [150,151]. Biomedical datasets belonging to the study of rare diseases or events are often severely imbalanced and most ML algorithms are not appropriate in such cases [152–154].

Despite the pervasiveness of imbalance in class distribution in real-world datasets, most ML classifiers including SVM, RF, and artificial neural networks (ANN) assume balance class distribution. This assumption means that the number of samples from each group or class is approximately the same (all categories are equally represented) [152,153]. Therefore, these classifiers overestimate the majority class and potentially ignore the minority class completely. Ironically, in most cases, minority class is the target class, e.g., a rare disease sub-type. A classifier trained on a rare disease dataset with 10,000 samples from the control group and 100 samples from the disease group can achieve 99% accuracy by predicting everything belonging to the majority class, without even detecting rare disease [155]. To tackle this problem, ML methods which are aware of the skewness in data or class imbalance learning (CIL) methods have been proposed. Broadly, CIL methods are divided into three categories; data sampling, algorithm modification and ensemble learning. Data sampling methods are frequently employed in biomedical domains because of its simplicity [145,147,149,156–158]. Data sampling approaches tackle class imbalance by balancing the dataset prior to applying the ML classifier. The majority class can be undersampled by removing some of the samples randomly, i.e., random undersampling (RUS) or informatively using one-sided selection [159]. New minority class samples can be synthetically created using the synthetic minority oversampling technique (SMOTE) [154]. Recently, a combination of undersampling and oversampling is becoming popular to tackle the imbalance problem more effectively, by overcoming the limitations associated with individual data sampling approach [145,151].

Algorithm modification approaches modify the machine learning algorithm, while still using the original imbalanced dataset. For example, cost-sensitive learning methods apply higher misclassification weight (cost) to minority class samples compared to majority class samples. Cost-sensitive weighting are frequently incorporated in SVM, ANN and boosting learning theory to tackle class imbalance [160–162]. Cost-sensitive learning approaches such as SVM_Weight [160] and WeightedELM (WELM) [163] are generally much more efficient than data sampling approaches, and hence attractive for big datasets [152]. However, they require theoretical understanding of the algorithm, as opposed to randomly undersampling the majority class [164]. Lastly, ensemble learning methods generally achieve better generalization performance than data sampling and cost-sensitive CIL methods [148,163,165,166]. In various clinical scenarios, it is a common practice to seek opinions of multiple doctors who are experts in the field. The final decision, for a particular treatment, is thus made by consulting a committee of experts and combining their opinions. In the context of ML, ensemble learning systems play a similar role [167,168]. The majority class is divided into several subsets (with or without replacement), each individual classifier in the ensemble is trained on all the minority class sample and a subset of majority class, and a final decision is based on aggregating the predictions from individual classifiers [139,142,148,163,167]. EasyEnsemble, Balanced Cascade, and ensemble WELM are some examples of ensemble methods for CIL [163,167]. It is important to mention that ensemble learning is a broad category of ML approaches that is not limited to class imbalance learning applications. For example, it has also been employed in integrated frameworks proposed for heterogenous and missing data [25,110].

Although many CIL methods exist for single omics studies, researchers have recently started developing imbalance-aware integrated omics analytical frameworks [69,169–172]. In [170], extensive simulations based on different integration algorithms and evaluation measures reveal that composite association network, relevance vector machine (RVM) and Ada-boost RVM were less influenced by class imbalance compared to other graph-based or kernel-based integration algorithms. A cross-organism PPI predive modelling was proposed based on tree-augmented naive Bayes (TAN) classifier (TAN relaxes the string independence assumption of NB) that integrated microarray expression and gene ontology (GO) values [173]. PPI data is highly imbalanced since the number of interacting proteins is much smaller than non-interacting protein pairs. Specifically, the imbalance ratio (IR) of non-interacting to interacting protein pairs was around 20. Dividing the imbalance dataset into 20 balanced datasets with the same positive samples produced better results as compared to imbalanced datasets. In [73], equal-class data sampling was performed to reduce the effects of class imbalance in identifying breast cancer sub-types through the integration of protein, methylation and gene expression data.

A PPI prediction method based on RF was proposed which not only considered affinity purification and mass spectrometry (APMS) data, but also various other indirect features including mRNA co-expression, gene ontologies and homologous protein [174]. This method, referred to as Spotlite, avoided the extreme imbalance in data, first by uniformly sampling the unknown interactions so that the IR is 10. Then, during the training of RF classifier, weights of 10 and 1 were assigned to known and unknown interaction classes, respectively. For automatic function prediction (APF), a cost-sensitive network integration approach unbalance-aware network integration and prediction of protein functions (UNIPred) [175] was proposed to integrate biological networks from different data sources. UNIPred addressed the imbalance between annotated and un-annotated proteins by building a consensus network from multiple protein networks derived from different omics data. MNet [176] builds a composite network by integrating multiple functional networks constructed from different proteomic sources to get a comprehensive view of proteins and predict their functions. The protein function prediction is an imbalanced classification problem and MNet addressed this problem by employing weighted functional labels (label represents distinct protein function), putting more emphasis on the labels that have fewer member proteins. A cost-sensitive SVM approach was proposed for diagnosing pancreatic cancer by integrating miRNA and mRNA expression data [177]. The dataset was imbalanced as there were 104 pancreatic ductal adenocarcinoma (PDAC) tissues and 17 benign pancreatic tissues. Therefore, class specific weights in SVM for cancer and normal samples were set to 1 and 6.117647 (104/17), respectively. Using their integrated approach, they were able to identify 705 multi-markers for 27 miRNAs and 289 genes as promising potential biomarkers for pancreatic cancer. The generalized simultaneous component analysis (GSCA) model, with GDP penalty, was proposed recently for the integrative analysis of gene expression and CNA [178]. This method was found to be more robust against class imbalance problem in CNA compared to iCluster+ method. In [179], authors showed that a simple ensemble learning method can work as well as state-of-the-art data integration methods such as kernel fusion. The ensemble comprised learners which were trained on different views of data and the predictions were combined using weighted majority voting (WMV). The weight was determined using F-score that considered the imbalance between gene classes.

Apart from data sampling, algorithm modification and ensemble learning based methods, some integration frameworks which perform model tuning based on CIL-specific evaluation measures were proposed recently. Traditional evaluation measures like overall accuracy are not appropriate for CIL [172]. The accuracy of the majority class (specificity) and the accuracy of the minority class (sensitivity) should be measured in a balanced way. Therefore, geometric mean (Gmean) of sensitivity and specificity is a commonly used evaluation measure for CIL [153]. Similarly, area under precision-recall curve (auPRC) provides more unbiased evaluation compared to the area under receiver operating characteristic (auROC). Matthews correlation coefficient (MCC) and F-scores also take into account imbalance in class sizes. F-score, which incorporates precision and recall, is a popular evaluation metric in information retrieval community [170]. MCC [14,145] considers true positives, true negatives, false positives and false negatives in its formula. It can have a value between -1 and 1; 1 means perfect prediction, 0 means random prediction and -1 means total disagreement. Balanced error rate (BER) calculates the average proportion of incorrectly classified samples in each class, weighted by the number of samples in each class. To address the imbalance problem in multi-omics predictive modelling, BER was incorporated as an evaluation measure for parameter tuning, through cross-validation, in data integration analysis for biomarker discovery using latent components (Diablo) [69,171]. Diablo is a multi-omics integrative framework which can identify biomarker panels that discriminate between different disease phenotypes. It transforms each omics dataset into latent components, and maximizes the correlations between these components and phenotype of interest. A novel neural network architecture incorporating cross-correlation between different modalities (e.g., gene expression and DNA methylation) was proposed in [172] to classify breast cancer patients. This method, referred to as super-layered neural network architecture (SNN), utilized MCC and F-scores to account for the imbalance in class sizes. In general, most methods and evaluation measures for CIL are proposed for binary class problems, i.e., there are only two categories in the dataset. However, multi-omics data analysis and hypothesis generation may involve more than two classes [66], with a varying degree of imbalance among them [172]. For example, instead of normal vs. disease samples, there can be different types or levels of diseases [72,157,163,172,180]. In recent years, researchers have started focusing on multi-class imbalance problems [152,163,181]. Fuzzy pattern random forest (FPRF) [181] employed multi-class version of F-score and Gmean for robust feature selection in the integrative analysis of an imbalanced Leukemia dataset.

Due to the inherent sparsity in various omics phenomena, rare events in diseases of interest and case-control imbalance in clinical studies, it is anticipated that integrated omics studies will present new challenges in predictive modelling and provide opportunities for researchers to propose specialized CIL algorithms. For example, beyond simple data sampling approaches, biomedical researchers can explore ensemble and algorithmic modification methods that generally have better theoretical foundations, natural scalability to multi-class classification, and lower risks of overfitting and information loss than data sampling approaches. Figure 4 shows categorization of class imbalance machine learning methods.

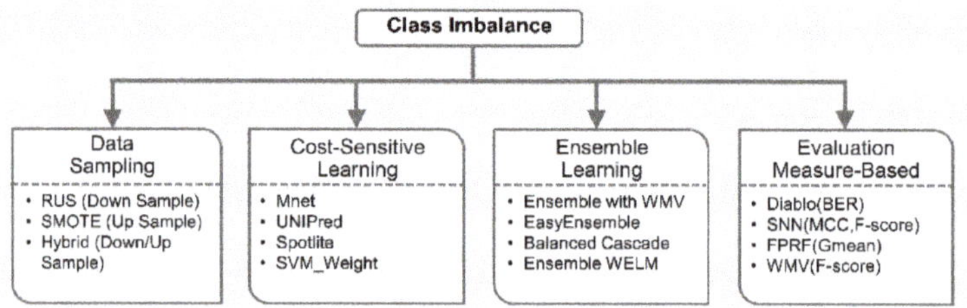

Figure 4. Machine learning with class imbalance. Class imbalance learning (CIL) methods are broadly classified into three types: data sampling, cost-sensitive learning and ensemble methods. Data sampling approaches balance the class distribution by either undersampling the majority class (e.g., random under sampling (RUS)), oversampling the minority class (e.g., synthetic minority oversampling technique (SMOTE)), or a combination of both (hybrid). Algorithm modification methods modify the learning algorithm generally by cost-sensitive weighting (e.g., Mnet, unbalance-aware network integration and prediction of protein functions (UNIPred), Spotlite and support vector machine (SVM)_weight). Cost-sensitive learning assigns a higher misclassification cost to minority class samples compared to majority class samples. Ensemble learning approaches like ensemble with weighted majority voting, EasyEnsemble, Balanced Cascade, and ensemble weighted extreme learning machine (WELM) train multiple classifiers, and aggregate their results to get the final output. Many existing integrative methods tackle imbalance by tuning models based on imbalance-aware evaluation measures. For example, data integration analysis for biomarker discovery using latent components (Diablo), super-layered neural network architecture (SNN), fuzzy pattern random forest (FPRF), and weighted majority voting (WMV) employ one or more CIL-specific evaluation measures like F-score, balanced error rate (BER), geometric mean (Gmean), Matthews correlation coefficient (MCC), area under precision-recall curve (auPRC), etc., instead of classification accuracy, to account for the bias introduced by imbalance in the dataset.

6. Big Data Scalability

Machine learning algorithms build data driven models whose performance generally gets better with the availability of more data. However, machine learning from big data acquired via multiple high-throughput omics platforms may raise scalability challenges. Implementation of multi-omics analytical workflows based on ML methods is increasingly becoming infeasible on a single computer. However, with the advancement in optimization algorithms for big data, online ML, parallelization of ML algorithms, and cloud computing, large-scale analysis can be performed efficiently on high-dimensional omics datasets. For example, a feed-forward neural network with multiple hidden layers can now be trained to accurately differentiate non-coding RNA types, i.e., circular RNAs (cirRNAs) from long non-coding RNAs (lncRNAs) in just a few hours on a single computer while the MKL method would take four days [182]. This is possible due to the development of computationally efficient training algorithms for neural networks [183–185]. Biomedical researchers can achieve large-scale machine learning by leveraging the computational approaches discussed below, as shown in Figure 5.

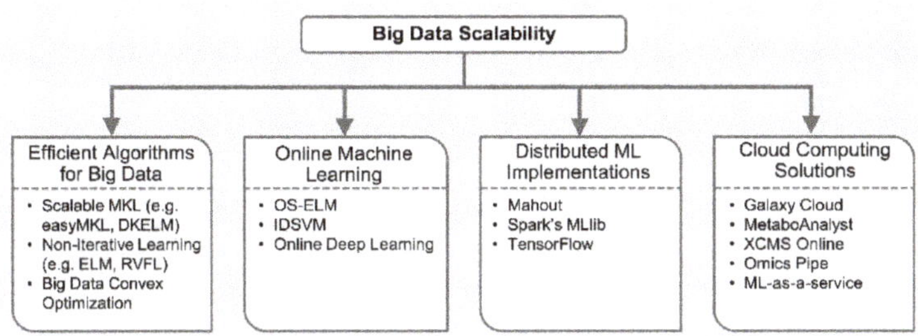

Figure 5. Large-scale machine learning. ML-based integrative analysis can be performed at large-scale by utilizing computationally efficient algorithms proposed for big data, online training algorithms, distributed data processing and computing frameworks, or cloud computing-based solutions. Efficient computational approaches tailored for big data include non-iterative neural networks (e.g., extreme learning machine (ELM) and random vector functional link (RVFL)), scalable multiple kernel learning (MKL) methods (e.g., easyMKL and dual-layer kernel ELM (DKELM)), convex optimization for big data, etc. Online machine learning algorithms including online sequential extreme learning machine (OS-ELM), incremental decremental support vector machine (IDSVM), and online deep learning are attractive for big data applications as they incrementally update the model with small chunks of data, instead of loading entire data in memory and learning all at once. In addition, ML algorithms can now be massively parallelized over a cluster of CPUs or graphics processing units (GPUs) using Spark's MLlib, Apache Mahout, and Google's TensorFlow programming frameworks. Cloud computing-based bioinformatics platforms including Galaxy Cloud, MetaboAnalyst, XCMS online, and Omics pipe are useful resources for multi-omics exploratory data analysis (EDA) and ML. Moreover, machine learning-as-a-service is being offered by leading commercial cloud service providers like Amazon, Google, Microsoft and IBM, which can be utilized for implementing ML-based analytical pipelines in large-scale multi-omics studies.

Various ML methods including ANN, SVM and DT estimate model parameters through iterative procedures; thus, they may not be easily scalable to big data applications. In recent years, there have been many efforts to optimize algorithms for training ML models efficiently on large datasets [183,184,186,187]. For example, non-iterative training algorithms are becoming popular for big data applications [187]. ANN can be trained in a single step without iterative tuning of hidden node parameters, as opposed to a back-propagation (BP) algorithm which is time-consuming, converges slowly, and can be stuck at local minima [183]. Non-iterative solutions for ANN include extreme learning machine (ELM) [188], random vector functional link (RVFL) [189,190], liquid state machine [191], echo state network [192], etc. In most of these methods, weights connecting input layer to hidden layer are randomly assigned, and output weights connecting hidden layer to output layer are determined analytically. Therefore, computational complexity of non-iterative methods is much lower than traditional BP methods for ANN. Furthermore, a highly parallel implementation of ELM for big data has been proposed by employing large-scale optimization [186]. Specifically, convex optimization, a key competent in training many ML and statistical models, is being reinvented for scalability and parallelism in the wake of big data [193]. Recently, methods based on ELM theory have been employed in single omics studies [163,182,194–197] and may be extended to multi-omics for efficient integrative analyses. Moreover, scalable MKL methods like dual-layer kernel extreme learning machine (DKELM) [198] and easyMKL [199] can be employed in multi-omics integrative analysis since MKL, a popular approach for integrating multiple omics datasets, can be computationally very expensive for large datasets.

Online algorithms are also useful in big data applications, especially when it is computationally infeasible to train models on the entire dataset all at once [200]. They are extremely popular in data stream analytics where the training samples arrive over time, e.g., in online prediction of glucose concentration in Type I diabetes [201]. Instead of retraining the model with the entire dataset every time new samples are received, online learning methods incrementally update the earlier learnt model only with the new samples. Previously learnt samples need not be stored in memory. On the other hand, batch ML algorithms would perform intensive training iterations over the entire dataset every time new samples arrive. In addition, batch learning requires complete datasets to be available in the memory prior to training, which may not be feasible in large-scale applications. Recursive least squares, a sequential (online) implementation of least squares method, is the building block of many online learning algorithms. For example, online sequential extreme learning machine (OS-ELM) [202] is a family of algorithms based on recursive least squares formulation for online training of single hidden layer feedforward networks (SLFNs). OS-ELM based algorithms can learn data one sample at a time or as chunks of samples, and have been employed for nonlinear classification and regression applications. Stochastic gradient decent (SGD), a variant of BP algorithm, is also a popular online optimization algorithm for training ML models [203]. SVM-based online learning algorithms such as incremental decremental SVM (IDSVM) and cost-sensitive learning-based online SVM [204,205] were proposed to address scalability issues in big data applications. Recently, multi-layer or deep online learning methods were proposed for better representation learning with high-dimensional datasets. These deep learning approaches are memory efficient as entire datasets need not be stored in memory, making them attractive for large-scale multi-omics analysis [206,207]. Online learning algorithms are now available for common ML tasks such as classification, regression, feature extraction, clustering, deep learning, etc.

Institutions can also leverage distributed implementations of ML algorithms, on a cluster of computers, when standalone commodity PCs lack the computational power required to learn from big data. For example, the MapReduce [208,209] programming framework provides a distributed platform to process big data in a fault tolerant way and can facilitate the scalability of ML algorithms on large biomedical datasets. Simply put, distributed frameworks like MapReduce and its open-source implementation Hadoop [210] divide the training data into many subsets such that each subset is processed by a single machine or slave. Slave machines perform operations in parallel and results are combined by a centralized master server. MapReduce is a good candidate for scaling those learning algorithms which can be expressed as computing sums of function of training data. Recently, a clustering algorithm KAymeans for MIxed LArge data (KAMILA) [211] was implemented on very large dataset using Hadoop [212]. KAMILA can be useful in multi-omics analysis since it was proposed for mixed-type data (combination of continuous and categorical data) clustering. From the original MapReduce framework, various computational platforms have arisen which are suitable for large-scale ML, such as Apache Spark [213]. These cluster computing platforms efficiently perform multiple iterations of matrix inversions and multiplications which are associated with many ML algorithms. Spark's MLlib [214] is a suite of scalable algorithms, providing distributed implementations of popular ML methods including regression models, PCA, *k*-means clustering, DT, Naïve Bayes, SVM, etc. Another open-source project that allows distributed implementation of ML algorithms for big data is Apache Mahout [215]. Mahout was successfully employed for scalable feature selection, data sampling and classification in protein structure prediction problems [140]. In addition, Google's TensorFlow programming model [216] allows parallelism of deep leaning approaches [31] such as convolutional neural networks (CNN) and long short-term memory (LSTM) algorithms, by distributed implementation on many CPUs or graphics processing units (GPUs) for large-scale analysis.

If memory and computational resources required for integrative analysis is beyond what is available in the cluster of a research lab or institution, cloud computing is an attractive option. Galaxy Cloud [217,218] allows users to run a private Galaxy installation on Amazon Web Services (AWS) elastic compute cloud (EC2) with the same functionalities as the main site using a virtual machine model. Omics pipe [219], an open source Python framework for automating multi-omics data analysis, is also available as Amazon virtual machine. XCMS online [220] is a cloud-based metabolomics data processing platform for predictive pathway analysis and enables multi-omics data analysis by integrating gene and protein data with metabolic pathways. MetaboAnalyst [221] is another cloud-based platform for integrative metabolomics analysis. It incorporates modules for multi-omics data integration through knowledge-based network analysis and various ML-based clustering, feature selection and classification algorithms. In addition to cloud-based bioinformatics platforms, machine learning-as-a-service is being offered by leading commercial cloud service providers like Amazon, Google, Microsoft and IBM. ML-as-a-service makes implementation of complex ML algorithms on large-scale datasets convenient for biomedical researchers [222]. It is apparent that the future of multi-omics integrative analysis is reliant on ML algorithms, and cloud-based solutions provide feasible options to implement them at large-scale.

7. Conclusions and Future Perspectives

High-throughput omics technologies are generating large volumes of multi-omics data at an unprecedented rate. Simultaneous analysis of data obtained from different platforms, for the same biological specimen, captures a holistic view of the complex biological interactions. For single-omics studies, traditional machine learning (ML) algorithms have been very successful in automatically identifying complex patterns from big data. However, multi-omics integrative analysis poses new computational challenges and amplifies the ones associated with single-omics studies. In this paper, we focused on five computational problems frequently encountered in integrative multi-omics data analysis, including the curse of dimensionality, data heterogeneity, missing data, rarity and class imbalance, and scalability issues. We reviewed some novel ML-based approaches recently applied to integrative analysis of multi-omics datasets, under each of the five problem categories. Furthermore, we also discussed state-of-the-art computational methods which have the potential to address these problems in multi-omics analysis. This article will help bioinformatics researchers in exploring modern computational approaches to tackle evolving challenges in integrative analysis. It also bridges the gap between problems in multi-omics integrative analysis, and novel machine learning approaches from the computer science community as potential solutions to these problems. Although this article addressed some key issues in integrative data analysis, there are other challenges that require attention in future studies. For example, specialized ML-based approaches need to be developed for multi-omics analysis in personalized medicine where cohort size can be very small (e.g., 100 patients or less) [223]. Moreover, additional machine learning frameworks which leverage prior knowledge of biological networks to integrate omics datasets should be proposed as they are vital for robust biomarker modelling [224–226]. In the integrative analysis of omics data and electronic health records (EHR) [227], or observational data and biomedical literature, sophisticated text mining and natural language processing approaches may play key roles to simultaneously handle structured and unstructured data [228–230]. However, the privacy and security of patient data should be ensured when developing ML approaches with EHRs and multi-omics. Integrative studies must comply with standards like Health Insurance Portability and Accountability Act (HIPPA) and any prediction or outcome from ML analysis must not compromise patient confidentiality. Collaborative studies can greatly benefit from privacy-preserving machine learning frameworks as institutions can jointly train accurate ML models without sharing sensitive patient data [231,232]. Finally, there is a need to benchmark ML methods for multi-omics analysis as numerous methods are available to solve the same problem. Although there are ongoing efforts to benchmark machine learning algorithms [233], benchmarking specific to multi-omics is required.

Author Contributions: Conceptualization, B.M., W.W., P.P.; Original draft preparation, B.M., W.W., J.W., H.C., N.C.C., P.P.; Review and editing, B.M., W.W., N.C.C., P.P.; Supervision, W.W., P.P.

Funding: This project is supported by the US National Institutes of Health funding through R35-HL135772 and U54-GM114833 (Peipei Ping).

Conflicts of Interest: The authors declare no conflict of interest.

References

1. Strobel, E.J.; Angela, M.Y.; Lucks, J.B. High-throughput determination of RNA structures. *Nat. Rev. Genet.* **2018**, *19*, 615–634. [CrossRef] [PubMed]

2. Hwang, B.; Lee, J.H.; Bang, D. Single-cell RNA sequencing technologies and bioinformatics pipelines. *Exp. Mol. Med.* **2018**, *50*, 96. [CrossRef] [PubMed]

3. Sedlazeck, F.J.; Lee, H.; Darby, C.A.; Schatz, M.C. Piercing the dark matter: Bioinformatics of long-range sequencing and mapping. *Nat. Rev. Genet.* **2018**, *19*, 329–346. [CrossRef] [PubMed]

4. Aebersold, R.; Mann, M. Mass spectrometry-based proteomics. *Nature* **2003**, *422*, 198. [CrossRef] [PubMed]

5. Dettmer, K.; Aronov, P.A.; Hammock, B.D. Mass spectrometry-based metabolomics. *Mass Spectrom. Rev.* **2007**, *26*, 51–78. [CrossRef] [PubMed]

6. Friedman, J.; Hastie, T.; Tibshirani, R. *The Elements of Statistical Learning*; Springer: New York, NY, USA, 2001.

7. Domingos, P. A few useful things to know about machine learning. *Commun. ACM* **2012**, *55*, 78–87. [CrossRef]

8. Cortes, C.; Vapnik, V. Support-vector networks. *Mach. Learn.* **1995**, *20*, 273–297. [CrossRef]

9. Rumelhart, D.E.; Hinton, G.E.; Williams, R.J. Learning representations by back-propagating errors. *Nature* **1986**, *323*, 533. [CrossRef]

10. Breiman, L. Statistical modeling: The two cultures (with comments and a rejoinder by the author). *Stat. Sci.* **2001**, *16*, 199–231. [CrossRef]

11. Obermeyer, Z.; Emanuel, E.J. Predicting the future—Big data, machine learning, and clinical medicine. *N. Engl. J. Med.* **2016**, *375*, 1216. [CrossRef]

12. Libbrecht, M.W.; Noble, W.S. Machine learning applications in genetics and genomics. *Nat. Rev. Genet.* **2015**, *16*, 321. [CrossRef] [PubMed]

13. Rohrback, S.; April, C.; Kaper, F.; Rivera, R.R.; Liu, C.S.; Siddoway, B.; Chun, J. Submegabase copy number variations arise during cerebral cortical neurogenesis as revealed by single-cell whole-genome sequencing. *Proc. Natl. Acad. Sci. USA* **2018**, *115*, 10804–10809. [CrossRef] [PubMed]

14. Wang, D.; Li, J.-R.; Zhang, Y.-H.; Chen, L.; Huang, T.; Cai, Y.-D. Identification of Differentially Expressed Genes between Original Breast Cancer and Xenograft Using Machine Learning Algorithms. *Genes* **2018**, *9*, 155. [CrossRef] [PubMed]

15. Kerepesi, C.; Daróczy, B.; Sturm, Á.; Vellai, T.; Benczúr, A. Prediction and characterization of human ageing-related proteins by using machine learning. *Sci. Rep.* **2018**, *8*, 4094. [CrossRef] [PubMed]

16. Bourdon, A.K.; Spano, G.M.; Marshall, W.; Bellesi, M.; Tononi, G.; Serra, P.A.; Baghdoyan, H.A.; Lydic, R.; Campagna, S.R.; Cirelli, C. Metabolomic analysis of mouse prefrontal cortex reveals upregulated analytes during wakefulness compared to sleep. *Sci. Rep.* **2018**, *8*, 11225. [CrossRef] [PubMed]

17. Zheng, P.-Z.; Wang, K.-K.; Zhang, Q.-Y.; Huang, Q.-H.; Du, Y.-Z.; Zhang, Q.-H.; Xiao, D.-K.; Shen, S.-H.; Imbeaud, S.; Eveno, E. Systems analysis of transcriptome and proteome in retinoic acid/arsenic trioxide-induced cell differentiation/apoptosis of promyelocytic leukemia. *Proc. Natl. Acad. Sci. USA* **2005**, *102*, 7653–7658. [CrossRef] [PubMed]

18. Azimzadeh, O.; Sievert, W.; Sarioglu, H.; Merl-Pham, J.; Yentrapalli, R.; Bakshi, M.V.; Janik, D.; Ueffing, M.; Atkinson, M.J.; Multhoff, G. Integrative proteomics and targeted transcriptomics analyses in cardiac endothelial cells unravel mechanisms of long-term radiation-induced vascular dysfunction. *J. Proteome Res.* **2015**, *14*, 1203–1219. [CrossRef] [PubMed]

19. Gerling, I.C.; Singh, S.; Lenchik, N.I.; Marshall, D.R.; Wu, J. New data analysis and mining approaches identify unique proteome and transcriptome markers of susceptibility to autoimmune diabetes. *Mol. Cell. Proteom.* **2006**, *5*, 293–305. [CrossRef]

20. Ryan, C.J.; Cimermančič, P.; Szpiech, Z.A.; Sali, A.; Hernandez, R.D.; Krogan, N.J. High-resolution network biology: Connecting sequence with function. *Nat. Rev. Genet.* **2013**, *14*, 865. [CrossRef] [PubMed]

21. Hoadley, K.A.; Yau, C.; Wolf, D.M.; Cherniack, A.D.; Tamborero, D.; Ng, S.; Leiserson, M.D.; Niu, B.; McLellan, M.D.; Uzunangelov, V. Multiplatform analysis of 12 cancer types reveals molecular classification within and across tissues of origin. *Cell* **2014**, *158*, 929–944. [CrossRef] [PubMed]

22. De Cecco, L.; Giannoccaro, M.; Marchesi, E.; Bossi, P.; Favales, F.; Locati, L.D.; Licitra, L.; Pilotti, S.; Canevari, S. Integrative miRNA-gene expression analysis enables refinement of associated biology and prediction of response to cetuximab in head and neck squamous cell cancer. *Genes* **2017**, *8*, 35. [CrossRef] [PubMed]

23. Argelaguet, R.; Velten, B.; Arnol, D.; Dietrich, S.; Zenz, T.; Marioni, J.C.; Buettner, F.; Huber, W.; Stegle, O. Multi-Omics Factor Analysis—A framework for unsupervised integration of multi-omics data sets. *Mol. Syst. Biol.* **2018**, *14*, e8124. [CrossRef] [PubMed]

24. Oberbach, A.; Blüher, M.; Wirth, H.; Till, H.; Kovacs, P.; Kullnick, Y.; Schlichting, N.; Tomm, J.M.; Rolle-Kampczyk, U.; Murugaiyan, J. Combined proteomic and metabolomic profiling of serum reveals association of the complement system with obesity and identifies novel markers of body fat mass changes. *J. Proteome Res.* **2011**, *10*, 4769–4788. [CrossRef] [PubMed]

25. Costello, J.C.; Heiser, L.M.; Georgii, E.; Gönen, M.; Menden, M.P.; Wang, N.J.; Bansal, M.; Hintsanen, P.; Khan, S.A.; Mpindi, J.-P. A community effort to assess and improve drug sensitivity prediction algorithms. *Nat. Biotechnol.* **2014**, *32*, 1202. [CrossRef] [PubMed]

26. Joyce, A.R.; Palsson, B.Ø. The model organism as a system: Integrating 'omics' data sets. *Nat. Rev. Mol. Cell Biol.* **2006**, *7*, 198. [CrossRef] [PubMed]

27. Cavill, R.; Jennen, D.; Kleinjans, J.; Briedé, J.J. Transcriptomic and metabolomic data integration. *Brief Bioinform.* **2015**, *17*, 891–901. [CrossRef] [PubMed]

28. Shen, R.; Olshen, A.B.; Ladanyi, M. Integrative clustering of multiple genomic data types using a joint latent variable model with application to breast and lung cancer subtype analysis. *Bioinformatics* **2009**, *25*, 2906–2912. [CrossRef]

29. Wang, B.; Mezlini, A.M.; Demir, F.; Fiume, M.; Tu, Z.; Brudno, M.; Haibe-Kains, B.; Goldenberg, A. Similarity network fusion for aggregating data types on a genomic scale. *Nat. Methods* **2014**, *11*, 333–337. [CrossRef]

30. LeCun, Y.; Bengio, Y.; Hinton, G. Deep learning. *Nature* **2015**, *521*, 436. [CrossRef]

31. Min, S.; Lee, B.; Yoon, S. Deep learning in bioinformatics. *Brief. Bioinform.* **2017**, *18*, 851–869. [CrossRef]

32. Kim, M.; Oh, I.; Ahn, J. An Improved Method for Prediction of Cancer Prognosis by Network Learning. *Genes* **2018**, *9*, 478. [CrossRef] [PubMed]

33. De Meulder, B.; Lefaudeux, D.; Bansal, A.T.; Mazein, A.; Chaiboonchoe, A.; Ahmed, H.; Balaur, I.; Saqi, M.; Pellet, J.; Ballereau, S. A computational framework for complex disease stratification from multiple large-scale datasets. *BMC Syst. Biol.* **2018**, *12*, 60. [CrossRef] [PubMed]

34. Wang, L.; Wang, Y.; Chang, Q. Feature selection methods for big data bioinformatics: A survey from the search perspective. *Methods* **2016**, *111*, 21–31. [CrossRef] [PubMed]

35. Hira, Z.M.; Gillies, D.F. A review of feature selection and feature extraction methods applied on microarray data. *Adv. Bioinform.* **2015**, *2015*. [CrossRef] [PubMed]

36. Guyon, I.; Elisseeff, A. An introduction to variable and feature selection. *J. Mach. Learn. Res.* **2003**, *3*, 1157–1182.

37. Van der Maaten, L.; Hinton, G. Visualizing data using t-SNE. *J. Mach. Learn. Res.* **2008**, *9*, 2579–2605.

38. Hinton, G.E.; Salakhutdinov, R.R. Reducing the dimensionality of data with neural networks. *Science* **2006**, *313*, 504–507. [CrossRef]

39. Wang, Y.; Yao, H.; Zhao, S. Auto-encoder based dimensionality reduction. *Neurocomputing* **2016**, *184*, 232–242. [CrossRef]

40. Meng, C.; Zeleznik, O.A.; Thallinger, G.G.; Kuster, B.; Gholami, A.M.; Culhane, A.C. Dimension reduction techniques for the integrative analysis of multi-omics data. *Brief. Bioinform.* **2016**, *17*, 628–641. [CrossRef]

41. Lock, E.F.; Hoadley, K.A.; Marron, J.S.; Nobel, A.B. Joint and individual variation explained (JIVE) for integrated analysis of multiple data types. *Ann. Appl. Stat.* **2013**, *7*, 523. [CrossRef]

42. Meng, C.; Kuster, B.; Culhane, A.C.; Gholami, A.M. A multivariate approach to the integration of multi-omics datasets. *BMC Bioinform.* **2014**, *15*, 162. [CrossRef] [PubMed]

43. Zhang, S.; Liu, C.-C.; Li, W.; Shen, H.; Laird, P.W.; Zhou, X.J. Discovery of multi-dimensional modules by integrative analysis of cancer genomic data. *Nucleic Acids Res.* **2012**, *40*, 9379–9391. [CrossRef] [PubMed]

44. Chalise, P.; Fridley, B.L. Integrative clustering of multi-level 'omic data based on non-negative matrix factorization algorithm. *PLoS ONE* **2017**, *12*, e0176278. [CrossRef] [PubMed]

45. Yang, Z.; Michailidis, G. A non-negative matrix factorization method for detecting modules in heterogeneous omics multi-modal data. *Bioinformatics* **2015**, *32*, 1–8. [CrossRef] [PubMed]

46. Lake, B.B.; Chen, S.; Sos, B.C.; Fan, J.; Kaeser, G.E.; Yung, Y.C.; Duong, T.E.; Gao, D.; Chun, J.; Kharchenko, P.V. Integrative single-cell analysis of transcriptional and epigenetic states in the human adult brain. *Nat. Biotechnol.* **2018**, *36*, 70–80. [CrossRef] [PubMed]

47. Butler, A.; Hoffman, P.; Smibert, P.; Papalexi, E.; Satija, R. Integrating single-cell transcriptomic data across different conditions, technologies, and species. *Nat. Biotechnol.* **2018**, *36*, 411–420. [CrossRef] [PubMed]

48. Ding, M.Q.; Chen, L.; Cooper, G.F.; Young, J.D.; Lu, X. Precision oncology beyond targeted therapy: Combining omics data with machine learning matches the majority of cancer cells to effective therapeutics. *Mol. Cancer Res.* **2018**, *16*, 269–278. [CrossRef] [PubMed]

49. Bengio, Y.; Courville, A.; Vincent, P. Representation learning: A review and new perspectives. *IEEE Trans. Pattern Anal. Mach. Intell.* **2013**, *35*, 1798–1828. [CrossRef] [PubMed]

50. Alshahrani, M.; Khan, M.A.; Maddouri, O.; Kinjo, A.R.; Queralt-Rosinach, N.; Hoehndorf, R. Neuro-symbolic representation learning on biological knowledge graphs. *Bioinformatics* **2017**, *33*, 2723–2730. [CrossRef] [PubMed]

51. Ma, T.; Zhang, A. Multi-view Factorization AutoEncoder with Network Constraints for Multi-omic Integrative Analysis. *arXiv*, 2018; arXiv:180901772.

52. Xu, Q.; Chen, J.; Ni, S.; Tan, C.; Xu, M.; Dong, L.; Yuan, L.; Wang, Q.; Du, X. Pan-cancer transcriptome analysis reveals a gene expression signature for the identification of tumor tissue origin. *Mod. Pathol.* **2016**, *29*, 546–556. [CrossRef] [PubMed]

53. Whalen, S.; Truty, R.M.; Pollard, K.S. Enhancer–promoter interactions are encoded by complex genomic signatures on looping chromatin. *Nat. Genet.* **2016**, *48*, 488–496. [CrossRef] [PubMed]

54. Kim, S.; Jhong, J.-H.; Lee, J.; Koo, J.-Y. Meta-analytic support vector machine for integrating multiple omics data. *BioData Min.* **2017**, *10*, 2. [CrossRef] [PubMed]

55. Liu, Z.; Sun, F.; McGovern, D.P. Sparse generalized linear model with L 0 approximation for feature selection and prediction with big omics data. *BioData Min.* **2017**, *10*, 39. [CrossRef] [PubMed]

56. Ding, C.; Peng, H. Minimum redundancy feature selection from microarray gene expression data. *J. Bioinform. Comput. Biol.* **2005**, *3*, 185–205. [CrossRef] [PubMed]

57. Sánchez-Maroño, N.; Alonso-Betanzos, A.; Tombilla-Sanromán, M. Filter methods for feature selection—A comparative study. In Proceedings of the International Conference on Intelligent Data Engineering and Automated Learning, Birmingham, UK, 16–19 December 2007; pp. 178–187.

58. Guyon, I.; Weston, J.; Barnhill, S.; Vapnik, V. Gene selection for cancer classification using support vector machines. *Mach. Learn.* **2002**, *46*, 389–422. [CrossRef]

59. Kursa, M.B.; Rudnicki, W.R. Feature selection with the Boruta package. *J. Stat. Softw.* **2010**, *36*, 1–13. [CrossRef]

60. Chung, N.C.; Storey, J.D. Statistical significance of variables driving systematic variation in high-dimensional data. *Bioinformatics* **2014**, *31*, 545–554. [CrossRef] [PubMed]

61. Meinshausen, N.; Bühlmann, P. Stability selection. *J. R. Stat. Soc. Ser. B Stat. Methodol.* **2010**, *72*, 417–473. [CrossRef]

62. Sill, M.; Saadati, M.; Benner, A. Applying stability selection to consistently estimate sparse principal components in high-dimensional molecular data. *Bioinformatics* **2015**, *31*, 2683–2690. [CrossRef]

63. Haury, A.-C.; Mordelet, F.; Vera-Licona, P.; Vert, J.-P. TIGRESS: Trustful inference of gene regulation using stability selection. *BMC Syst. Biol.* **2012**, *6*, 145. [CrossRef]

64. Zou, H.; Hastie, T. Regularization and variable selection via the elastic net. *J. R. Stat. Soc. Ser. B Stat. Methodol.* **2005**, *67*, 301–320. [CrossRef]

65. Pineda, S.; Real, F.X.; Kogevinas, M.; Carrato, A.; Chanock, S.J.; Malats, N.; Van Steen, K. Integration analysis of three omics data using penalized regression methods: An application to bladder cancer. *PLoS Genet.* **2015**, *11*, e1005689. [CrossRef] [PubMed]

66. Li, Y.; Wu, F.-X.; Ngom, A. A review on machine learning principles for multi-view biological data integration. *Brief. Bioinform.* **2016**, *19*, 325–340. [CrossRef] [PubMed]

67. Tini, G.; Marchetti, L.; Priami, C.; Scott-Boyer, M.-P. Multi-omics integration—A comparison of unsupervised clustering methodologies. *Brief Bioinform.* **2017**. [CrossRef] [PubMed]
68. Kim, S.; Oesterreich, S.; Kim, S.; Park, Y.; Tseng, G.C. Integrative clustering of multi-level omics data for disease subtype discovery using sequential double regularization. *Biostatistics* **2017**, *18*, 165–179. [CrossRef] [PubMed]
69. Rohart, F.; Gautier, B.; Singh, A.; Le Cao, K.-A. mixOmics: An R package for 'omics feature selection and multiple data integration. *PLoS Comput. Biol.* **2017**, *13*, e1005752. [CrossRef] [PubMed]
70. Mallik, S.; Bhadra, T.; Maulik, U. Identifying epigenetic biomarkers using maximal relevance and minimal redundancy based feature selection for multi-omics data. *IEEE Trans. Nanobiosci.* **2017**, *16*, 3–10. [CrossRef] [PubMed]
71. Liu, C.; Wang, X.; Genchev, G.Z.; Lu, H. Multi-omics facilitated variable selection in Cox-regression model for cancer prognosis prediction. *Methods* **2017**, *124*, 100–107. [CrossRef] [PubMed]
72. Poruthoor, A.; Phan, J.H.; Kothari, S.; Wang, M.D. Exploration of genomic, proteomic, and histopathological image data integration methods for clinical prediction. In Proceedings of the IEEE China Summit & International Conference on Signal and Information Processing, IEEE China Summit & International Conference on Signal and Information Processing, Beijing, China, 6–10 July 2013; p. 259.
73. Narvaez-Bandera, I.; Sanchez, F. Integration of Multi Omics Data for Breast Cancer Subtype Classification. In *IIE Annual Conference Proceedings*; Institute of Industrial and Systems Engineers (IISE): Norcross, GA, USA, 2017; pp. 1314–1319.
74. Chen, Q.; Meng, Z.; Liu, X.; Jin, Q.; Su, R. Decision Variants for the Automatic Determination of Optimal Feature Subset in RF-RFE. *Genes* **2018**, *9*, 301. [CrossRef] [PubMed]
75. Mo, Q.; Wang, S.; Seshan, V.E.; Olshen, A.B.; Schultz, N.; Sander, C.; Powers, R.S.; Ladanyi, M.; Shen, R. Pattern discovery and cancer gene identification in integrated cancer genomic data. *Proc. Natl. Acad. Sci. USA* **2013**. [CrossRef] [PubMed]
76. Kim, M.; Rai, N.; Zorraquino, V.; Tagkopoulos, I. Multi-omics integration accurately predicts cellular state in unexplored conditions for Escherichia coli. *Nat. Commun.* **2016**, *7*, 13090. [CrossRef] [PubMed]
77. Zhang, Y.; Li, A.; Peng, C.; Wang, M. Improve glioblastoma multiforme prognosis prediction by using feature selection and multiple kernel learning. *IEEE ACM Trans. Comput. Biol. Bioinform. TCBB* **2016**, *13*, 825–835. [CrossRef] [PubMed]
78. Liaw, A.; Wiener, M. Classification and regression by randomForest. *R News* **2002**, *2*, 18–22.
79. Barretina, J.; Caponigro, G.; Stransky, N.; Venkatesan, K.; Margolin, A.A.; Kim, S.; Wilson, C.J.; Lehár, J.; Kryukov, G.V.; Sonkin, D. The Cancer Cell Line Encyclopedia enables predictive modelling of anticancer drug sensitivity. *Nature* **2012**, *483*, 603. [CrossRef] [PubMed]
80. Spicker, J.S.; Brunak, S.; Frederiksen, K.S.; Toft, H. Integration of clinical chemistry, expression, and metabolite data leads to better toxicological class separation. *Toxicol. Sci.* **2008**, *102*, 444–454. [CrossRef] [PubMed]
81. Aben, N.; Vis, D.J.; Michaut, M.; Wessels, L.F. TANDEM: A two-stage approach to maximize interpretability of drug response models based on multiple molecular data types. *Bioinformatics* **2016**, *32*, i413–i420. [CrossRef] [PubMed]
82. Gönen, M.; Alpaydın, E. Multiple kernel learning algorithms. *J. Mach. Learn. Res.* **2011**, *12*, 2211–2268.
83. Rakotomamonjy, A.; Bach, F.R.; Canu, S.; Grandvalet, Y. SimpleMKL. *J. Mach. Learn. Res.* **2008**, *9*, 2491–2521.
84. Speicher, N.K.; Pfeifer, N. Integrating different data types by regularized unsupervised multiple kernel learning with application to cancer subtype discovery. *Bioinformatics* **2015**, *31*, i268–i275. [CrossRef] [PubMed]
85. Le, D.-H.; Pham, V.-H. Drug Response Prediction by Globally Capturing Drug and Cell Line Information in a Heterogeneous Network. *J. Mol. Biol.* **2018**, *18*, 2993–3004. [CrossRef]
86. Koller, D.; Friedman, N. *Probabilistic Graphical Models: Principles and Techniques*; MIT Press: Cambridge, MA, USA, 2009; ISBN 0-262-01319-3.
87. Davies, S.; Moore, A. Mix-nets: Factored mixtures of gaussians in Bayesian networks with mixed continuous and discrete variables. In *Proceedings of the Sixteenth Conference on Uncertainty in Artificial Intelligence*; Morgan Kaufmann Publishers Inc.: Burlington, MA, USA, 2000; pp. 168–175.
88. Wahl, S.; Vogt, S.; Stückler, F.; Krumsiek, J.; Bartel, J.; Kacprowski, T.; Schramm, K.; Carstensen, M.; Rathmann, W.; Roden, M. Multi-omic signature of body weight change: Results from a population-based cohort study. *BMC Med.* **2015**, *13*, 48. [CrossRef] [PubMed]

89. Langfelder, P.; Horvath, S. WGCNA: An R package for weighted correlation network analysis. *BMC Bioinform.* **2008**, *9*, 559. [CrossRef] [PubMed]

90. Krumsiek, J.; Suhre, K.; Illig, T.; Adamski, J.; Theis, F.J. Gaussian graphical modeling reconstructs pathway reactions from high-throughput metabolomics data. *BMC Syst. Biol.* **2011**, *5*, 21. [CrossRef] [PubMed]

91. Vaske, C.J.; Benz, S.C.; Sanborn, J.Z.; Earl, D.; Szeto, C.; Zhu, J.; Haussler, D.; Stuart, J.M. Inference of patient-specific pathway activities from multi-dimensional cancer genomics data using PARADIGM. *Bioinformatics* **2010**, *26*, i237–i245. [CrossRef] [PubMed]

92. Cheng, W.; Shi, Y.; Zhang, X.; Wang, W. Fast and robust group-wise eQTL mapping using sparse graphical models. *BMC Bioinform.* **2015**, *16*, 2. [CrossRef] [PubMed]

93. Dimitrakopoulos, C.; Hindupur, S.K.; Häfliger, L.; Behr, J.; Montazeri, H.; Hall, M.N.; Beerenwinkel, N. Network-based integration of multi-omics data for prioritizing cancer genes. *Bioinformatics* **2018**, *34*, 2441–2448. [CrossRef] [PubMed]

94. Shi, C.; Li, Y.; Zhang, J.; Sun, Y.; Philip, S.Y. A survey of heterogeneous information network analysis. *IEEE Trans. Knowl. Data Eng.* **2017**, *29*, 17–37. [CrossRef]

95. Tsuyuzaki, K.; Nikaido, I. Biological Systems as Heterogeneous Information Networks: A Mini-review and Perspectives. *arXiv*, 2017; arXiv:171208865.

96. Hosseini, A.; Chen, T.; Wu, W.; Sun, Y.; Sarrafzadeh, M. HeteroMed: Heterogeneous Information Network for Medical Diagnosis. In Proceedings of the 27th ACM International Conference on Information and Knowledge Management, Torino, Italy, 22–26 October 2018; pp. 763–772.

97. Ge, S.-G.; Xia, J.; Sha, W.; Zheng, C.-H. Cancer subtype discovery based on integrative model of multigenomic data. *IEEE/ACM Trans. Comput. Biol. Bioinform.* **2017**, *14*, 1115–1121. [CrossRef] [PubMed]

98. Nguyen, T.D.; Tran, T.; Phung, D.; Venkatesh, S. Latent patient profile modelling and applications with mixed-variate restricted Boltzmann machine. In Proceedings of the Pacific-Asia Conference on Knowledge Discovery and Data Mining, Gold Coast, Australia, 14–17 April 2013; pp. 123–135.

99. Frey, B.J.; Dueck, D. Clustering by passing messages between data points. *Science* **2007**, *315*, 972–976. [CrossRef] [PubMed]

100. Liang, M.; Li, Z.; Chen, T.; Zeng, J. Integrative data analysis of multi-platform cancer data with a multimodal deep learning approach. *IEEE ACM Trans. Comput. Biol. Bioinform. TCBB* **2015**, *12*, 928–937. [CrossRef] [PubMed]

101. Srivastava, N.; Salakhutdinov, R.R. Multimodal learning with deep boltzmann machines. In Proceedings of the Advances in Neural Information Processing Systems, Lake Tahoe, NV, USA, 3–6 December 2012; pp. 2222–2230.

102. Choi, J.; Park, S.; Yoon, Y.; Ahn, J. Improved prediction of breast cancer outcome by identifying heterogeneous biomarkers. *Bioinformatics* **2017**, *33*, 3619–3626. [CrossRef] [PubMed]

103. Sun, D.; Wang, M.; Li, A. A multimodal deep neural network for human breast cancer prognosis prediction by integrating multi-dimensional data. *IEEE/ACM Trans. Comput. Biol. Bioinform.* **2018**. [CrossRef] [PubMed]

104. Chaudhary, K.; Poirion, O.B.; Lu, L.; Garmire, L.X. Deep Learning–Based Multi-Omics Integration Robustly Predicts Survival in Liver Cancer. *Clin. Cancer Res.* **2018**, *24*, 1248–1259. [CrossRef] [PubMed]

105. Zhang, T.; Zhang, L.; Payne, P.R.; Li, F. Synergistic Drug Combination Prediction by Integrating Multi-omics Data in Deep Learning Models. *arXiv*, 2018; arXiv:181107054.

106. Choi, H.; Pavelka, N. When one and one gives more than two: Challenges and opportunities of integrative omics. *Front. Genet.* **2012**, *2*, 105. [CrossRef] [PubMed]

107. Torres-García, W.; Zhang, W.; Runger, G.C.; Johnson, R.H.; Meldrum, D.R. Integrative analysis of transcriptomic and proteomic data of Desulfovibrio vulgaris: A non-linear model to predict abundance of undetected proteins. *Bioinformatics* **2009**, *25*, 1905–1914. [CrossRef] [PubMed]

108. Misra, B.B.; Langefeld, C.D.; Olivier, M.; Cox, L.A. Integrated Omics: Tools, Advances, and Future Approaches. *J. Mol. Endocrinol.* **2018**. [CrossRef] [PubMed]

109. Rouillard, A.D.; Wang, Z.; Ma'ayan, A. Abstraction for data integration: Fusing mammalian molecular, cellular and phenotype big datasets for better knowledge extraction. *Comput. Biol. Chem.* **2015**, *58*, 104. [CrossRef] [PubMed]

110. Lin, D.; Zhang, J.; Li, J.; Xu, C.; Deng, H.-W.; Wang, Y.-P. An integrative imputation method based on multi-omics datasets. *BMC Bioinform.* **2016**, *17*, 247. [CrossRef] [PubMed]

111. Rubin, D.B. Inference and missing data. *Biometrika* **1976**, *63*, 581–592. [CrossRef]

112. Allison, P.D. Estimation of linear models with incomplete data. *Sociol. Methodol.* **1987**, 71–103. [CrossRef]
113. Allison, P.D. *Missing Data*; Sage Publications: Thousand Oaks, CA, USA, 2001; Volume 136, ISBN 1-4522-0790-9.
114. Allison, P.D. Handling missing data by maximum likelihood. In Proceedings of the SAS Global Forum, Statistical Horizons, Havenford, PA, USA, 22–25 April 2012.
115. Mias, G.I.; Yusufaly, T.; Roushangar, R.; Brooks, L.R.; Singh, V.V.; Christou, C. MathIOmica: An integrative platform for dynamic omics. *Sci. Rep.* **2016**, *6*, 37237. [CrossRef] [PubMed]
116. Kohl, M.; Megger, D.A.; Trippler, M.; Meckel, H.; Ahrens, M.; Bracht, T.; Weber, F.; Hoffmann, A.-C.; Baba, H.A.; Sitek, B. A practical data processing workflow for multi-OMICS projects. *Biochim. Biophys. Acta BBA-Proteins Proteom.* **2014**, *1844*, 52–62. [CrossRef] [PubMed]
117. Newgard, C.D.; Lewis, R.J. Missing data: How to best account for what is not known. *Jama* **2015**, *314*, 940–941. [CrossRef] [PubMed]
118. Schafer, J.L. *Analysis of Incomplete Multivariate Data*; Chapman and Hall/CRC: Boca Raton, FL, USA, 1997; ISBN 1-4398-2186-0.
119. Van Buuren, S.; Brand, J.P.; Groothuis-Oudshoorn, C.G.; Rubin, D.B. Fully conditional specification in multivariate imputation. *J. Stat. Comput. Simul.* **2006**, *76*, 1049–1064. [CrossRef]
120. Honaker, J.; King, G.; Blackwell, M. Amelia II: A program for missing data. *J. Stat. Softw.* **2011**, *45*, 1–47. [CrossRef]
121. Morris, T.P.; White, I.R.; Royston, P. Tuning multiple imputation by predictive mean matching and local residual draws. *BMC Med. Res. Methodol.* **2014**, *14*, 75. [CrossRef] [PubMed]
122. Rubin, D.B. *Multiple Imputation for Nonresponse in Surveys*; John Wiley & Sons: Hoboken, NJ, USA, 2004; Volume 81, ISBN 0-471-65574-0.
123. Voillet, V.; Besse, P.; Liaubet, L.; San Cristobal, M.; González, I. Handling missing rows in multi-omics data integration: Multiple imputation in multiple factor analysis framework. *BMC Bioinform.* **2016**, *17*, 402. [CrossRef] [PubMed]
124. Graham, J.W. Missing data analysis: Making it work in the real world. *Annu. Rev. Psychol.* **2009**, *60*, 549–576. [CrossRef] [PubMed]
125. Carpenter, J.; Kenward, M. *Multiple Imputation and Its Application*; John Wiley & Sons: Hoboken, NJ, USA, 2012; ISBN 1-119-94227-6.
126. Yadav, M.L.; Roychoudhury, B. Handling Missing Values: A study of Popular Imputation Packages in R. *Knowl.-Based Syst.* **2018**, *160*, 104–118. [CrossRef]
127. Sovilj, D.; Eirola, E.; Miche, Y.; Björk, K.-M.; Nian, R.; Akusok, A.; Lendasse, A. Extreme learning machine for missing data using multiple imputations. *Neurocomputing* **2016**, *174*, 220–231. [CrossRef]
128. Shah, A.D.; Bartlett, J.W.; Carpenter, J.; Nicholas, O.; Hemingway, H. Comparison of random forest and parametric imputation models for imputing missing data using MICE: A CALIBER study. *Am. J. Epidemiol.* **2014**, *179*, 764–774. [CrossRef] [PubMed]
129. Beaulieu-Jones, B.K.; Moore, J.H. Missing data imputation in the electronic health record using deeply learned autoencoders. In Proceedings of the Pacific Symposium on Biocomputing, Kohala Coast, HI, USA, 3–7 January 2017; pp. 207–218.
130. Gondara, L.; Wang, K. Mida: Multiple imputation using denoising autoencoders. In Proceedings of the Pacific-Asia Conference on Knowledge Discovery and Data Mining, Melbourne, VIC, Australia, 3–6 June 2018; pp. 260–272.
131. Gondara, L.; Wang, K. Recovering loss to followup information using denoising autoencoders. *arXiv*, 2018; arXiv:180204664.
132. Talwar, D.; Mongia, A.; Sengupta, D.; Majumdar, A. AutoImpute: Autoencoder based imputation of single-cell RNA-seq data. *Sci. Rep.* **2018**, *8*, 16329. [CrossRef] [PubMed]
133. Linderman, G.C.; Zhao, J.; Kluger, Y. Zero-preserving imputation of scRNA-seq data using low-rank approximation. *bioRxiv* **2018**. [CrossRef]
134. Troyanskaya, O.; Cantor, M.; Sherlock, G.; Brown, P.; Hastie, T.; Tibshirani, R.; Botstein, D.; Altman, R.B. Missing value estimation methods for DNA microarrays. *Bioinformatics* **2001**, *17*, 520–525. [CrossRef] [PubMed]

135. Jiang, B.; Ma, S.; Causey, J.; Qiao, L.; Hardin, M.P.; Bitts, I.; Johnson, D.; Zhang, S.; Huang, X. SparRec: An effective matrix completion framework of missing data imputation for GWAS. *Sci. Rep.* **2016**, *6*, 35534. [CrossRef] [PubMed]

136. Davies, R.W.; Flint, J.; Myers, S.; Mott, R. Rapid genotype imputation from sequence without reference panels. *Nat. Genet.* **2016**, *48*, 965. [CrossRef] [PubMed]

137. Liu, X.; Zhu, X.; Li, M.; Wang, L.; Tang, C.; Yin, J.; Shen, D.; Wang, H.; Gao, W. Late Fusion Incomplete Multi-view Clustering. *IEEE Trans. Pattern Anal. Mach. Intell.* **2018**. [CrossRef]

138. Yu, H.; Sun, C.; Yang, W.; Xu, S.; Dan, Y. A Review of Class Imbalance Learning Methods in Bioinformatics. *Curr. Bioinform.* **2015**, *10*, 360–369. [CrossRef]

139. Kleftogiannis, D.; Kalnis, P.; Bajic, V.B. DEEP: A general computational framework for predicting enhancers. *Nucleic Acids Res.* **2014**, *43*, e6. [CrossRef]

140. Triguero, I.; del Río, S.; López, V.; Bacardit, J.; Benítez, J.M.; Herrera, F. ROSEFW-RF: The winner algorithm for the ECBDL'14 big data competition: An extremely imbalanced big data bioinformatics problem. *Knowl.-Based Syst.* **2015**, *87*, 69–79. [CrossRef]

141. Aledo, J.C.; Cantón, F.R.; Veredas, F.J. A machine learning approach for predicting methionine oxidation sites. *BMC Bioinform.* **2017**, *18*, 430. [CrossRef] [PubMed]

142. Hu, J.; Li, Y.; Zhang, M.; Yang, X.; Shen, H.-B.; Yu, D.-J. Predicting protein-DNA binding residues by weightedly combining sequence-based features and boosting multiple SVMs. *IEEE/ACM Trans. Comput. Biol. Bioinform.* **2017**, *14*, 1389–1398. [CrossRef] [PubMed]

143. Ding, J.; Zhou, S.; Guan, J. MiRenSVM: Towards better prediction of microRNA precursors using an ensemble SVM classifier with multi-loop features. *BMC Bioinform.* **2010**, *11*, S11. [CrossRef] [PubMed]

144. Fernández-Martínez, J.L.; de Andrés-Galiana, E.J.; Sonis, S.T. Genomic data integration in chronic lymphocytic leukemia. *J. Gene Med.* **2017**, *19*, e2936. [CrossRef] [PubMed]

145. Liu, Z.; Xiao, X.; Qiu, W.-R.; Chou, K.-C. iDNA-Methyl: Identifying DNA methylation sites via pseudo trinucleotide composition. *Anal. Biochem.* **2015**, *474*, 69–77. [CrossRef] [PubMed]

146. Zhang, W.; Spector, T.D.; Deloukas, P.; Bell, J.T.; Engelhardt, B.E. Predicting genome-wide DNA methylation using methylation marks, genomic position, and DNA regulatory elements. *Genome Biol.* **2015**, *16*, 14. [CrossRef] [PubMed]

147. Wei, Z.-S.; Yang, J.-Y.; Shen, H.-B.; Yu, D.-J. A cascade random forests algorithm for predicting protein-protein interaction sites. *IEEE Trans. Nanobioscience* **2015**, *14*, 746–760. [CrossRef]

148. Wei, Z.-S.; Han, K.; Yang, J.-Y.; Shen, H.-B.; Yu, D.-J. Protein–protein interaction sites prediction by ensembling SVM and sample-weighted random forests. *Neurocomputing* **2016**, *193*, 201–212. [CrossRef]

149. Lin, W.; Xu, D. Imbalanced multi-label learning for identifying antimicrobial peptides and their functional types. *Bioinformatics* **2016**, *32*, 3745–3752. [CrossRef]

150. Troisi, J.; Sarno, L.; Martinelli, P.; Di Carlo, C.; Landolfi, A.; Scala, G.; Rinaldi, M.; D'Alessandro, P.; Ciccone, C.; Guida, M. A metabolomics-based approach for non-invasive diagnosis of chromosomal anomalies. *Metabolomics* **2017**, *13*, 140. [CrossRef]

151. Dubey, R.; Zhou, J.; Wang, Y.; Thompson, P.M.; Ye, J.; Initiative, A.D.N. Analysis of sampling techniques for imbalanced data: An n= 648 ADNI study. *NeuroImage* **2014**, *87*, 220–241. [CrossRef] [PubMed]

152. Haixiang, G.; Yijing, L.; Shang, J.; Mingyun, G.; Yuanyue, H.; Bing, G. Learning from class-imbalanced data: Review of methods and applications. *Expert Syst. Appl.* **2017**, *73*, 220–239. [CrossRef]

153. He, H.; Garcia, E.A. Learning from imbalanced data. *IEEE Trans. Knowl. Data Eng.* **2008**, 1263–1284.

154. Chawla, N.V.; Bowyer, K.W.; Hall, L.O.; Kegelmeyer, W.P. SMOTE: Synthetic minority over-sampling technique. *J. Artif. Intell. Res.* **2002**, *16*, 321–357. [CrossRef]

155. Lin, W.-J.; Chen, J.J. Class-imbalanced classifiers for high-dimensional data. *Brief. Bioinform.* **2012**, *14*, 13–26. [CrossRef] [PubMed]

156. Huang, C.-C.; Chang, C.-C.; Chen, C.-W.; Ho, S.; Chang, H.-P.; Chu, Y.-W. PClass: Protein Quaternary Structure Classification by Using Bootstrapping Strategy as Model Selection. *Genes* **2018**, *9*, 91. [CrossRef] [PubMed]

157. Zhang, X.; Yan, L.-F.; Hu, Y.-C.; Li, G.; Yang, Y.; Han, Y.; Sun, Y.-Z.; Liu, Z.-C.; Tian, Q.; Han, Z.-Y. Optimizing a machine learning based glioma grading system using multi-parametric MRI histogram and texture features. *Oncotarget* **2017**, *8*, 47816. [CrossRef]

158. Bach, M.; Werner, A.; Żywiec, J.; Pluskiewicz, W. The study of under-and over-sampling methods' utility in analysis of highly imbalanced data on osteoporosis. *Inf. Sci.* **2017**, *384*, 174–190. [CrossRef]

159. Kubat, M.; Matwin, S. Addressing the curse of imbalanced training sets: One-sided selection. In Proceedings of the ICML, Nashville, TN, USA, 8–12 July 1997; pp. 179–186.

160. Veropoulos, K.; Campbell, C.; Cristianini, N. Controlling the sensitivity of support vector machines. In Proceedings of the International Joint Conference on AI, Stockholm, Sweden, 31 July–6 August 1999; p. 60.

161. Bao, F.; Deng, Y.; Zhao, Y.; Suo, J.; Dai, Q. Bosco: Boosting corrections for genome-wide association studies with imbalanced samples. *IEEE Trans. Nanobiosci.* **2017**, *16*, 69–77. [CrossRef]

162. Martina, F.; Beccuti, M.; Balbo, G.; Cordero, F. Peculiar Genes Selection: A new features selection method to improve classification performances in imbalanced data sets. *PLoS ONE* **2017**, *12*, e0177475. [CrossRef]

163. Liu, Z.; Tang, D.; Cai, Y.; Wang, R.; Chen, F. A hybrid method based on ensemble WELM for handling multi class imbalance in cancer microarray data. *Neurocomputing* **2017**, *266*, 641–650. [CrossRef]

164. Liu, G.-H.; Shen, H.-B.; Yu, D.-J. Prediction of protein–protein interaction sites with machine-learning-based data-cleaning and post-filtering procedures. *J. Membr. Biol.* **2016**, *249*, 141–153. [CrossRef] [PubMed]

165. Mirza, B.; Lin, Z.; Liu, N. Ensemble of subset online sequential extreme learning machine for class imbalance and concept drift. *Neurocomputing* **2015**, *149*, 316–329. [CrossRef]

166. Chen, L.; Jin, P.; Qin, Z.S. DIVAN: Accurate identification of non-coding disease-specific risk variants using multi-omics profiles. *Genome Biol.* **2016**, *17*, 252. [CrossRef] [PubMed]

167. Liu, X.-Y.; Wu, J.; Zhou, Z.-H. Exploratory undersampling for class-imbalance learning. *IEEE Trans. Syst. Man Cybern. Part B Cybern.* **2009**, *39*, 539–550.

168. Yang, P.; Hwa Yang, Y.; Zhou, B.B.; Zomaya, A.Y. A review of ensemble methods in bioinformatics. *Curr. Bioinform.* **2010**, *5*, 296–308. [CrossRef]

169. Li, C.-X.; Wheelock, C.E.; Sköld, C.M.; Wheelock, Å.M. Integration of multi-omics datasets enables molecular classification of COPD. *Eur. Respir. J.* **2018**, 1701930. [CrossRef]

170. Yan, K.K.; Zhao, H.; Pang, H. A comparison of graph-and kernel-based–omics data integration algorithms for classifying complex traits. *BMC Bioinform.* **2017**, *18*, 539. [CrossRef]

171. Singh, A.; Gautier, B.; Shannon, C.P.; Rohart, F.; Vacher, M.; Tebutt, S.J.; Le Cao, K.-A. DIABLO: From multi-omics assays to biomarker discovery, an integrative approach. *bioRxiv* **2018**. [CrossRef]

172. Bica, I.; Velickovic, P.; Xiao, H.; Li, P. Multi-omics data integration using cross-modal neural networks. In Proceedings of the 26th European Symposium on Artificial Neural Networks, Computational Intelligence and Machine Learning (ESANN 2018), Bruges, Belgium, 25–27 April 2018.

173. Lin, X.; Chen, X. Heterogeneous data integration by tree-augmented naïve B ayes for protein–protein interactions prediction. *Proteomics* **2013**, *13*, 261–268. [CrossRef]

174. Goldfarb, D.; Hast, B.; Wang, W.; Major, M.B. An Improved Algorithm and Web Application for Predicting Co-Complexed Proteins from Affinity Purification–Mass Spectrometry Data. *J. Proteome Res.* **2014**, *13*, 5944. [CrossRef] [PubMed]

175. Frasca, M.; Bertoni, A.; Valentini, G. UNIPred: Unbalance-aware Network Integration and Prediction of protein functions. *J. Comput. Biol.* **2015**, *22*, 1057–1074. [CrossRef] [PubMed]

176. Yu, G.; Zhu, H.; Domeniconi, C.; Guo, M. Integrating multiple networks for protein function prediction. In *Proceedings of the BMC Systems Biology*; BioMed Central: London, UK, 2015; Volume 9, p. S3.

177. Kwon, M.-S.; Kim, Y.; Lee, S.; Namkung, J.; Yun, T.; Yi, S.G.; Han, S.; Kang, M.; Kim, S.W.; Jang, J.-Y. Integrative analysis of multi-omics data for identifying multi-markers for diagnosing pancreatic cancer. *BMC Genom.* **2015**, *16*, S4. [CrossRef] [PubMed]

178. Song, Y.; Westerhuis, J.A.; Aben, N.; Wessels, L.F.; Groenen, P.J.; Smilde, A.K. Generalized Simultaneous Component Analysis of Binary and Quantitative data. *arXiv*, 2018; arXiv:180704982.

179. Re, M.; Valentini, G. Simple ensemble methods are competitive with state-of-the-art data integration methods for gene function prediction. In Proceedings of the MLSB, PMLR, Ljubljana, Slovenia, 5–6 September 2009; Volume 8, pp. 98–111.

180. Yu, H.; Hong, S.; Yang, X.; Ni, J.; Dan, Y.; Qin, B. Recognition of multiple imbalanced cancer types based on DNA microarray data using ensemble classifiers. *BioMed Res. Int.* **2013**, *2013*, 239628. [CrossRef] [PubMed]

181. Fortino, V.; Kinaret, P.; Fyhrquist, N.; Alenius, H.; Greco, D. A robust and accurate method for feature selection and prioritization from multi-class OMICs data. *PLoS ONE* **2014**, *9*, e107801. [CrossRef] [PubMed]

182. Chen, L.; Zhang, Y.-H.; Huang, G.; Pan, X.; Wang, S.; Huang, T.; Cai, Y.-D. Discriminating cirRNAs from other lncRNAs using a hierarchical extreme learning machine (H-ELM) algorithm with feature selection. *Mol. Genet. Genom.* **2018**, *293*, 137–149. [CrossRef] [PubMed]

183. Zhang, L.; Suganthan, P.N. A survey of randomized algorithms for training neural networks. *Inf. Sci.* **2016**, *364*, 146–155. [CrossRef]

184. Cao, W.; Wang, X.; Ming, Z.; Gao, J. A review on neural networks with random weights. *Neurocomputing* **2018**, *275*, 278–287. [CrossRef]

185. Tang, J.; Deng, C.; Huang, G.-B. Extreme learning machine for multilayer perceptron. *IEEE Trans. Neural Netw. Learn. Syst.* **2016**, *27*, 809–821. [CrossRef]

186. Lai, X.; Cao, J.; Lin, Z. A Novel Relaxed ADMM with Highly Parallel Implementation for Extreme Learning Machine. In Proceedings of the 2018 IEEE International Symposium on Circuits and Systems (ISCAS), Florence, Italy, 27–30 May 2018; pp. 1–5.

187. Wang, X.; Cao, W. Non-Iterative Approaches in Training Feed-Forward Neural Networks and Their Applications. *Soft Comput.* **2018**, *22*, 3473–3476. [CrossRef]

188. Huang, G.-B.; Zhou, H.; Ding, X.; Zhang, R. Extreme learning machine for regression and multiclass classification. *IEEE Trans. Syst. Man Cybern. Part B Cybern.* **2012**, *42*, 513–529. [CrossRef] [PubMed]

189. Pao, Y.-H.; Takefuji, Y. Functional-link net computing: Theory, system architecture, and functionalities. *Computer* **1992**, *25*, 76–79. [CrossRef]

190. Zhang, L.; Suganthan, P.N. A comprehensive evaluation of random vector functional link networks. *Inf. Sci.* **2016**, *367*, 1094–1105. [CrossRef]

191. Maass, W.; Natschläger, T.; Markram, H. Real-time computing without stable states: A new framework for neural computation based on perturbations. *Neural Comput.* **2002**, *14*, 2531–2560. [CrossRef] [PubMed]

192. Jaeger, H. Adaptive nonlinear system identification with echo state networks. In *Proceedings of the Advances in Neural Information Processing Systems*; MIT Press: Cambridge, MA, USA, 2003; Volume 15, pp. 593–600.

193. Cevher, V.; Becker, S.; Schmidt, M. Convex optimization for big data: Scalable, randomized, and parallel algorithms for big data analytics. *IEEE Signal Process. Mag.* **2014**, *31*, 32–43. [CrossRef]

194. Rubiolo, M.; Milone, D.H.; Stegmayer, G. Extreme learning machines for reverse engineering of gene regulatory networks from expression time series. *Bioinformatics* **2017**, *34*, 1253–1260. [CrossRef] [PubMed]

195. Lei, H.; Wen, Y.; Elazab, A.; Tan, E.-L.; Zhao, Y.; Lei, B. Protein-protein Interactions Prediction via Multimodal Deep Polynomial Network and Regularized Extreme Learning Machine. *IEEE J. Biomed. Health Inform.* **2018**. [CrossRef]

196. Belciug, S.; Gorunescu, F. Learning a single-hidden layer feedforward neural network using a rank correlation-based strategy with application to high dimensional gene expression and proteomic spectra datasets in cancer detection. *J. Biomed. Inform.* **2018**, *83*, 159–166. [CrossRef] [PubMed]

197. Pian, C.; Zhang, G.; Chen, Z.; Chen, Y.; Zhang, J.; Yang, T.; Zhang, L. LncRNApred: Classification of long non-coding RNAs and protein-coding transcripts by the ensemble algorithm with a new hybrid feature. *PLoS ONE* **2016**, *11*, e0154567. [CrossRef] [PubMed]

198. Nguyen, T.V.; Mirza, B. Dual-layer kernel extreme learning machine for action recognition. *Neurocomputing* **2017**, *260*, 123–130. [CrossRef]

199. Aiolli, F.; Donini, M. EasyMKL: A scalable multiple kernel learning algorithm. *Neurocomputing* **2015**, *169*, 215–224. [CrossRef]

200. Hoi, S.C.; Sahoo, D.; Lu, J.; Zhao, P. Online Learning: A Comprehensive Survey. *arXiv* **2018**, arXiv:180202871.

201. Georga, E.I.; Protopappas, V.C.; Polyzos, D.; Fotiadis, D.I. Online prediction of glucose concentration in type 1 diabetes using extreme learning machines. In Proceedings of the 2015 37th Annual International Conference of the IEEE Engineering in Medicine and Biology Society (EMBC), Milan, Italy, 25–29 August 2015; pp. 3262–3265.

202. Liang, N.-Y.; Huang, G.-B.; Saratchandran, P.; Sundararajan, N. A fast and accurate online sequential learning algorithm for feedforward networks. *IEEE Trans. Neural Netw.* **2006**, *17*, 1411–1423. [CrossRef] [PubMed]

203. LeCun, Y.A.; Bottou, L.; Orr, G.B.; Müller, K.-R. Efficient backprop. In *Neural Networks: Tricks of the Trade*; Springer: Berlin, Germany, 2012; pp. 9–48.

204. Cauwenberghs, G.; Poggio, T. Incremental and decremental support vector machine learning. In *Advances in Neural Information Processing Systems*; MIT Press: Cambridge, MA, USA, 2001; Volume 13, pp. 409–415.

205. Gu, B.; Quan, X.; Gu, Y.; Sheng, V.S. Chunk Incremental Learning for Cost-Sensitive Hinge Loss Support Vector Machine. *Pattern Recognit.* **2018**, *83*, 196–208. [CrossRef]
206. Mirza, B.; Kok, S.; Dong, F. Multi-layer online sequential extreme learning machine for image classification. In *Proceedings of ELM-2015*; Springer: Berlin, Germany, 2016; Volume 1, pp. 39–49.
207. Sahoo, D.; Pham, Q.; Lu, J.; Hoi, S.C. Online deep learning: Learning deep neural networks on the fly. *arXiv* **2017**, arXiv:171103705.
208. Dean, J.; Ghemawat, S. MapReduce: Simplified data processing on large clusters. *Commun. ACM* **2008**, *51*, 107–113. [CrossRef]
209. Zou, Q.; Li, X.-B.; Jiang, W.-R.; Lin, Z.-Y.; Li, G.-L.; Chen, K. Survey of MapReduce frame operation in bioinformatics. *Brief. Bioinform.* **2013**, *15*, 637–647. [CrossRef] [PubMed]
210. White, T. *Hadoop: The Definitive Guide*; O'Reilly Media, Inc.: Sebastopol, CA, USA, 2012; ISBN 1-4493-1152-0.
211. Foss, A.; Markatou, M.; Ray, B.; Heching, A. A semiparametric method for clustering mixed data. *Mach. Learn.* **2016**, *105*, 419–458. [CrossRef]
212. Foss, A.H.; Markatou, M. kamila: Clustering Mixed-Type Data in R and Hadoop. *J. Stat. Softw.* **2018**, *83*, 1–44. [CrossRef]
213. Zaharia, M.; Xin, R.S.; Wendell, P.; Das, T.; Armbrust, M.; Dave, A.; Meng, X.; Rosen, J.; Venkataraman, S.; Franklin, M.J. Apache spark: A unified engine for big data processing. *Commun. ACM* **2016**, *59*, 56–65. [CrossRef]
214. Meng, X.; Bradley, J.; Yavuz, B.; Sparks, E.; Venkataraman, S.; Liu, D.; Freeman, J.; Tsai, D.B.; Amde, M.; Owen, S. Mllib: Machine learning in apache spark. *J. Mach. Learn. Res.* **2016**, *17*, 1235–1241.
215. Owen, S.; Anil, R.; Dunning, T.; Friedman, E. *Mahout in Action*; Manning Publications Co.: Shelter Island, NY, USA, 2011; ISBN 1-935182-68-4.
216. Abadi, M.; Barham, P.; Chen, J.; Chen, Z.; Davis, A.; Dean, J.; Devin, M.; Ghemawat, S.; Irving, G.; Isard, M. Tensorflow: A system for large-scale machine learning. In Proceedings of the 12th USENIX Symposium on Operating Systems Design and Implementation (OSDI), Savannah, GA, USA, 2–4 November 2016; USENIX Association: Berkeley, CA, USA; Volume 16, pp. 265–283.
217. Afgan, E.; Baker, D.; Batut, B.; van den Beek, M.; Bouvier, D.; Čech, M.; Chilton, J.; Clements, D.; Coraor, N.; Grüning, B.A. The Galaxy platform for accessible, reproducible and collaborative biomedical analyses: 2018 update. *Nucleic Acids Res.* **2018**, *46*, W537–W544. [CrossRef]
218. Afgan, E.; Baker, D.; Coraor, N.; Goto, H.; Paul, I.M.; Makova, K.D.; Nekrutenko, A.; Taylor, J. Harnessing cloud computing with Galaxy Cloud. *Nat. Biotechnol.* **2011**, *29*, 972–974. [CrossRef] [PubMed]
219. Fisch, K.M.; Meißner, T.; Gioia, L.; Ducom, J.-C.; Carland, T.M.; Loguercio, S.; Su, A.I. Omics Pipe: A community-based framework for reproducible multi-omics data analysis. *Bioinformatics* **2015**, *31*, 1724–1728. [CrossRef] [PubMed]
220. Forsberg, E.M.; Huan, T.; Rinehart, D.; Benton, H.P.; Warth, B.; Hilmers, B.; Siuzdak, G. Data processing, multi-omic pathway mapping, and metabolite activity analysis using XCMS Online. *Nat. Protoc.* **2018**, *13*, 633–651. [CrossRef] [PubMed]
221. Chong, J.; Soufan, O.; Li, C.; Caraus, I.; Li, S.; Bourque, G.; Wishart, D.S.; Xia, J. MetaboAnalyst 4.0: Towards more transparent and integrative metabolomics analysis. *Nucleic Acids Res.* **2018**, *46*, W486–W494. [CrossRef] [PubMed]
222. Tafti, A.P.; LaRose, E.; Badger, J.C.; Kleiman, R.; Peissig, P. Machine learning-as-a-service and its application to medical informatics. In Proceedings of the International Conference on Machine Learning and Data Mining in Pattern Recognition, New York, NY, USA, 15–20 July 2017; pp. 206–219.
223. Price, N.D.; Magis, A.T.; Earls, J.C.; Glusman, G.; Levy, R.; Lausted, C.; McDonald, D.T.; Kusebauch, U.; Moss, C.L.; Zhou, Y. A wellness study of 108 individuals using personal, dense, dynamic data clouds. *Nat. Biotechnol.* **2017**, *35*, 747. [CrossRef] [PubMed]
224. Glaab, E. Using prior knowledge from cellular pathways and molecular networks for diagnostic specimen classification. *Brief. Bioinform.* **2015**, *17*, 440–452. [CrossRef] [PubMed]
225. Greene, C.S.; Krishnan, A.; Wong, A.K.; Ricciotti, E.; Zelaya, R.A.; Himmelstein, D.S.; Zhang, R.; Hartmann, B.M.; Zaslavsky, E.; Sealfon, S.C. Understanding multicellular function and disease with human tissue-specific networks. *Nat. Genet.* **2015**, *47*, 569. [CrossRef] [PubMed]

226. Yao, V.; Kaletsky, R.; Keyes, W.; Mor, D.E.; Wong, A.K.; Sohrabi, S.; Murphy, C.T.; Troyanskaya, O.G. An integrative tissue-network approach to identify and test human disease genes. *Nat. Biotechnol.* **2018**, *36*, 1091–1099. [CrossRef] [PubMed]

227. Li, J.; Pan, C.; Zhang, S.; Spin, J.M.; Deng, A.; Leung, L.L.; Dalman, R.L.; Tsao, P.S.; Snyder, M. Decoding the Genomics of Abdominal Aortic Aneurysm. *Cell* **2018**, *174*, 1361–1372. [CrossRef] [PubMed]

228. Ritchie, M.D. Large-Scale Analysis of Genetic and Clinical Patient Data. *Annu. Rev. Biomed. Data Sci.* **2018**, *1*, 263–274. [CrossRef]

229. Liem, D.A.; Murali, S.; Sigdel, D.; Shi, Y.; Wang, X.; Shen, J.; Choi, H.; Caufield, J.H.; Wang, W.; Ping, P. Phrase Mining of Textual Data to Analyze Extracellular Matrix Protein Patterns Across Cardiovascular Disease. *Am. J. Physiol.-Heart Circ. Physiol.* **2018**. [CrossRef] [PubMed]

230. Tao, F.; Zhuang, H.; Yu, C.W.; Wang, Q.; Cassidy, T.; Kaplan, L.R.; Voss, C.R.; Han, J. Multi-Dimensional, Phrase-Based Summarization in Text Cubes. *IEEE Data Eng. Bull.* **2016**, *39*, 74–84.

231. Shokri, R.; Shmatikov, V. Privacy-preserving deep learning. In Proceedings of the 22nd ACM SIGSAC Conference on Computer and Communications Security, Denver, CO, USA, 12–16 October 2015; pp. 1310–1321.

232. Beaulieu-Jones, B.K.; Wu, Z.S.; Williams, C.; Greene, C.S. Privacy-preserving generative deep neural networks support clinical data sharing. *BioRxiv* **2017**. [CrossRef]

233. Olson, R.S.; La Cava, W.; Orzechowski, P.; Urbanowicz, R.J.; Moore, J.H. PMLB: A large benchmark suite for machine learning evaluation and comparison. *BioData Min.* **2017**, *10*, 36. [CrossRef] [PubMed]

 *genes*

Review

From Genotype to Phenotype: Through Chromatin

Julia Romanowska [1,2] **and Anagha Joshi** [2,*]

1   Department of Global Public Health and Primary Care, University of Bergen, 5018 Bergen, Norway;
    Julia.Romanowska@uib.no
2   Computational Biology Unit, Department of Clinical Science, University of Bergen, 5021 Bergen, Norway
*   Correspondence: Anagha.Joshi@uib.no; Tel.: +47-55-58-54-35

Received: 19 December 2018; Accepted: 21 January 2019; Published: 23 January 2019

Abstract: Advances in sequencing technologies have enabled the exploration of the genetic basis for several clinical disorders by allowing identification of causal mutations in rare genetic diseases. Sequencing technology has also facilitated genome-wide association studies to gather single nucleotide polymorphisms in common diseases including cancer and diabetes. Sequencing has therefore become common in the clinic for both prognostics and diagnostics. The success in follow-up steps, i.e., mapping mutations to causal genes and therapeutic targets to further the development of novel therapies, has nevertheless been very limited. This is because most mutations associated with diseases lie in inter-genic regions including the so-called regulatory genome. Additionally, no genetic causes are apparent for many diseases including neurodegenerative disorders. A complementary approach is therefore gaining interest, namely to focus on *epigenetic* control of the disease to generate more complete functional genomic maps. To this end, several recent studies have generated large-scale epigenetic datasets in a disease context to form a link between genotype and phenotype. We focus DNA methylation and important histone marks, where recent advances have been made thanks to technology improvements, cost effectiveness, and large meta-scale epigenome consortia efforts. We summarize recent studies unravelling the mechanistic understanding of epigenetic processes in disease development and progression. Moreover, we show how methodology advancements enable causal relationships to be established, and we pinpoint the most important issues to be addressed by future research.

Keywords: epigenetics; chromatin modification; sequencing; regulatory genomics; disease variants

1. Introduction

1.1. Definition of Epigenetics

The human body consists of hundreds of different tissues and cell types, each with its characteristic well-defined function. For example, myosin is produced by muscle cells while hemoglobin is produced by red blood cells to facilitate cell type specific functions. Despite the diversity of functional molecules in an individual cell type, nearly all cell types in an organism contain the same genetic information or genome. To explain how this diversity of cell types can be achieved from a single cell or zygote, Conrad Waddington proposed the concept of *"epigenesis"* in 1956, where pluripotent cells have the "potential" to generate all other cell types of restricted potential, in which they gradually lose this "potential" during differentiation, famously depicted by the Waddington landscape [1]. This so-called potential was later associated with a physical phenomenon, the methylation of DNA [2], which is a methyl group added to position 5 on the cytosine ring. In mammals, it is mainly $5'$—C—phosphate—G—$3'$ dinucleotide (CpG) that is subjected to methylation. Originally, methylation was found to act as a silencing mark. Accordingly, in embryonic stem cells, the majority of promoters have un-methylated DNA, and some of them become methylated during differentiation, assisting the acquisition of their final cell identity [3]. Over the years, many other epigenetic and transcription control mechanisms responsible for establishing unique gene expression profiles characteristic for different cell and tissue types during embryonic development have been studied in detail [4,5]. Gene regulatory elements receive and execute transcriptional signals, dependent on their epigenetic state and chromatin accessibility, controlling the expression of key developmental factors [6]. Chromatin dynamics are regulated through two main mechanisms: methylation of DNA and post-translational modifications of histone tails [7] (Figure 1). Histone modifications include, among others, phosphorylation, acetylation, methylation, and ubiquitylation, with methylation at specific residues as one of the most important posttranslational modifications regulating nuclear function, including transcriptional regulation, epigenetic inheritance, and maintenance of genome integrity [8]. Recently, it has become evident that histone modifications act together and a term "histone code" was coined to refer to a scheme of gene control exhibited by the complex interactions of histone modifications [9,10]. Accordingly, specific functions can be associated to a group of histone modifications, such as H3K27ac and H3K4me1, and are associated with enhancer regions. Several reviews written over the years focus on state-of-the-art studies providing structure function associations of histone modifications and successive layers of chromatin structure in mammalian genomes [11–13].

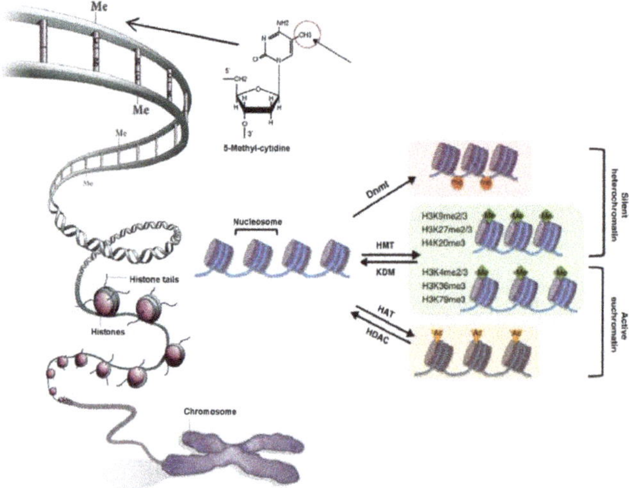

Figure 1. Diagrammatic representation of epigenetic mechanisms namely DNA methylation and chromatin modifications [14].

1.2. Broadening the Definition of Epigenetics

Epigenetics are widely understood as any mechanism by which heritable changes in gene expression occur without changing the DNA sequence, but the precise definition has evolved over the years. Apart from the above mechanisms, the role of non-coding RNAs (ncRNAs) is becoming evident in epigenetic control (reviewed in [15]). In short, ncRNAs are transcribed from the genome sequence without producing a functional protein, are highly cell type specific and regulate epigenetic patterning by establishing epigenetic modifications (DNA methylation and chromatin modifications). For example, *Xist* is an ncRNA expressed from the X chromosome that silences the other X chromosome in females. Non-coding RNAs can function as a guide or tethers, and may be the molecules of choice for epigenetic regulation of DNA methylation [16]. Some authors therefore now include ncRNAs in their definition of epigenetics. Nevertheless, we will stick to the classical definition and discuss only DNA methylation and chromatin modifications in this review.

1.3. Epigenetic Mechanisms Regulate Gene Expression Using Environmental Cues

Epigenetic mechanisms are thought to act as a memory of a cell and might be the key process by which the environment interacts with the genome [17]. DNA methylation plays a crucial role during early development including active demethylation of paternal genome before the first cleavage and subsequent demethylation of maternal genome [18]. Furthermore, environmental factors also affect gene expression via epigenetic mechanisms during embryonic development, which can manifest into adulthood or even old age. Cigarette smoking is an environmental factor, associated with dose- and time-dependent changes in the DNA methylation signature, which manifests in gene and protein expression leading to an increased vulnerability to other forms of complex illnesses [19,20]. Harmful environmental factors need not be substances. Trauma and stress also influence gene expression through epigenetic mechanisms, and furthermore these epigenetic modifications can be passed over the generations [21].

2. Chromatin Modifications and the Genome Organization

2.1. Chromatin's Structure Defines Its Function

To understand epigenetic control mechanisms, we will begin with the structure of chromatin. DNA is wrapped around the core histone proteins, forming a structure named nucleosome (two copies of H2A, H2B, H3, H4, and 147 base pairs (bps) of DNA around them). This is further compacted, with the assistance of assembly and packaging related proteins, to form a higher-order chromatin structure [22], with two distinct chromatin states "euchromatin" and "heterochromatin" (Figure 1). A more open chromatin environment, euchromatin, is where the majority of active genes localize, while heterochromatin is characterized by a more compact environment where inactive genes, non-coding DNA and repeat elements reside [8]. Heterochromatin can be further separated into two groups, facultative and constitutive. Facultative heterochromatin includes regions that consist of genes that are highly differentially expressed during development. Constitutive heterochromatin on the other hand is gene poor, rich in repeat elements, mainly found in centromeres and telomeres, and silenced indefinitely [23]. These chromatin states are marked by distinct epigenetic factors [24] (Figure 1), and in euchromatin, the histone modification density correlates with the density of TF binding sites [25]. However, neither euchromatin nor heterochromatin is marked uniformly with epigenetic and transcriptional signals. Chromatin is further organized into so-called topologically associated domains (TADs), (first described by Dixon et al. (2012) [26]), regions spanning several hundred kilobases. Topologically associated domains are organized hierarchically and are highly enriched for insulating factor CCCTC-binding factor (*CTCF*) binding and histone marks at the boundaries [27]. Intra-chromosomal interactions are particularly enriched within TADs and accordingly genes within a TAD show highly correlated gene expression. The chromatin structure allows manifestation of genetic information in a cellular context, and mutations in chromatin organization genes lead to developmental pathologies [28,29]. Understanding of cell type specific 3D genome organization is therefore highly valuable in a disease context [30,31], where by disruption of TADs can result in chromatin interaction changes leading to mis-regulation of oncogenic or tumor suppressor genes [32].

2.2. Chromatin Structure is Dynamic and Marked by Histone Modifications

The chromatin structure is organized with the help of DNA sequence and epigenetic modifications, including histone modifications, and a cross-talk between them is potentially facilitated through histone amino (N)-terminal tails interacting with neighboring nucleosomes [33]. Of various histone modifications, the most well-studied types are methylation, acetylation, phosphorylation, and ubiquitination [8]. Histone modifications influence chromatin mainly in two ways. The first mode of the modifications affects directly the structure of the chromatin over a long or short distance by the recruitment of DNA binding proteins and chromatin remodelers affecting nucleosome location. Hence, nucleosome removal could open the chromatin and a possible transcription factor binding motif could be revealed, or otherwise, newly recruited nucleosomes could conceal a binding motif, hindering transcriptional machinery recruitment at the locus [34]. The second mode of histone modification is carried out by three sets of enzymes named "writer", "reader", and "eraser", based on the function of each enzyme related to each histone modification. For example, COMPASS family members maintain H3K4m3 modification, while polycomb family members maintain H3K27me3 modification. Both activating (H3K4me3) and repressing (H3K27me3) modifications are indeed present simultaneously at promoters enriched for developmental genes and have a distinct sequence signature [35]. Histone modifications also work jointly with DNA methylation, for repression of gene loci [36].

3. Epigenetics in Disease Context

3.1. Genome-Wide Studies Are Not Enough

Monogenic diseases are caused by the malfunctioning of only a single gene. For example, fragile x syndrome is caused by epigenetic changes in the *FMR1* gene. The silenced promoter of *FMR1* in disease shows heterochromatin markers, including DNA hypermethylation and histone deacetylation. This can be treated by pharmacological reactivation of gene transcription, particularly through the use of DNA demethylating agents or inhibitors of histone deacetylases [37]. Unfortunately, the vast majority of common diseases are not caused by mutations in a single gene, but rather by a large number of single nucleotide variations (SNPs) spread throughout the genome. These diseases are therefore called complex diseases. Complex diseases including cancer, diabetes, and neurodegenerative disorders such as Alzheimer's and Parkinson's disease are common and therefore form a global health burden. Though a large number of genetic variants have been identified (and will be identified) that increase the risk for these diseases, most explain only a small fraction of risk. Moreover, despite the fact that over 1000 genetic loci are associated with susceptibility to common diseases in human [38], only a handful of these loci have resulted in the identification of causal genes or pathways for potential therapeutic applications [39]. It is becoming clear that understanding of only genetic variation will not be sufficient to get a complete understanding of disease, and the role of epigenetic alterations in gene regulation is becoming evident in many diseases, including cancer. Understanding how a genotype influences human health and disease now requires characterization of the epigenome as well. For example, copy number aberrations of genes responsible for writing, reading, and removing H3K9 methylation were identified in medulloblastoma, demonstrating that defective control of the histone code contributes to the pathogenesis of medulloblastoma [40]. Large studies have therefore been designed to unravel epigenetic malfunctionalities in diverse diseases (Table 1). It is important to note that another major challenge in interpreting genome-wide data in a clinical context is the fact that the vast majority of genetic and epigenetic modifications lie in non-coding genomic regions, particularly [41] where the disease-associated variants in enhancers explain a greater proportion of the disease heritability [42].

Table 1. A collection of epigenetic studies (excluding DNA methylation) in disease context including the data type, number of samples, disease type, and publication reference.

Num.	Data Type	Disease	Available data	# of Samples	Reference
1	ATAC-seq	23 cancer types	Genotype, ATAC-seq, RNA-seq	410	[43]
2	ChIP-seq	Prostate cancer	H3K27ac, H3K4me3, H3K27me3	100	GSE120738
3	ChIP-seq	Breast cancer	H3K4me1, TFs	-	[44]
4	ChIP-seq	Adenocarcinoma	H3K27ac, H3K4me3, H3K4me1	94	[45]
5	ChIP-seq	Acute myeloid leukemia	H3K9me3	108	[46]
6	ChIP-seq	Glioma	Multiple	-	[47]
7	ChIP-on-chip	Acute myeloid leukemia	H3	73	[48]
8	ChIP-on-chip	Acute promyelocytic leukemia	H3, H3K9me3, H3K4me3	372	[49]
9	ChIP-seq	Acute myeloid leukemia	H3K9me2	16	[50]
10	ChIP-seq	Hepatocarcinoma	Multiple	5	[51]
11	ATAC-seq, ChIP-seq	Colorectal cancer	Multiple	4	[52]
12	FAIRE-seq, ChIP-seq	Ovarian cancer	H3K27ac, H3K4me1	5	[53]

3.2. Largescale Epigenetic Studies in Cancer

3.2.1. Epigenetic Mechanisms Are Major Drivers in Cancer

The studies exploring mutational landscapes of cancer have highlighted frequent mutations in genes encoding chromatin-associated proteins. The exploration of functional mechanisms behind these mutations have improved our understanding of oncogenic mechanisms at different levels of chromatin organization and regulation (reviewed in Valencia et al. (2019) [54]). DNA methylation remains by far the most studied epigenetic mechanism in cancer where inactivation of tumor-suppressor genes occurs as a consequence of hypermethylation of the gene promoters. Numerous studies have identified a broad range of genes silenced by DNA methylation in different cancer types [55]. Importantly, different cancer subtypes show characteristic DNA methylation signatures [56], which can be translated in clinical medicine by using hypermethylated promoters as biomarkers. Human pluripotent stem cells were found to have more hypermethylated DNA than fibroblast cells [57]. Similarly, oncogenesis is thought to modify the cell state into a stem or progenitor epigenetic state. In cancer, mutations in key transcription factors lead to changes in DNA methylation, such that the number of genes with gene expression changes explained by DNA methylation are 10-fold higher than those explained by genetic mutations. Over 75% of DNA hypermethylated genes are marked by polycomb repressor components forming bivalent chromatin [58]. Wang et al. [59] pointed to one molecular mechanism to explain the role of *MLL3* mutations in cancer pathogenesis by examining changes in histone modification and gene expression after depletion of Polycomb or COMPASS family members. Next, they proposed a potential therapeutic strategy for cancers harboring COMPASS mutations which will allow resetting the epigenetically (Polycomb/COMPASS) balanced state of gene expression.

3.2.2. Epigenetic Mechanisms in Hematopoietic Malignancies and Their Therapeutic Implications

Epigenetic changes in cancer are possibly reversible making them precious targets for cancer therapy. Indeed, DNA methylation biomarkers with diagnostic, prognostic, and predictive power are already in clinical trials or in a clinical setting [60]. DNA methyltransferase inhibitors have been approved for the treatment of several hematopoietic malignancies, including myelodysplastic syndromes, chronic myelomonocytic leukemia, and acute myelogenous leukemia (AML) [61]. Other epigenetic regulatory mechanisms also play a critical role in the pathogenesis of AML. Epigenome-wide analyses of histone H3 acetylation identified that epigenetic silencing of PRDX2, a growth suppressor, contributed to the malignant phenotype in AML [48]. A combination of the H3K9me3 signature with established clinical prognostic markers outperformed prognosis prediction based on clinical parameters alone in AML [46]. Epigenetic control is systematically studied in other hematopoietic malignancies as well. For example, the translocation t (15;17) forming a chimeric PML–RARα transcription factor is the initiating event of acute promyelocytic leukemia. PML-RARα regulates key cancer related genes and pathways by inducing a repressed chromatin at its target genes [49]. The PML–RARα binding universally led to histone deacetylase (HDAC) recruitment, loss of histone H3 acetylation, and increased H3K9me3. Accordingly, several anticancer drugs acting as inhibitors of *HDAC* or bromodomain and extra-terminal proteins (*BET*) were designed, tested, and in clinical trial. The use of these inhibitors is not limited to hematopoietic malignancies. The *HDAC* inhibitors have been used in glioblastomas, where mutations in tumor suppressors such as *IDH1* induce epigenetic changes that drive the development of gliomas [47]. Both *HDAC* and *BET* inhibitors work synergistically, primarily by suppressing super-enhancers, the regulatory regions driving cancer phenotype through epigenetic reprogramming. Indeed, adenocarcinoma super-enhancers classified according to their somatic alteration status display distinct epigenetic, transcriptional and pathway enrichments and are enriched in genetic risk SNPs associated with cancer predisposition [45].

3.2.3. Epigenetic Targets for Cancer Therapy

Unfortunately, the current cancer drugs targeting epigenetic mechanisms are unspecific and can often have serious side effects. Understanding other epigenetic changes in cancer is therefore highly urgent to open up avenues for new therapies. The pharmaceutical industry is therefore focused on identifying new compounds that target the reader, writer, and eraser mechanisms of histone modifications. To this end, functional genomics studies in disease are gaining pace. A recent large study generated ATAC-seq data, a proxy for mapping genome-wide open chromatin, in over 400 tumors across 23 cancer types from The Cancer Genome Atlas project [43]. The authors further identified enhancer–promoter interactions in different cancer types by integrating it with RNA-seq data and validated some of their predictions through CRISPR-Cas9 assays [43].

3.3. Largescale Epigenetic Studies in Other Diseases

The potential of epigenetic therapies for cancer treatment has influenced an increase in studies investigating epigenetic control across a wide range of other diseases. Such efforts have generated knowledge about the combinatorial effects of genetic mutations and epigenetics on the phenotype. For example, the interaction of genetic variants and DNA methylation of the interleukin-4 receptor gene increases the risk of asthma [62], and a genetic/epigenetic interaction in the reduced folate carrier (RFC1) gene locus influence fetal predisposition to autism [63]. The study of epigenetic mechanisms is highly relevant to some diseases. One of the major concerns of the aging world population today are neurodegenerative disorders. There is no cure for many of the neuropathies and the majority of the cases have no genetic basis. Many compounds function via epigenetic mechanisms, and epidrugs (discussed above) developed for cancer treatment have been submitted to clinical trials for the treatment of Alzheimer's and Parkinson's diseases [64]. For example, HDAC inhibitors change the epigenetic state and expression of *FXN* in the neurodegenerative disease Friedreich ataxia, making it highly effective in an in vitro disease model and also showing promising results in a patient study [65]. In summary, understanding of epigenomic landscape of neurodegenerative and other disorders will likely provide a possibility of early detection and intervention of pre-symptomatic pathological events. This will allow development and implementation of novel strategies or treatments to halt pathological progress. It is important to stress that it is the putative reversibility of epigenetic aberrations that enables pharmacological interventions (epidrugs) as potential novel candidates for successful treatments of multifactorial disorders [64].

4. Computational Approaches towards Epigenetic Data Analysis and Integration

4.1. Epigenetic Data Integration to Understand the "Epigenetic Code"

Several studies have connected specific combinations of histone modifications and DNA methylation to the presence or absence of transcriptional activity and genomic functional elements. For instance, H3K4me3 is highly enriched at the promoters of actively transcribed genes [25], H3K36me3 is found on the gene body of genes under transcription and high levels of H3K9me3 are associated with facultative heterochromatin [23]. ChIP sequencing technology has allowed to generate a genome-wide high-resolution map of the distribution and co-localization of histone marks. Large initiatives have focused on unravelling the human epigenetic landscape. The Roadmap Epigenomics consortium has collected 111 reference human epigenomes by profiling histone modification patterns, DNA accessibility, DNA methylation, and RNA expression to define global maps of regulatory elements, regulatory modules of coordinated activity, and their likely activators and repressors [41]. They further used a method based on Hidden Markov Models (HMMs) to derive a minimal informative set of epigenetic modifications for differentiating between cell types, tissues and development stages, as well as between healthy and diseased cells. Increasingly, epigenetic data is generated in clinical settings, for a move towards precision medicine. For example, Polak et al. [66] were able to pinpoint differences in the mutational landscape between cancers based on their cell type of origin. In their work, a random forest based approach was used to predict mutation densities using 424 predictor variables. When gene expression is available, together with DNA methylation levels and genotypes, one could construct a network of interactions between these features, as introduced by Hou et al. [67]. Such an approach is useful in prognosis of various cancers. This was also demonstrated by Zhu et al. [68], who tested a kernel machine learning method on various omics data and clinical factors to predict prognosis in 14 cancer types. They found that the prognostic power of copy number and somatic mutations was quite low compared to expression profiles. Moreover, they demonstrated that incorporating omics data to predictions based on clinical variables can improve the results, as it may account for the absence of unknown or unmeasured clinical features.

The Function of Epigenetic Modifications Still Remains Understudied

Sekhon et al. [69] integrated five different histone modification datasets to predict gene expression levels with the use of deep neural networks. Hlady et al. [51] performed integrative analysis of multiple epigenetic modifications in hepatic cancer to identify epigenetic driver loci, and further demonstrated that two loci, *COMT* and *FMO3*, increase apoptosis and decrease cell viability in a liver-derived cancer cell line. There is an effort to integrate more and more epigenetic phenomena in such studies, but the large number of histone modifications possible at histone tails increases the combinatorial complexity of the histone code. Furthermore, histone modifications or the histone status varies during development [70]. The histone code is therefore complex and dynamic. More importantly, the causal relationship between histone modifications and transcription activity has not yet been deciphered. For example, H3K4me1 is present at regulatory elements called enhancers, and is widely used to predict enhancer elements [71]. However, whether H3K4me1 controls or simply correlates with enhancer activity and function has remained unclear. Recent studies suggest that H3K4me1 might fine-tune, rather than tightly control, enhancer activity and function [72].

4.2. Linking Epigenetic Mechanisms to Phenotypes: Epigenetic Epidemiology

4.2.1. More Data Equals More Challenges

The success of genome-wide association studies (GWAS) in identifying genetic loci associated with common diseases have facilitated exploration of epigenetic loci associated with diseases, also known as the epigenome-wide association studies (EWAS). Much focus in the EWAS-type analysis has been on genome-wide DNA methylation studies, where a statistical framework is developed to identify statistically significant association between the methylation level of each CpG site and the trait of interest (reviewed in References [73,74]). However, as the technologies constantly improve to make data from other epigenetic markers available, more and more researchers integrate this data, together with genetic information to improve predictions and risk assessment [75–78]. The integration of data from diverse sources is generally a daunting task. This challenge can be simplified with the help of new experimental methods such as assay for transposase-accessible chromatin using sequencing (ATAC-seq) allow for extracting information about different epigenetic phenomena from a single experiment [79]. Moreover, one can use existing databases that enable visualization of publicly available datasets, sometimes also giving the possibility to overlay user's data [80].

4.2.2. New Data Integration Opportunities

The most widely applied method in epigenetic epidemiology is to use a regression model to check associations between variations in the data and the trait, as in standard epidemiology. This methodology is used by various studies where principal components (PCs) [81], level of methylation [82,83], or association score from EWAS analysis [84] are used to represent the variation. In order to facilitate the interpretation of the results from such an analysis, one typically uses bioinformatics databases to search for possible biological explanations for connections between the significant genomic regions and the trait of interest. This can be done, for example, in the Cistrome database [85] that gathers published gene regulatory data, and enables interactive visual analysis. Another easy-to-use tool is HaploReg [86]. Although the output is less intuitive, the database provides rich information about possible regulatory functions of SNPs or genomic regions of interest. Having found a set of genes that contain differential epigenetic modifications allows to perform a gene enrichment analysis, for example, with the help of LAGO (https://go.princeton.edu/cgi-bin/LAGO), STRING [87] or Reactome [88]. Another interesting possibility is to infer disease–gene connections by accounting for associations between different types of data, as implemented in Hetionet [89]. This tool integrates around 30 different databases, creating a heterogeneous network from information such as expression data, differential gene regulation, GWAS gene-trait associations, drug banks, etc. The implementation of the database in a neo4j network service allows for a quick online querying and visually appealing output that can inform on hidden connections between, for example, influence of vitamin intake, genes, and a disease [90].

4.2.3. Epigenome-Wide Association Studies Analyses Are Informative Only about an Association and Not Causality

In classical epidemiology Mendelian randomization (MR) is widely used to infer causality whenever a standard randomized trial is impossible to perform. It is based on an assumption that the underlying genotype is randomly assigned to each individual and is the cause of the measured exposure (e.g., body mass index (BMI)), not vice versa. This method has been recently adapted to DNA methylation data [91,92]. However, since DNA methylation can be both an inducer and the outcome of the disease, MR with epigenetic data needs to be used with caution [93]. Nevertheless, the remarkably simple idea behind the MR allows researchers to make very interesting claims, studying causality between the epigenetic marks and a wide range of outcomes, from blood lipid levels [94] through features such as physical aggression [95]. Used together with EWAS and GWAS analyses, MR gives us the possibility to propose biomarker loci or targets for therapies for patients [82].

Many more methods have been developed recently to infer causality from epigenetic data. For example, Howey et al. [96] fit a Bayesian network to the most significant findings from their linear regression modeling to show the directions of influence between DNA methylation and blood lipid levels. In another study, structural equation modeling (SEM) was used to search for the pathways by which the genetic variants lead to a disease [97]. With this method, one can establish significant interactions between all the different measurements (here, blood lipid levels, variant allele in the chosen SNP, and methylation levels on the nearby CpGs) and importantly, the model predicts the directionality of these interactions.

4.2.4. Causality Inference from Translational Studies

To test whether a change in the levels of epigenetic modifications is the cause or consequence of a disease, one can conduct a translational study, following patients over a specific time. Such time-dependent information can then be used to check whether a certain locus displays epigenetic changes, e.g., DNA methylation, before or after a certain event; disease onset. Using this concept, a computational approach GATE [98] has been implemented as a two-layer model, where one layer categorizes the spatial characteristics of the chromatin, and the other layer focuses on transitions between different chromatin states. This allows to create a model of transitions between different epigenetic states of a cell. Another recent method, ChromTime, uses the raw signal from data generated by CHiP-seq and similar techniques to track temporal changes in the peaks [99]. It not only detects diminishing or appearing peaks, but also asymmetrical changes in peak shapes. The authors further demonstrate that ChromTime can be applied on ATAC-seq, CHiP-seq, and DNase-seq data to infer on gene expression levels and TF binding.

4.3. Combining Levels of Epigenetic Marks within Genomic Regions

One of the important shortcomings of many methods is that they consider each epigenetic locus independently of other loci to evaluate its significance for association with a certain trait. For example, the majority of studies focus on methylation level of one CpG at a time even when they integrate it with several other data sources. Recent studies summarized methylation level within a region [81,83], though this is not yet widely used despite the fact that changes in DNA methylation of only one CpG site would likely not lead to big changes in TF binding affinity to this site, unless it is followed by coordinated changes on neighboring CpG sites [100]. To this end, we developed a statistical framework that integrates DNA methylation and genetic information to identify statistically significant interactions between an SNP and methylation level within a group of neighboring CpGs [101]. The CpGs are grouped based on whether the CpGs are assigned to a promoter, enhancer or a gene body, to facilitate the downstream analysis for the biological interpretations.

The ultimate goal is to understand how genetic and epigenetic variations manifest in a phenotype under certain environmental conditions (Figure 2). To this end, an ideal computational approach would take into account the genotype and several epigenetic modifications at the same time, to explain a phenotype or perhaps a proxy such as transcriptomic data. There is already a huge amount of such data in the public domain, and tools and resources such as Omics Discovery Index web service (https://www.omicsdi.org/) to search for datasets. There is a need to take the advantage of this enormous amount of data, test ideas, and to develop tools to maximize the information extraction from the data.

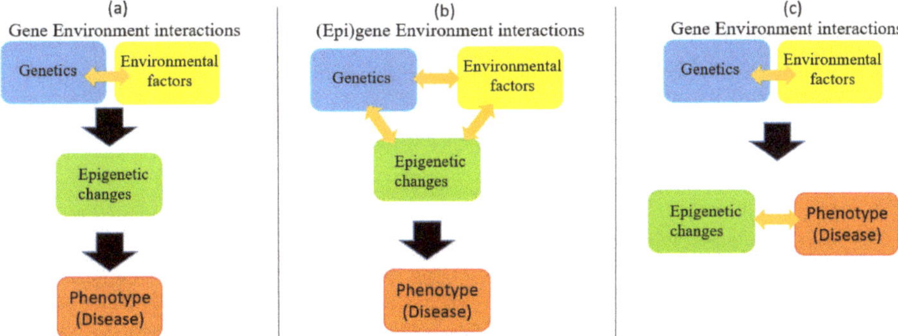

Figure 2. The figure depicts three likely scenarios where epigenetics might fit with from the genotype to phenotype (gene expression) information flow: (**a**) epigenetic changes are downstream of gene environment interactions and determine the phenotype; (**b**) genome sequence, environment, and epigenetic modification work together to establish the phenotype; and (**c**) epigenetic landscape and phenotype are both determined and established by gene–environment interactions. SNP: single nucleotide variations.

5. Conclusions

5.1. Possible Scenarios Linking Epigenetics, Genetics, and Phenotype

Hundreds of human cell types have a unique gene expression signature despite sharing the same genome sequence, largely due to tight control by epigenetic modifications of the non-coding genome in a cell type specific manner. Epigenetic aberrations are thought to result in complex diseases such as cancers. The vast majority of genetic variants found associated with common diseases by genome-wide association studies are indeed located in the non-coding genome. Only in a very limited number of cases such as for Crohn's disease or rheumatoid arthritis, have these associations led to the successful identification of causal genes with a potential of being therapeutic targets. However, most disease-associated variants have no known biological context to disease, limiting their utility for prognosis or treatment. Human epidemiological studies provide evidence for prenatal and early postnatal environmental factors influencing adult risk of developing various chronic diseases, such as cancer, cardiovascular disease, diabetes, obesity, and behavioral disorders such as schizophrenia [17]. Some of these environmental factors can be linked directly to alterations of the epigenetic landscape that affect gene regulation and finally the disease. Though the association is proven in many cases, the chain of causality remains to be established. This leads to three possible scenarios of how epigenetic mechanisms control genes and influence disease occurrence (Figure 2). The first scenario is where environmental factors alter epigenetic modifications, which in turn alter the phenotype (Figure 2A). This scenario is supported by mouse experiments where maternal methyl-donor supplementation during pregnancy with folic acid, vitamin B12, choline, and betaine was shown to affect the phenotype of the Avy (viable yellow agouti) offspring by directly altering the epigenome [102]. The second possibility is that gene–environment interactions affect both epigenetic status and transcription read-out, as their correlation does not imply causality (Figure 2B). Indeed, as most of the epigenetic modifications are "lost" during the gametogenesis, this scenario is assumed to be true for many cases. Careful research has nevertheless identified that at least some epigenetic modifications are passed on to the next generation [103]. This leads to a third scenario where epigenetic modifications are not downstream of but work together with gene environment interactions to result in a phenotype (Figure 2C). The relative abundance of the three scenarios and the molecular mechanisms controlling them need to be understood. Over the coming years, research should be focused not only on identifying epigenetic phenomena affecting gene regulation to find epigenetic biomarkers for disease and environmental exposure, but also on establishing the causal relationship between the three components (gene–environment, epigenetics, and phenotype). Only by understanding causal relations can we develop new epigenetic interventions to truly revolutionize medicine to move towards preventive medicine.

5.2. New Approaches and Technologies Must Aim on Establishing a Causal Link between Epigenetics and Disease

The most important challenge in precision medicine is thus to link genetic variation within the non-coding genome to candidate causal gene(s) or pathways for disease or other physiological phenotypes. It is urgent not only to identify the regulatory regions but also the spatial organization of DNA to understand how these regulatory regions interact to manifest into a phenotype. It is now accepted that a large number of possible regulatory interactions are potentially pathogenic and might be unique to tumors [43]. Although experimental techniques such as chromosome conformation capture (3C, 4C) [104] combined with next generation sequencing (Hi-C) show a great promise [105,106], their time and cost will limit the availability of comprehensive, experimentally verified 3D chromatin landscapes to a tiny fraction of the hundreds of different human cell types in the foreseeable future. The development of novel cost-effective high-throughput experimental methods is ongoing. Meanwhile, computational tools to predict enhancer–promoter interactions will be essential to model the effects of non-coding genetic variation on epigenetic modifications and downstream gene expression programs in human health and disease. Though a regulatory region is associated to its proximal promoter, the integration of known or putative enhancer promoter interactions in GWAS analysis has a potential to identify novel disease associated genes and pathways [107]. This will require a significant leap beyond studies which have only used correlations between epigenetic states of enhancers with promoter expression [108]. We have recently performed preliminary work to establish causality using regulatory information [109]. More computational approaches to systematically combine epigenetic information into causal network models are needed.

5.3. Epigenetic Studies and Therapies Have an Important Role in Shaping the Future of Medicine

Finally, segregating patients based on different factors into more coherent groups for better treatment is the foundation of precision medicine, but many factors used to stratify patients have no known functional mechanisms. For example, sexual differences in cancer risk and survival are well studied, with males having an increased risk and poorer survival for most cancers [110]. The understanding of functional mechanisms behind these sex differences is gathering pace. For example, male breast cancer is rare, poorly characterized and resistant to hormonal treatment. An integrative epigenetic and transcriptomic analysis revealed a gender-selective and genomic location-specific hormone receptor action associated with survival in male breast cancer [39]. Epigenetics therefore has a big role to play in the foundations of the precision medicine.

Author Contributions: Conceptualization, A.J.; data curation, J.R. and A.J.; writing—original draft preparation, J.R. and A.J.; writing—review and editing, J.R. and A.J.; visualization, A.J.; funding acquisition, A.J.

Funding: AJ is supported by the Bergen Research Foundation Grant no. BFS2017TMT01. The APC was funded by open access fund of University of Bergen.

Acknowledgments: We thank Tom Michoel for useful feedback.

Conflicts of Interest: The authors declare no conflict of interest.

References

1. Nicoglou, A. Waddington's epigenetics or the pictorial meetings of development and genetics. *Hist. Philos. Life Sci.* **2018**, *40*, 61. [CrossRef] [PubMed]
2. Bird, A.P.; Wolffe, A.P. Methylation-induced repression—Belts, braces, and chromatin. *Cell* **1999**, *99*, 451–454. [CrossRef]
3. Mohn, F.; Weber, M.; Rebhan, M.; Roloff, T.C.; Richter, J.; Stadler, M.B.; Bibel, M.; Schübeler, D. Lineage-specific polycomb targets and de novo DNA methylation define restriction and potential of neuronal progenitors. *Mol. Cell* **2008**, *30*, 755–766. [CrossRef] [PubMed]
4. Bird, A. DNA methylation patterns and epigenetic memory. *Genes Dev.* **2002**, *16*, 6–21. [CrossRef] [PubMed]
5. Morgan, H.D.; Santos, F.; Green, K.; Dean, W.; Reik, W. Epigenetic reprogramming in mammals. *Hum. Mol. Genet.* **2005**, *14*, R47–R58. [CrossRef] [PubMed]
6. Reik, W. Stability and flexibility of epigenetic gene regulation in mammalian development. *Nature* **2007**, *447*, 425–432. [CrossRef] [PubMed]
7. Turner, B.M. Defining an epigenetic code. *Nat. Cell Biol.* **2007**, *9*, 2–6. [CrossRef] [PubMed]
8. Bannister, A.J.; Kouzarides, T. Regulation of chromatin by histone modifications. *Cell Res.* **2011**, *21*, 381–395. [CrossRef]
9. Jenuwein, T.; Allis, C.D. Translating the histone code. *Science* **2001**, *293*, 1074–1080. [CrossRef]
10. Strahl, B.D.; Allis, C.D. The language of covalent histone modifications. *Nature* **2000**, *403*, 41–45. [CrossRef]
11. Zhou, V.W.; Goren, A.; Bernstein, B.E. Charting histone modifications and the functional organization of mammalian genomes. *Nat. Rev. Genet.* **2011**, *12*, 7–18. [CrossRef] [PubMed]
12. Zentner, G.E.; Henikoff, S. Regulation of nucleosome dynamics by histone modifications. *Nat. Struct. Mol. Biol.* **2013**, *20*, 259–266. [CrossRef]
13. Allis, C.D.; Jenuwein, T. The molecular hallmarks of epigenetic control. *Nat. Rev. Genet.* **2016**, *17*, 487–500. [CrossRef] [PubMed]
14. Maleszewska, M.; Kaminska, B. Is Glioblastoma an Epigenetic Malignancy? *Cancers* **2013**, *5*, 1120–1139. [CrossRef] [PubMed]
15. Frías-Lasserre, D.; Villagra, C.A. The Importance of ncRNAs as Epigenetic Mechanisms in Phenotypic Variation and Organic Evolution. *Front. Microbiol.* **2017**, *8*, 2483. [CrossRef] [PubMed]
16. Lee, J.T. Lessons from X-chromosome inactivation: Long ncRNA as guides and tethers to the epigenome. *Genes Dev.* **2009**, *23*, 1831–1842. [CrossRef] [PubMed]
17. Jirtle, R.L.; Skinner, M.K. Environmental epigenomics and disease susceptibility. *Nat. Rev. Genet.* **2007**, *8*, 253–262. [CrossRef]
18. Okano, M.; Bell, D.W.; Haber, D.A.; Li, E. DNA methyltransferases Dnmt3a and Dnmt3b are essential for de novo methylation and mammalian development. *Cell* **1999**, *99*, 247–257. [CrossRef]
19. Philibert, R.A.; Beach, S.R.H.; Brody, G.H. The DNA methylation signature of smoking: An archetype for the identification of biomarkers for behavioral illness. *Neb. Symp. Motiv.* **2014**, *61*, 109–127.
20. Joubert, B.R.; Felix, J.F.; Yousefi, P.; Bakulski, K.M.; Just, A.C.; Breton, C.; Reese, S.E.; Markunas, C.A.; Richmond, R.C.; Xu, C.-J.; et al. DNA Methylation in Newborns and Maternal Smoking in Pregnancy: Genome-wide Consortium Meta-analysis. *Am. J. Hum. Genet.* **2016**, *98*, 680–696. [CrossRef]

21. Keleher, M.R.; Zaidi, R.; Shah, S.; Oakley, M.E.; Pavlatos, C.; El Idrissi, S.; Xing, X.; Li, D.; Wang, T.; Cheverud, J.M. Maternal high-fat diet associated with altered gene expression, DNA methylation, and obesity risk in mouse offspring. *PLoS ONE* **2018**, *13*, e0192606. [CrossRef] [PubMed]

22. Li, E. Chromatin modification and epigenetic reprogramming in mammalian development. *Nat. Rev. Genet.* **2002**, *3*, 662–673. [CrossRef] [PubMed]

23. Trojer, P.; Reinberg, D. Facultative heterochromatin: Is there a distinctive molecular signature? *Mol. Cell* **2007**, *28*, 1–13. [CrossRef] [PubMed]

24. Filion, G.J.; van Bemmel, J.G.; Braunschweig, U.; Talhout, W.; Kind, J.; Ward, L.D.; Brugman, W.; de Castro, I.J.; Kerkhoven, R.M.; Bussemaker, H.J.; et al. Systematic Protein Location Mapping Reveals Five Principal Chromatin Types in *Drosophila* Cells. *Cell* **2010**, *143*, 212–224. [CrossRef]

25. Barski, A.; Cuddapah, S.; Cui, K.; Roh, T.-Y.; Schones, D.E.; Wang, Z.; Wei, G.; Chepelev, I.; Zhao, K. High-resolution profiling of histone methylations in the human genome. *Cell* **2007**, *129*, 823–837. [CrossRef]

26. Dixon, J.R.; Selvaraj, S.; Yue, F.; Kim, A.; Li, Y.; Shen, Y.; Hu, M.; Liu, J.S.; Ren, B. Topological domains in mammalian genomes identified by analysis of chromatin interactions. *Nature* **2012**, *485*, 376–380. [CrossRef]

27. Filippova, D.; Patro, R.; Duggal, G.; Kingsford, C. Identification of alternative topological domains in chromatin. *Algorithms Mol. Biol. AMB* **2014**, *9*, 14. [CrossRef]

28. Hnisz, D.; Weintraub, A.S.; Day, D.S.; Valton, A.-L.; Bak, R.O.; Li, C.H.; Goldmann, J.; Lajoie, B.R.; Fan, Z.P.; Sigova, A.A.; et al. Activation of proto-oncogenes by disruption of chromosome neighborhoods. *Science* **2016**, *351*, 1454–1458. [CrossRef]

29. Kaiser, V.B.; Taylor, M.S.; Semple, C.A. Mutational Biases Drive Elevated Rates of Substitution at Regulatory Sites across Cancer Types. *PLoS Genet.* **2016**, *12*, e1006207. [CrossRef]

30. Mifsud, B.; Tavares-Cadete, F.; Young, A.N.; Sugar, R.; Schoenfelder, S.; Ferreira, L.; Wingett, S.W.; Andrews, S.; Grey, W.; Ewels, P.A.; et al. Mapping long-range promoter contacts in human cells with high-resolution capture Hi-C. *Nat. Genet.* **2015**, *47*, 598–606. [CrossRef]

31. Corces, M.R.; Corces, V.G. The three-dimensional cancer genome. *Curr. Opin. Genet. Dev.* **2016**, *36*, 1–7. [CrossRef] [PubMed]

32. Achinger-Kawecka, J.; Clark, S.J. Disruption of the 3D cancer genome blueprint. *Epigenomics* **2017**, *9*, 47–55. [CrossRef] [PubMed]

33. Kouzarides, T. SnapShot: Histone-modifying enzymes. *Cell* **2007**, *131*, 822. [CrossRef] [PubMed]

34. Margueron, R.; Trojer, P.; Reinberg, D. The key to development: Interpreting the histone code? *Curr. Opin. Genet. Dev.* **2005**, *15*, 163–176. [CrossRef] [PubMed]

35. Mantsoki, A.; Devailly, G.; Joshi, A. CpG island erosion, polycomb occupancy and sequence motif enrichment at bivalent promoters in mammalian embryonic stem cells. *Sci. Rep.* **2015**, *5*, 16791. [CrossRef] [PubMed]

36. Bartke, T.; Vermeulen, M.; Xhemalce, B.; Robson, S.C.; Mann, M.; Kouzarides, T. Nucleosome-interacting proteins regulated by DNA and histone methylation. *Cell* **2010**, *143*, 470–484. [CrossRef] [PubMed]

37. Tabolacci, E.; Chiurazzi, P. Epigenetics, fragile X syndrome and transcriptional therapy. *Am. J. Med. Genet. A.* **2013**, *161A*, 2797–2808. [CrossRef]

38. Welter, D.; MacArthur, J.; Morales, J.; Burdett, T.; Hall, P.; Junkins, H.; Klemm, A.; Flicek, P.; Manolio, T.; Hindorff, L.; et al. The NHGRI GWAS Catalog, a curated resource of SNP-trait associations. *Nucleic Acids Res.* **2014**, *42*, D1001–D1006. [CrossRef]

39. Visscher, P.M.; Brown, M.A.; McCarthy, M.I.; Yang, J. Five years of GWAS discovery. *Am. J. Hum. Genet.* **2012**, *90*, 7–24. [CrossRef]

40. Northcott, P.A.; Nakahara, Y.; Wu, X.; Feuk, L.; Ellison, D.W.; Croul, S.; Mack, S.; Kongkham, P.N.; Peacock, J.; Dubuc, A.; et al. Multiple recurrent genetic events converge on control of histone lysine methylation in medulloblastoma. *Nat. Genet.* **2009**, *41*, 465–472. [CrossRef]

41. Roadmap Epigenomics Consortium; Kundaje, A.; Meuleman, W.; Ernst, J.; Bilenky, M.; Yen, A.; Heravi-Moussavi, A.; Kheradpour, P.; Zhang, Z.; Wang, J.; et al. Integrative analysis of 111 reference human epigenomes. *Nature* **2015**, *518*, 317–330. [CrossRef] [PubMed]

42. Corradin, O.; Scacheri, P.C. Enhancer variants: Evaluating functions in common disease. *Genome Med.* **2014**, *6*, 85. [CrossRef] [PubMed]

43. Corces, M.R.; Granja, J.M.; Shams, S.; Louie, B.H.; Seoane, J.A.; Zhou, W.; Silva, T.C.; Groeneveld, C.; Wong, C.K.; Cho, S.W.; et al. The chromatin accessibility landscape of primary human cancers. *Science* **2018**, *362*, eaav1898. [CrossRef] [PubMed]

44. Severson, T.M.; Kim, Y.; Joosten, S.E.P.; Schuurman, K.; van der Groep, P.; Moelans, C.B.; Ter Hoeve, N.D.; Manson, Q.F.; Martens, J.W.; van Deurzen, C.H.M.; et al. Characterizing steroid hormone receptor chromatin binding landscapes in male and female breast cancer. *Nat. Commun.* **2018**, *9*, 482. [CrossRef] [PubMed]

45. Ooi, W.F.; Xing, M.; Xu, C.; Yao, X.; Ramlee, M.K.; Lim, M.C.; Cao, F.; Lim, K.; Babu, D.; Poon, L.-F.; et al. Epigenomic profiling of primary gastric adenocarcinoma reveals super-enhancer heterogeneity. *Nat. Commun.* **2016**, *7*, 12983. [CrossRef] [PubMed]

46. Müller-Tidow, C.; Klein, H.-U.; Hascher, A.; Isken, F.; Tickenbrock, L.; Thoennissen, N.; Agrawal-Singh, S.; Tschanter, P.; Disselhoff, C.; Wang, Y.; et al. Profiling of histone H3 lysine 9 trimethylation levels predicts transcription factor activity and survival in acute myeloid leukemia. *Blood* **2010**, *116*, 3564–3571. [CrossRef] [PubMed]

47. Turcan, S.; Makarov, V.; Taranda, J.; Wang, Y.; Fabius, A.W.M.; Wu, W.; Zheng, Y.; El-Amine, N.; Haddock, S.; Nanjangud, G.; et al. Mutant-IDH1-dependent chromatin state reprogramming, reversibility, and persistence. *Nat. Genet.* **2018**, *50*, 62–72. [CrossRef]

48. Agrawal-Singh, S.; Isken, F.; Agelopoulos, K.; Klein, H.-U.; Thoennissen, N.H.; Koehler, G.; Hascher, A.; Bäumer, N.; Berdel, W.E.; Thiede, C.; et al. Genome-wide analysis of histone H3 acetylation patterns in AML identifies PRDX2 as an epigenetically silenced tumor suppressor gene. *Blood* **2012**, *119*, 2346–2357. [CrossRef]

49. Hoemme, C.; Peerzada, A.; Behre, G.; Wang, Y.; McClelland, M.; Nieselt, K.; Zschunke, M.; Disselhoff, C.; Agrawal, S.; Isken, F.; et al. Chromatin modifications induced by PML-RARalpha repress critical targets in leukemogenesis as analyzed by ChIP-Chip. *Blood* **2008**, *111*, 2887–2895. [CrossRef]

50. Salzberg, A.C.; Harris-Becker, A.; Popova, E.Y.; Keasey, N.; Loughran, T.P.; Claxton, D.F.; Grigoryev, S.A. Genome-wide mapping of histone H3K9me2 in acute myeloid leukemia reveals large chromosomal domains associated with massive gene silencing and sites of genome instability. *PLoS ONE* **2017**, *12*, e0173723. [CrossRef]

51. Hlady, R.A.; Sathyanarayan, A.; Thompson, J.J.; Zhou, D.; Wu, Q.; Pham, K.; Lee, J.H.; Liu, C.; Robertson, K.D. Integrating the Epigenome to Identify Novel Drivers of Hepatocellular Carcinoma. *Hepatol. Baltim. Md* **2018**.

52. Kelso, T.W.R.; Porter, D.K.; Amaral, M.L.; Shokhirev, M.N.; Benner, C.; Hargreaves, D.C. Chromatin accessibility underlies synthetic lethality of SWI/SNF subunits in ARID1A-mutant cancers. *eLife* **2017**, *6*, e30506. [CrossRef] [PubMed]

53. Coetzee, S.G.; Shen, H.C.; Hazelett, D.J.; Lawrenson, K.; Kuchenbaecker, K.; Tyrer, J.; Rhie, S.K.; Levanon, K.; Karst, A.; Drapkin, R.; et al. Cell-type-specific enrichment of risk-associated regulatory elements at ovarian cancer susceptibility loci. *Hum. Mol. Genet.* **2015**, *24*, 3595–3607. [CrossRef] [PubMed]

54. Valencia, A.M.; Kadoch, C. Chromatin regulatory mechanisms and therapeutic opportunities in cancer. *Nat. Cell Biol.* **2019**. [CrossRef] [PubMed]

55. Kulis, M.; Esteller, M. DNA Methylation and Cancer. *Adv. Genet.* **2010**. [CrossRef]

56. Bernhart, S.H.; Kretzmer, H.; Holdt, L.M.; Jühling, F.; Ammerpohl, O.; Bergmann, A.K.; Northoff, B.H.; Doose, G.; Siebert, R.; Stadler, P.F.; et al. Changes of bivalent chromatin coincide with increased expression of developmental genes in cancer. *Sci. Rep.* **2016**, *6*, 37393. [CrossRef] [PubMed]

57. Deng, J.; Shoemaker, R.; Xie, B.; Gore, A.; LeProust, E.M.; Antosiewicz-Bourget, J.; Egli, D.; Maherali, N.; Park, I.-H.; Yu, J.; et al. Targeted bisulfite sequencing reveals changes in DNA methylation associated with nuclear reprogramming. *Nat. Biotechnol.* **2009**, *27*, 353–360. [CrossRef] [PubMed]

58. Easwaran, H.; Johnstone, S.E.; Van Neste, L.; Ohm, J.; Mosbruger, T.; Wang, Q.; Aryee, M.J.; Joyce, P.; Ahuja, N.; Weisenberger, D.; et al. A DNA hypermethylation module for the stem/progenitor cell signature of cancer. *Genome Res.* **2012**, *22*, 837–849. [CrossRef] [PubMed]

59. Wang, L.; Zhao, Z.; Ozark, P.A.; Fantini, D.; Marshall, S.A.; Rendleman, E.J.; Cozzolino, K.A.; Louis, N.; He, X.; Morgan, M.A.; et al. Resetting the epigenetic balance of Polycomb and COMPASS function at enhancers for cancer therapy. *Nat. Med.* **2018**, *24*, 758–769. [CrossRef] [PubMed]

60. Mikeska, T.; Craig, J.M. DNA methylation biomarkers: Cancer and beyond. *Genes* **2014**, *5*, 821–864. [CrossRef]

61. Da Costa, E.M.; McInnes, G.; Beaudry, A.; Raynal, N.J.-M. DNA Methylation–Targeted Drugs. *Cancer J.* **2017**, *23*, 270–276. [CrossRef] [PubMed]

62. Soto-Ramírez, N.; Arshad, S.H.; Holloway, J.W.; Zhang, H.; Schauberger, E.; Ewart, S.; Patil, V.; Karmaus, W. The interaction of genetic variants and DNA methylation of the interleukin-4 receptor gene increase the risk of asthma at age 18 years. *Clin. Epigenet.* **2013**, *5*, 1. [CrossRef] [PubMed]

63. James, S.J.; Melnyk, S.; Jernigan, S.; Pavliv, O.; Trusty, T.; Lehman, S.; Seidel, L.; Gaylor, D.W.; Cleves, M.A. A functional polymorphism in the reduced folate carrier gene and DNA hypomethylation in mothers of children with autism. *Am. J. Med. Genet. Part B Neuropsychiatr. Genet.* **2010**, *153B*, 1209–1220. [CrossRef] [PubMed]

64. Teijido, O.; Cacabelos, R. Pharmacoepigenomic Interventions as Novel Potential Treatments for Alzheimer's and Parkinson's Diseases. *Int. J. Mol. Sci.* **2018**, *19*, 3199. [CrossRef] [PubMed]

65. Soragni, E.; Miao, W.; Iudicello, M.; Jacoby, D.; De Mercanti, S.; Clerico, M.; Longo, F.; Piga, A.; Ku, S.; Campau, E.; et al. Epigenetic therapy for Friedreich ataxia. *Ann. Neurol.* **2014**, *76*, 489–508. [CrossRef] [PubMed]

66. Polak, P.; Karlic, R.; Koren, A.; Thurman, R.; Sandstrom, R.; Lawrence, M.S.; Reynolds, A.; Rynes, E.; Vlahovicek, K.; Stamatoyannopoulos, J.A.; et al. Cell-of-origin chromatin organization shapes the mutational landscape of cancer. *Nature* **2015**, *518*, 360–364. [CrossRef] [PubMed]

67. Hou, X.; He, X.; Wang, K.; Hou, N.; Fu, J.; Jia, G.; Zuo, X.; Xiong, H.; Pang, M. Genome-Wide Network-Based Analysis of Colorectal Cancer Identifies Novel Prognostic Factors and an Integrative Prognostic Index. *Cell. Physiol. Biochem.* **2018**, *49*, 1703–1716. [CrossRef] [PubMed]

68. Zhu, B.; Song, N.; Shen, R.; Arora, A.; Machiela, M.J.; Song, L.; Landi, M.T.; Ghosh, D.; Chatterjee, N.; Baladandayuthapani, V.; et al. Integrating Clinical and Multiple Omics Data for Prognostic Assessment across Human Cancers. *Sci. Rep.* **2017**, *7*, 16954. [CrossRef] [PubMed]

69. Sekhon, A.; Singh, R.; Qi, Y. DeepDiff: DEEP-learning for predicting DIFFerential gene expression from histone modifications. *Bioinformatics* **2018**, *34*, i891–i900. [CrossRef] [PubMed]

70. Wu, J.; Xu, J.; Liu, B.; Yao, G.; Wang, P.; Lin, Z.; Huang, B.; Wang, X.; Li, T.; Shi, S.; et al. Chromatin analysis in human early development reveals epigenetic transition during ZGA. *Nature* **2018**, *557*, 256–260. [CrossRef] [PubMed]

71. Hon, G.C.; Hawkins, R.D.; Ren, B. Predictive chromatin signatures in the mammalian genome. *Hum. Mol. Genet.* **2009**, *18*, R195–R201. [CrossRef]

72. Rada-Iglesias, A. Is H3K4me1 at enhancers correlative or causative? *Nat. Genet.* **2018**, *50*, 4–5. [CrossRef]

73. Laird, P.W. Principles and challenges of genomewide DNA methylation analysis. *Nat. Rev. Genet.* **2010**, *11*, 191–203. [CrossRef] [PubMed]

74. Rakyan, V.K.; Down, T.A.; Balding, D.J.; Beck, S. Epigenome-wide association studies for common human diseases. *Nat. Rev. Genet.* **2012**, *12*, 529–541. [CrossRef] [PubMed]

75. Jiang, S.; Mortazavi, A. Integrating ChIP-seq with other functional genomics data. *Brief. Funct. Genom.* **2018**, *17*, 104–115. [CrossRef]

76. Lappalainen, T.; Greally, J.M. Associating cellular epigenetic models with human phenotypes. *Nat. Rev. Genet.* **2017**, *18*, 441–451. [CrossRef]

77. Ritchie, M.D.; Holzinger, E.R.; Li, R.; Pendergrass, S.A.; Kim, D. Methods of integrating data to uncover genotype-phenotype interactions. *Nat. Rev. Genet.* **2015**, *16*, 85–97. [CrossRef] [PubMed]

78. Auerbach, J.; Howey, R.; Jiang, L.; Justice, A.; Li, L.; Oualkacha, K.; Sayols-Baixeras, S.; Aslibekyan, S.W. Causal modeling in a multi-omic setting: Insights from GAW20. *BMC Genet.* **2018**, *19*, 74. [CrossRef]

79. Buenrostro, J.D.; Giresi, P.G.; Zaba, L.C.; Chang, H.Y.; Greenleaf, W.J. Transposition of native chromatin for fast and sensitive epigenomic profiling of open chromatin, DNA-binding proteins and nucleosome position. *Nat. Methods* **2013**, *10*, 1213–1218. [CrossRef] [PubMed]

80. Devailly, G.; Mantsoki, A.; Joshi, A. Heat*seq: An interactive web tool for high-throughput sequencing experiment comparison with public data. *Bioinforma. Oxf. Engl.* **2016**, *32*, 3354–3356. [CrossRef] [PubMed]

81. Romanescu, R.G.; Espin-Garcia, O.; Ma, J.; Bull, S.B. Integrating epigenetic, genetic, and phenotypic data to uncover gene-region associations with triglycerides in the GOLDN study 06 Biological Sciences 0604 Genetics. *BMC Proc.* **2018**, *12*, 57. [CrossRef] [PubMed]

82. Liang, L.; Willis-Owen, S.A.G.; Laprise, C.; Wong, K.C.C.; Davies, G.A.; Hudson, T.J.; Binia, A.; Hopkin, J.M.; Yang, I.V.; Grundberg, E.; et al. An epigenome-wide association study of total serum immunoglobulin e concentration. *Nature* **2015**, *520*, 670–674. [CrossRef] [PubMed]

83. Wang, B.; DeStefano, A.L.; Lin, H. Integrative methylation score to identify epigenetic modifications associated with lipid changes resulting from fenofibrate treatment in families. *BMC Proc.* **2018**, *12*, 28. [CrossRef] [PubMed]

84. Shah, S.; Bonder, M.J.; Marioni, R.E.; Zhu, Z.; McRae, A.F.; Zhernakova, A.; Harris, S.E.; Liewald, D.; Henders, A.K.; Mendelson, M.M.; et al. Improving Phenotypic Prediction by Combining Genetic and Epigenetic Associations. *Am. J. Hum. Genet.* **2015**, *97*, 75–85. [CrossRef] [PubMed]

85. Zheng, R.; Wan, C.; Mei, S.; Qin, Q.; Wu, Q.; Sun, H.; Chen, C.-H.; Brown, M.; Zhang, X.; Meyer, C.A.; et al. Cistrome Data Browser: Expanded datasets and new tools for gene regulatory analysis. *Nucleic Acids Res.* **2018**, 1–7. [CrossRef]

86. Ward, L.D.; Kellis, M. HaploReg v4: Systematic mining of putative causal variants, cell types, regulators and target genes for human complex traits and disease. *Nucleic Acids Res.* **2016**, *44*, D877–D881. [CrossRef]

87. Szklarczyk, D.; Franceschini, A.; Wyder, S.; Forslund, K.; Heller, D.; Huerta-Cepas, J.; Simonovic, M.; Roth, A.; Santos, A.; Tsafou, K.P.; et al. STRING v10: Protein-protein interaction networks, integrated over the tree of life. *Nucleic Acids Res.* **2015**, *43*, D447–D452. [CrossRef]

88. Fabregat, A.; Jupe, S.; Matthews, L.; Sidiropoulos, K.; Gillespie, M.; Garapati, P.; Haw, R.; Jassal, B.; Korninger, F.; May, B.; et al. The Reactome Pathway Knowledgebase. *Nucleic Acids Res.* **2018**, *46*, D649–D655. [CrossRef]

89. Himmelstein, D.S.; Lizee, A.; Hessler, C.; Brueggeman, L.; Chen, S.L.; Hadley, D.; Green, A.; Khankhanian, P.; Baranzini, S.E. Systematic integration of biomedical knowledge prioritizes drugs for repurposing. *eLife* **2017**, *6*, e26726. [CrossRef]

90. Haaland, Ø.A.; Lie, R.T.; Romanowska, J.; Gjerdevik, M.; Gjessing, H.K.; Jugessur, A. A Genome-Wide Search for Gene-Environment Effects in Isolated Cleft Lip with or without Cleft Palate Triads Points to an Interaction between Maternal Periconceptional Vitamin Use and Variants in ESRRG. *Front. Genet.* **2018**, *9*, 1–16. [CrossRef]

91. Lawlor, D.A.; Harbord, R.M.; Sterne, J.A.C.; Timpson, N.; Davey Smith, G. Mendelian randomization: Using genes as instruments for making causal inferences in epidemiology. *Stat. Med.* **2008**, *27*, 1133–1163. [CrossRef] [PubMed]

92. Relton, C.L.; Davey Smith, G. Two-step epigenetic mendelian randomization: A strategy for establishing the causal role of epigenetic processes in pathways to disease. *Int. J. Epidemiol.* **2012**, *41*, 161–176. [CrossRef] [PubMed]

93. Latvala, A.; Ollikainen, M. Mendelian randomization in (epi)genetic epidemiology: An effective tool to be handled with care. *Genome Biol.* **2016**, *17*, 156. [CrossRef] [PubMed]

94. Dekkers, K.F.; van Iterson, M.; Slieker, R.C.; Moed, M.H.; Bonder, M.J.; van Galen, M.; Mei, H.; Zhernakova, D.V.; van den Berg, L.H.; Deelen, J.; et al. Blood lipids influence DNA methylation in circulating cells. *Genome Biol.* **2016**, *17*, 138. [CrossRef] [PubMed]

95. Cecil, C.A.M.; Walton, E.; Pingault, J.-B.; Provençal, N.; Pappa, I.; Vitaro, F.; Côté, S.; Szyf, M.; Tremblay, R.E.; Tiemeier, H.; et al. DRD4 methylation as a potential biomarker for physical aggression: An epigenome-wide, cross-tissue investigation. *Am. J. Med. Genet. B Neuropsychiatr. Genet.* **2018**, *177*, 746–764. [CrossRef] [PubMed]

96. Howey, R.A.J.; Cordell, H.J. Application of Bayesian networks to GAW20 genetic and blood lipid data. *BMC Proc.* **2018**, *12*, 19. [CrossRef] [PubMed]

97. Justice, A.E.; Howard, A.G.; Fernández-Rhodes, L.; Graff, M.; Tao, R.; North, K.E. Direct and indirect genetic effects on triglycerides through omics and correlated phenotypes. *BMC Proc.* **2018**, *12*, 22. [CrossRef] [PubMed]

98. Yu, P.; Xiao, S.; Xin, X.; Song, C.-X.; Huang, W.; McDee, D.; Tanaka, T.; Wang, T.; He, C.; Zhong, S. Spatiotemporal clustering of the epigenome reveals rules of dynamic gene regulation. *Genome Res.* **2013**, *23*, 352–364. [CrossRef]

99. Fiziev, P.; Ernst, J. ChromTime: Modeling spatio-temporal dynamics of chromatin marks. *Genome Biol.* **2018**, *19*, 109. [CrossRef]

100. Guo, S.; Diep, D.; Plongthongkum, N.; Fung, H.L.; Zhang, K.; Zhang, K. Identification of methylation haplotype blocks AIDS in deconvolution of heterogeneous tissue samples and tumor tissue-of-origin mapping from plasma DNA. *Nat. Genet.* **2017**, *49*, 635–642. [CrossRef]

101. Romanowska, J.; Haaland, Ø.A.; Jugessur, A.; Gjerdevik, M.; Xu, Z.; Taylor, J.; Wilcox, A.J.; Jonassen, I.; Lie, R.T.; Håkon, K. Gjessing Integrating genome-wide methylation and genotype data to elucidate how region-wise methylation level might influence allele-defined relative risks. Submitted. 2019.

102. Dolinoy, D.C.; Weidman, J.R.; Waterland, R.A.; Jirtle, R.L. Maternal genistein alters coat color and protects Avy mouse offspring from obesity by modifying the fetal epigenome. *Environ. Health Perspect.* **2006**, *114*, 567–572. [CrossRef] [PubMed]

103. Briffa, J.F.; Wlodek, M.E.; Moritz, K.M. Transgenerational programming of nephron deficits and hypertension. *Semin. Cell Dev. Biol.* **2018**. [CrossRef]

104. Dekker, J.; Rippe, K.; Dekker, M.; Kleckner, N. Capturing chromosome conformation. *Science* **2002**, *295*, 1306–1311. [CrossRef] [PubMed]

105. Lieberman-Aiden, E.; van Berkum, N.L.; Williams, L.; Imakaev, M.; Ragoczy, T.; Telling, A.; Amit, I.; Lajoie, B.R.; Sabo, P.J.; Dorschner, M.O.; et al. Comprehensive mapping of long-range interactions reveals folding principles of the human genome. *Science* **2009**, *326*, 289–293. [CrossRef]

106. Fraser, J.; Williamson, I.; Bickmore, W.A.; Dostie, J. An Overview of Genome Organization and How We Got There: From FISH to Hi-C. *Microbiol. Mol. Biol. Rev. MMBR* **2015**, *79*, 347–372. [CrossRef]

107. Wu, C.; Pan, W. Integration of Enhancer-Promoter Interactions with GWAS Summary Results Identifies Novel Schizophrenia-Associated Genes and Pathways. *Genetics* **2018**, *209*, 699–709. [CrossRef] [PubMed]

108. Thurman, R.E.; Rynes, E.; Humbert, R.; Vierstra, J.; Maurano, M.T.; Haugen, E.; Sheffield, N.C.; Stergachis, A.B.; Wang, H.; Vernot, B.; et al. The accessible chromatin landscape of the human genome. *Nature* **2012**, *489*, 75–82. [CrossRef] [PubMed]

109. Vipin, D.; Wang, L.; Devailly, G.; Michoel, T.; Joshi, A. Causal Transcription Regulatory Network Inference Using Enhancer Activity as a Causal Anchor. *Int. J. Mol. Sci.* **2018**, *19*, 3609. [CrossRef] [PubMed]

110. Radkiewicz, C.; Johansson, A.L.V.; Dickman, P.W.; Lambe, M.; Edgren, G. Sex differences in cancer risk and survival: A Swedish cohort study. *Eur. J. Cancer* **2017**, *84*, 130–140. [CrossRef] [PubMed]

 *genes*

Article

An Analytic Approach Using Candidate Gene Selection and Logic Forest to Identify Gene by Environment Interactions (G × E) for Systemic Lupus Erythematosus in African Americans

Bethany J. Wolf [1,*,†], Paula S. Ramos [1,2,†], J. Madison Hyer [1], Viswanathan Ramakrishnan [1], Gary S. Gilkeson [2], Gary Hardiman [1,3,4,5], Paul J. Nietert [1,‡] and Diane L. Kamen [2,‡]

[1] Department of Public Health Sciences, Medical University of South Carolina, Charleston, SC 29425, USA; ramosp@musc.edu (P.S.R.); madison.hyer@osumc.edu (J.M.H.); ramakris@musc.edu (V.R.); hardiman@musc.edu (G.H.); nieterpj@musc.edu (P.J.N.)

[2] Division of Rheumatology and Immunology, Department of Medicine, Medical Univeristy of South Carolina, Charleston, SC 29425, USA; gilkeson@musc.edu (G.S.G.); kamend@musc.edu (D.L.K.)

[3] Center for Genomic Medicine, Department of Medicine, Medical Univeristy of South Carolina, Charleston, SC 29425, USA

[4] Division of Nephrology, Department of Medicine, Medical Univeristy of South Carolina, Charleston, SC 29425, USA

[5] School of Biological Sciences & Institute for Global Food Security, Queens University Belfast, Stranmillis Road, Belfast BT9 5AG, UK

* Correspondence: wolfb@musc.edu; Tel.: +1-843-876-1940

† These authors share first authorship.

‡ These authors share senior authorship.

Received: 31 August 2018; Accepted: 3 October 2018; Published: 15 October 2018

Abstract: Development and progression of many human diseases, such as systemic lupus erythematosus (SLE), are hypothesized to result from interactions between genetic and environmental factors. Current approaches to identify and evaluate interactions are limited, most often focusing on main effects and two-way interactions. While higher order interactions associated with disease are documented, they are difficult to detect since expanding the search space to all possible interactions of p predictors means evaluating $2^p - 1$ terms. For example, data with 150 candidate predictors requires considering over 10^{45} main effects and interactions. In this study, we present an analytical approach involving selection of candidate single nucleotide polymorphisms (SNPs) and environmental and/or clinical factors and use of Logic Forest to identify predictors of disease, including higher order interactions, followed by confirmation of the association between those predictors and interactions identified with disease outcome using logistic regression. We applied this approach to a study investigating whether smoking and/or secondhand smoke exposure interacts with candidate SNPs resulting in elevated risk of SLE. The approach identified both genetic and environmental risk factors, with evidence suggesting potential interactions between exposure to secondhand smoke as a child and genetic variation in the *ITGAM* gene associated with increased risk of SLE.

Keywords: candidate genes; gene–environment interactions; logic forest; systemic lupus erythematosus

1. Introduction

Many complex human diseases have been hypothesized to be the result of interactions between genetic and environmental risk factors [1–9]. Research studies aimed at detecting potential gene by environment (G×E) interactions as risk factors for human disease most often take one of two approaches. The first approach, often applied in genome-wide association studies, evaluates all two-way interactions. However, higher order interactions would not be detected using the this approach since expanding the search space to include higher order interactions is prohibitively laborious and computationally intensive, as evaluating all possible main effects and interactions in a data set with p predictors would mean evaluating $2^p - 1$ terms [9,10]. A second approach is to identify a set of candidate factors and/or interactions between these factors. The selection of the "best" subset of genetic and environmental factors may be based on the marginal effects of each factor passing a specific statistical significance threshold. In this case, only those factors that have a strong marginal effect are selected for interaction screening, which will fail to identify those factors with minimal marginal effects but strong interaction effects [2,8,9]. Alternatively, a subset of candidate genetic and environmental factors may be selected *a priori* [10]. Selecting candidate single nucleotide polymorphisms (SNPs) from genome wide data coupled with the environmental exposures provides a sufficiently concise and targeted sample space to be thorough while computationally manageable.

Identification of candidate variants and exposures can be prioritized based on *a priori* knowledge (e.g., reported association, biomedical data from databases, involvement in relevant biological mechanisms or pathways) and can be facilitated through existing literature and databases. If a suitable subset of candidate genes and environmental exposures can be identified, the analytical approach to evaluate the possible interactions among these factors must be considered. Statistical approaches such as case-only studies have been proposed to improve the efficiency of interaction identification in such studies [11–13]. However, results from such designs may be misleading as there is an assumption of independence between factors, which if violated can lead to erroneous conclusions [12,14,15]. Additionally, such studies typically focus on two-way interactions as each interaction is evaluated individually, which can be a limitation if seeking to identify interactions with more than two terms [9,10,14,16]. For example, data with only 25 predictors still requires evaluating over 10^7 terms (predictors) while data with 150 predictors would require evaluating over 10^{45} terms. Machine learning methods such as artificial neural networks, support vector machines, and forest approaches offer flexibility in modeling outcomes and can incorporate complex relationships such as higher order interactions in modeling disease outcomes based on a large number of predictors [17–22]. However, analytic approaches should provide guidance for determining the subset of predictors and predictor interactions from among a larger set that are most relevant for determining outcome. Both random forest and Logic Forest provide quantitative importance measures for individual predictors allowing them to be ranked according to their relative importance in determining an outcome [17,22]. However, predictor importance for each variable represents the marginal effect of a predictor and if a set of predictors is associated with the outcome only through interactions effects, these marginal importance measures may mask such interaction effects [23]. Unlike random forest, Logic Forest also provides a quantitative measure of importance for interactions identified by the forest, which is advantageous in complex disease settings where interactions among genetic and environmental factors rather than main effects lead to disease. Despite the availability and usefulness of such tools, they have been under utilized. An ideal approach would combine identification of candidate factors based on prior knowledge with an efficient method for evaluating the space of possible interactions, including higher order interactions, among these candidate factors.

In this paper, we present an analytic approach to evaluate main effects and interactions between genetic and environmental factors associated with a disease outcome by coupling selection of relevant genetic and environmental factors based on available literature and public databases with a machine learning approach, Logic Forest. To illustrate this approach, we examine varying degrees of tobacco smoke exposure as environmental factors, disease-associated SNPs as genetic factors, and their individual and combined associations with the diagnosis of systemic lupus erythematosus (SLE) in a cohort from the Sea Island Gullah population of South Carolina. The Gullah population is a distinctive group of African Americans from the coastal Sea Islands of South Carolina and Georgia who are descendants of enslaved Africans from the African Rice Coast [24]. On many plantations, Africans vastly outnumbered Europeans, and the Gullah remained in the geographically isolated Sea

Islands until recent times [24–26]. This population is unique in that they have low non-African genetic admixture [25,26] and high ancestral homogeneity from their ancestral home, Sierra Leone [27–29], offering a unique opportunity to study genetic and environmental disease risk factors. SLE is a "prototype" autoimmune rheumatic disease with a well substantiated genetic etiology and many of the SNPs identified as increasing the risk for SLE are in genes that enhance immune reactivity [30–38]. Additionally, given that the concordance rate between monozygotic twins only ranges between 24% and 35% [31], epigenetic or environmental factors are likely to have an important role in SLE susceptibility. Known environmental triggers in SLE include ultraviolet (UV) light, silica dust, certain infections, and smoking [39]. We apply our proposed approach to evaluate associations between risk of SLE with genetic factors thought to amplify the inflammatory/immune response to tobacco smoke exposure, which has been implicated in earlier research [40]. Results of the analysis found evidence of both a main effect for smoke exposure and several interactions between genetic factors and smoke exposure, demonstrating the applicability of our approach.

2. Materials and Methods

We present an analytical approach for identifying main effects and interactions between genetic and environmental factors associated with a disease outcome. The approach involves selection of candidate genetic and/or environmental factors, use of a machine learning algorithm to identify important main effects and interactions in disease, followed by confirmation of the association between interactions identified by the algorithm using logistic regression. To give this theoretical approach context, it is applied to a study examining the association between SNPs and cigarette smoke exposure with risk of developing SLE as shown in Figure 1.

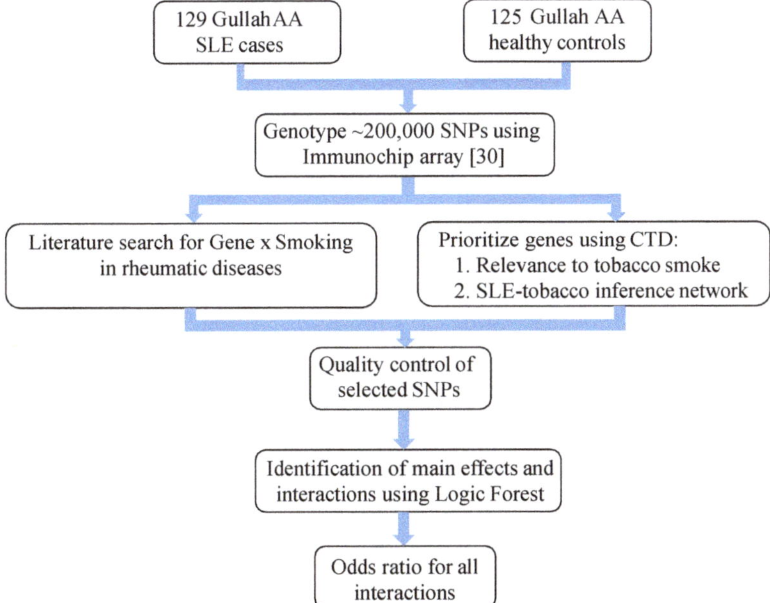

Figure 1. Flowchart of the proposed analytic approach. AA: African American; SLE: Systemic lupus erythematosus; CTD: Comparative Toxicogenomics Database, and SNP: Single nucleotide polymorphism.

2.1. Study Subjects and Design

The Gullah population is a distinctive group of African Americans from the coastal Sea Islands of South Carolina and Georgia they are descendants of enslaved Africans from the African Rice Coast [24] and thus represent a unique population of African Americans, which, while not a genetic isolate, is a more genetically homogeneous group relative to other African Americans [25–29,41]. Systemic lupus erythematosus is also known to have a high disease load in African Americans relative to Americans of European descent with an estimated prevalence in South Carolina of $\frac{1}{200}$ in African American women; the prevalence in the Gullah is unknown, but it is believed to be similar [41].

The SLE study used a case control design, and subjects were selected from people participating in the SLE in Gullah Health (SLEIGH) Study, which began recruitment in 2003 [42]. Systemic lupus erythematosus cases fulfilled the 1997 American College of Rheumatology classification criteria for "definite" SLE [43]. Race was self-reported and Gullah ancestry was self-identified as African American (AA) Gullah from the Sea Island region of South Carolina, with all known grandparents being of Gullah descent [42,44,45]. Unrelated non-SLE Gullah controls were also recruited by asking the cases to "bring a friend" of the same gender and community to the screening visit. As described in our recent manuscript [45], first-degree relatives were not considered for the analysis. These subjects received a clinical examination by a rheumatologist to ensure they did not meet criteria for any inflammatory rheumatologic disease before inclusion in the genetic studies as unaffected Gullah controls. This study was approved by the Medical University of South Carolina Institutional Review Board (Pro#00021985, approved 1/15/2013). All study participants provided written consent prior to study enrollment.

Genotypic data was available on 129 Gullah AA SLE cases and 125 AA unrelated controls genotyped on the Immunochip genotyping array [45]. Tobacco smoke exposure, including both secondhand smoke exposure as a child and current smoking status, was collected as a part of the SLEIGH study protocol. At baseline, each subject was asked the following questions as part of an in-person interview related to smoking: "Have you ever smoked cigarettes?" (If yes) "What was the maximum daily amount (packs per day) smoked?" "What is the total number of years you smoked?" "Are you currently smoking?" "If not, how many years since quitting?". Participants were also asked the following questions about secondhand smoke exposure: "Were you ever routinely exposed to passive smoke as an adult (at work or in the home)?" "Were you ever exposed to passive smoke as a child (before age 18)?". From responses to these questions, four binary variables were created for each case and control to indicate whether or not they (1) had ever been a smoker prior to SLE diagnosis (for cases) or prior to their study visit (for controls), (2) were current smokers at the time of SLE diagnosis (for cases) or at their baseline visit (for controls), (3) were ever regularly exposed to secondhand smoke, and (4) were ever regularly exposed to secondhand smoke as a child (<18 years old). Twenty participants were missing information on smoking and smoke exposure data and were excluded for analysis.

2.2. Prioritization of SNPs

2.2.1. Gene Selection

We searched the literature for reports of interactions between genetic variation and tobacco smoke in SLE and related rheumatic diseases. We identified genes with reported interactions with tobacco smoke in SLE (*NAT2*) [40] and rheumatoid arthritis (*HLA-DRB1* shared epitope [46], *PTPN22* [47], and *HMOX1* [48]). In addition to these candidate genes from the literature, we also used information compiled in the Comparative Toxicogenomics Database (CTD) [49], a database that contains curated scientific data describing relationships between chemicals/drugs, genes/proteins, diseases, phenotypes, pathways, and interaction modules. We used the CTD to prioritize genes relevant to tobacco smoke (*APOE*, *NFE2L2*, *IL6* and *CXCL8*) and genes in an inference network between tobacco smoke and SLE (*IRF5*, *ITGAM* and *ITGAX*; *IL6* is also part of this network).

2.2.2. Genotypic Dataset and Quality Control

Genotypic data on 129 Gullah AA SLE cases and 125 AA controls genotyped on the Immunochip array was subject to the following quality control (QC) filters: exclusion of individuals with missing genotypes, markers that did not statistically conform to Hardy–Weinberg Equilibrium (HWE) (at $p < 0.001$) in controls, markers with missing data, and markers with minor allele frequency (MAF) < 0.05. We used all the SNPs that met these QC thresholds in a region including ±5 kb around each gene. Most promoters are located within 1 kb of the transcription start site , a 5 kb flanking region around a gene is a common and reasonable choice. For the four genes with previously reported interactions with tobacco smoke (*NAT2*, *HLA-DRB1*, *PTPN22* and *HMOX*), we searched the 1000 Genomes and HapMap Projects for SNPs that tag the reported alleles (as defined by an r-squared > 0.4 in the YRI (Yoruba in Ibadan, Nigeria) population) that might have been genotyped and met QC in our dataset. Populations of African ancestry have decreased linkage disequilibrium (LD) and a rapid decay of LD with distance genome-wide relative to populations of European ancestry [45]. A threshold of r-squared > 0.4 is thus reasonable to identify proxy SNPs in our population. Finally, the genotypic cluster plots for each SNP were visually inspected, and SNPs with poor or questionable plots (without clear cluster separation) were excluded. After applying these QC filters, the following were available for further analyses: *NAT2* (4 SNPs), *HLA-DRB1* (6 SNPs), *APOE* (2 SNPs), *IL6* (17 SNPs), *CXCL8* (1 SNP), *IRF5* (20 SNPs), *ITGAM* (67 SNPs), and *ITGAX* (31 SNPs). Genotype frequencies for each of the SNPs discussed in the manuscript are listed in Supplemental Table S1. Thirty participants failed to meet quality control parameters and were excluded from the analysis.

2.3. Identification of Important Main Effects and Interactions

The primary goal of the SLE study was to identify potential gene×gene and gene×environment interactions associated with risk of SLE among the Gullah population. We used a binary classification algorithm to identify main effects and interactions among the candidate SNPs and smoke exposure for classifying individuals according to SLE status.

2.3.1. Logic Forest

Logic Forest (LF) is a machine learning algorithm designed to identify interactions among binary variables (for example, SNPs or smoking status) and quantify the importance of potential predictors and predictor interactions identified in the forest in terms of correctly classifying disease status [22]. Logic Forest does not require *a priori* specification of interactions as it iteratively evaluates the space of all possible interactions to identify the subset of interactions best able to classify disease status. The LF algorithm and methods for calculating LF model misclassification rate and predictor interaction importance have been previously described by [22] and detailed description of the algorithm can be found there. For completeness, we provide details of the algorithm here. Given data $\mathbf{W} = \{\mathbf{X}, \mathbf{y}\}$ where $\mathbf{X} = (x_1, x_2, \ldots, x_p)'$ is an $n \times p$ matrix of binary predictors and $\mathbf{y} = (y_1, y_2, \ldots, y_n)'$ is a binary vector indicating disease status for $i = 1, 2, \ldots, n$ subjects, an LF model consists of a collection of B logic regression trees constructed from B bootstrap samples from data $\mathbf{W}$ and is denoted as $LF(\mathbf{W}, B) = \{T^1, T^2, \ldots, T^B\} = \{T^b\}$. A single logic regression tree, T^b, represents the predictors and predictor interactions, referred to as "prime implicants", identified for the b-th bootstrap sample as being associated with having SLE. Trees in an LF model are allowed to grow up to maximum size of eight leaves. Thus, trees in the forest can explore interactions of up to eight variables. Figure 2 shows an example logic regression tree with three prime implicants identified as associated with SLE: (1) exposure to passive smoking as a child and having at least one copy of the major allele of rs2359661 (A) in *ITGAM*; (2) having two copies of the minor allele of rs4632147 (T) in *ITGAX*; and (3) having two copies of the minor allele of rs11761199 (G) in *IRF5*. When all predictor variables are categorical (e.g., SNPs), an interaction between two variables occurs when specific conditions for both variables must be met to confer additional risk of disease. For example, the first prime implicant for Figure 2 suggests that additional risk for SLE from having at least one copy of the major allele of rs4632147 occurs only if the subject also had passive smoke exposure as a child. For tree T_b in the forest, subjects are predicted to have disease if they meet any of the conditions defined by the tree.

Figure 2. Example of a logic regression tree. White boxes represent the predictor, in the case of SNPs, the recessive effect of the minor allele, and black boxes represent the complement of that predictor (e.g., for a SNP, this means the dominant effect of the major allele). There are three independent predictors/predictor interactions identified within the tree: (1) exposure to passive smoking as a child and having at least one copy of the major allele of rs2359661 (A) in *ITGAM*; (2) having two copies of the minor allele of rs4632147 (T) in *ITGAX*; and (3) having two copies of the minor allele of rs11761199 (G) in *IRF5*.

Predictions for the LF model of B trees is determined by the proportion of trees that predict the subject to have SLE. Each tree T^b in the LF has an associated out-of-bag (OOB) dataset, $OOB\left(T^b\right)$, comprised of those observations left out of the b-th bootstrap sample that can be used for an unbiased estimate of the model's prediction error (similar to internal bootstrap validation). The LF OOB prediction for observation y_i is determined by Equation (1) where $I\left(W_i \in OOB\left(T^b\right)\right)$ is the indicator of the i-th observations membership in $OOB\left(T^b\right)$.

$$\widehat{y_i^{OOB}}\left(\{T^b\},\mathbf{x}_i\right) = \begin{cases} 1, & \text{if} \dfrac{\sum_{b=1}^{B} \widehat{y}_i\left(T^b,\mathbf{x}_i\right) I\left(W_i \in OOB\left(T^b\right)\right)}{\sum_{b=1}^{B} I\left(W_i \in OOB\left(T^b\right)\right)} \geq 0.5, \\ 0, & \text{otherwise.} \end{cases} \tag{1}$$

Accordingly, the LF OOB misclassification rate is

$$MC^{OOB}\left(\{T^b\},\mathbf{y},\mathbf{X}\right) = \frac{1}{n}\sum_{i=1}^{n}\left(y_i - \widehat{y_i^{OOB}}\left(\{T^b\},\mathbf{x}_i\right)\right)^2. \tag{2}$$

Logic Forest also provides two quantitative measures of importance for all prime implicants identified in the forest. The first measure evaluates the change in classification error for each tree in the forest before and after permutation of the data. The misclassification rate for tree T^b is

$$MC^{OOB}\left(T^b,\mathbf{y},\mathbf{X}\right) = \frac{\sum_{i=1}^{n}\left(y_i - \widehat{y_i^{OOB}}\left(T^b,\mathbf{x}_i\right)\right)^2 I\left(W_i \in OOB\left(T^b\right)\right)}{\sum_{i=1}^{n} I\left(W_i \in OOB\left(T^b\right)\right)}. \tag{3}$$

Let $\mathbf{X}^{(j)}$ be the matrix of predictors with X_j randomly permuted, where X_j can be an individual predictor or more generally a prime implicant. The importance of prime implicant X_j is

$$VI_1\left(X_j\right) = \frac{1}{B}\sum_{b=1}^{B}\left[MC^{OOB}\left(T^b,\mathbf{y},\mathbf{X}^{(j)}\right) - MC^{OOB}\left(T^b,\mathbf{y},\mathbf{X}\right)\right]. \tag{4}$$

Values for Equation (4) range from -1 to 1 with positive values indicating a positive association between response **y** and prime implicant X_j. The second measure of prime implicant importance is the frequency with which the prime implicant occurs across trees in the forest and can be calculated according to Equation (5)

$$VI_2\left(X_j\right) = \frac{1}{B}\sum_{b=1}^{B} I\left(X_j \in T^b\right),\qquad(5)$$

where $I\left(X_j \in T^b\right)$ is an indicator of prime implicant X_j's inclusion in tree T^b. Permutation p-values for importance measures for each prime implicant X_j can be calculated by randomly permuting the outcome many times and fitting LF models to the data with the permuted outcome. The permutation p-value is the proportion of times LF models fitted to data with the outcome permuted yield an importance score for prime implicant X_j as large as or larger than the importance score from the original model.

For analysis of the SLE study, three LF models including 200 logic regression trees each were fit using (1) the recessive effect of the minor allele for each SNP (i.e., subjects have two copies of the minor allele); (2) the dominant effect of the minor allele for each SNP (i.e., subjects having at least one copy of the minor allele); and (3) the genotypic model with two indicators for of the number of copies of the minor allele (with 0 being a reference group). Demographic and environmental variables, namely gender, passive smoke exposure as a child, passive smoke exposure as an adult, and smoking status as an adult were also considered in each model. Permutation p-values for prime implicants identified by LF models were calculated based on 500 LF models fitted to the data with SLE case-control status randomly permuted. All analyses were conducted in R v. 3.2.5 using the *LogicForest* package [50,51].

2.3.2. Validation of Main Effects and Interactions

To further validate the association between prime implicants identified by the LF and response y, logistic regression models were also constructed to estimate odds ratios associated with each risk factor (i.e., main effects and interactions) identified using the LF approach.

3. Results

Twenty subjects were missing information on childhood and/or adult smoke exposure and 30 additional subjects had missing genotype information, thus the final study population included 204 participants with both genetic and environmental exposure data available, 100 of whom were diagnosed with SLE. There was no notable difference in sex or case/control status between subjects included in the final population compared to those who were excluded (data not shown). Participants included in the study were on average four years older than participants that were excluded ($p = 0.042$). A majority of the study participants were female (85.8%), consistent with the historical gender distribution for the disease. Participant demographic characteristics for cases and unrelated controls are shown in Table 1.

Table 1. Participant characteristics by SLE status.

Characteristic	Control ($n = 104$)	SLE ($n = 100$)	p-Value *
Age (Mean $\pm$ Std Dev)	42.6 ± 11.7	38.6 ± 13.4	0.022
Female (n, %)	87 (83.6)	88 (88.0)	0.491
Passive Smoke Exposure as a Child (n, %)	28 (26.9)	41 (41.0)	0.048
Passive Smoke Exposure as an Adult (n, %)	18 (17.3)	20 (20.0)	0.754
Ever Smoker (n, %)	24 (23.1)	24 (24.0)	1.000
Current Smoker (n, %)	13 (12.5)	17 (17.0)	0.478

* p-values reported in the table for the association with SLE status are based on a two-sample t-test for age and chi-square test for all categorical variables.

The results from the LF model that included the recessive effect of the minor allele and the environmental and demographic variables are presented, since the gene–environment interactions identified in this model showed the strongest relationship with SLE status. Logic Forest identified 426 unique prime implicants across the 200 trees in the model. Figure 3 is a plot of the number of trees in the model that include each predictor by the normalized importance scores for each predictor. Points shown in red represent those predictors that have the largest combination of predictor frequency and importance score. As seen in Figure 3, the LF model identified passive smoke exposure as a child as the most important predictor of SLE status (permutation $p < 0.01$). The SNPs rs11770589 (*IRF5*), rs58408589 (*ITGAX*), rs67898294 (*ITGAX*), rs11761199 (*IRF5*), and rs7190807 (*ITGAM*) had both a high predictor importance score and occurred frequently in the LF model (permutation $p < 0.01$ for all).

Predictor Importance

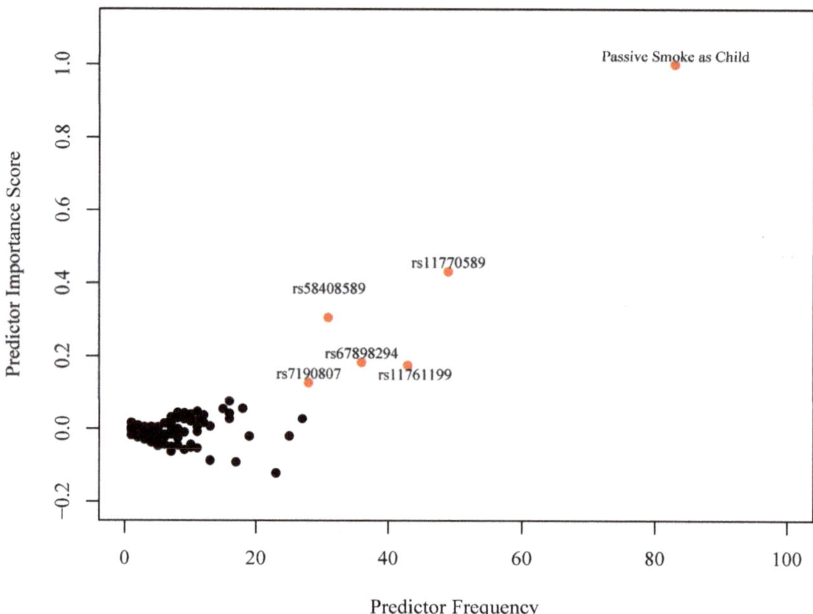

Figure 3. Predictor frequency by normalized predictor importance score for all predictors in the Logic Forest (LF) model. Points highlighted in red represent the predictors that have the largest combination of frequency and importance score.

Figure 4 shows the number of trees in the model that include each prime implicant by the normalized importance scores for all prime implicants that were identified in the forest. The most important and most frequent prime implicants identified in the forest were the main effects for passive smoke exposure as a child (permutation $p = 0.008$) and the following SNPs: rs4632147 (*ITGAX*), rs11761199 (*IRF5*), rs11770589 (*IRF5*), and rs58408589 (*ITGAX*) (permutation $p = 0.006$, 0.01, 0.01, and 0.028, respectively).

Interaction Importance

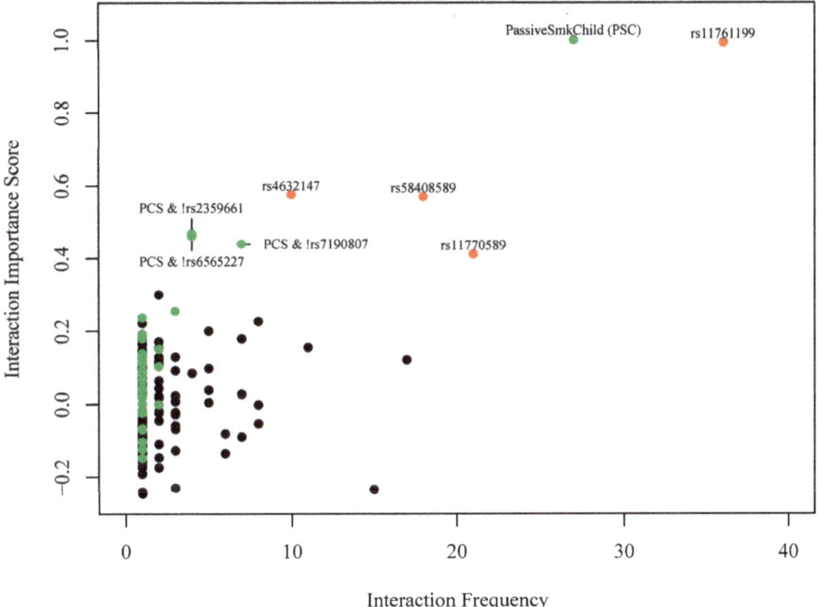

Figure 4. Interaction frequency by normalized interaction importance score for all interactions identified in the LF model. Points highlighted in red represent the interactions that have the largest combination of frequency and importance score. Points in green represent additional interaction terms identified in the forest that include passive smoke exposure as a child with at least one SNP.

There are three additional interaction terms that were ranked as highly important and occurred with some regularity that included SNPs in the *ITGAM* gene and passive smoke exposure as a child (permutation $p < 0.002$ for all three interactions). The points in Figure 4 highlighted in red represent the interactions that have the largest combination of frequency and importance score. Points in green represent interaction terms identified in the forest that include passive smoke exposure as a child with at least one SNP. Passive smoke exposure as a child occurred in 88 of the 200 trees, and in 27 of those instances it occurred as a main effect. In the remaining 61 instances, it occurred as an interaction with different SNPs. Although the main goal of this analysis is to identify potential gene–gene and gene–environment interactions; for completeness, we also examined the ability of the LF model to discriminate SLE cases from controls. The estimated prediction error rate for the final LF model is 43%, with an area under the receiver operating characteristic (ROC) curve of 0.54 (ROC curve for the final model is shown in Supplemental Figure S1).

The Logic Forest model identified four main effects and three interactions as the most important predictors in for determining SLE status based on the importance score. Separate logistic regression models for these seven predictors that had the largest importance scores from the LF model were fit by including an indicator variable for whether or not the subject had the combination of exposures in the interaction. Table 2 shows the odds ratios and associated p-values for these logistic regression models. The LF model included indicators for the recessive effect of the minor allele; however, if the model found an interaction with the complement of a recessive effect, this is equivalent to the interaction term including at least one copy of the major allele (i.e., dominant effect of the major allele as noted in the last three interactions shown in Table 2). These results generally agree with the results from the LF model in that a majority of the prime implicants reported in the table have a statistically significant association with being SLE positive. The only exception is rs11770589 in the *IRF5* gene, which has a *p*-value from the logistic regression model of 0.18.

Table 2. Odds ratios with 95% confidence intervals (CI) from a series of logistic regression models. The implied reference category for each odds ratio is the complement of the effect defined in the first column.

Effect	Gene	Odds Ratio (95% CI)	Unadjusted *p*-Value
Passive Smoke Exposure as Child (PSC)		1.88 (1.01, 3.55)	0.039
2 copies of the minor allele of rs4632147 (T)	*ITGAX*	3.09 (1.09, 10.1)	0.023
2 copies of the minor allele of rs58408589 (C)	*ITGAX*	2.96 (1.23, 7.75)	0.011
2 copies of the minor allele of rs11761199 (G)	*IRF5*	7.69 (1.01, 352)	0.033
2 copies of the minor allele of rs11770589 (A)	*IRF5*	1.65 (0.81, 3.42)	0.179
PSC & > 1 copy of the major allele of rs2359661 (A)	*ITGAM*	2.28 (1.18, 4.48)	0.009
PSC & > 1 copy of the major allele of rs7190807 (G)	*ITGAM*	2.46 (1.25, 4.92)	0.005
PSC & > 1 copy of the major allele of rs6565227 (T)	*ITGAM*	2.37 (1.23, 4.66)	0.006

4. Discussion

In this study, we demonstrate the utility of the proposed analytical approach to examine main effects and interactions between 148 SNPs, gender, and four different types of smoke exposure in a well-characterized cohort of Gullah African Americans participating in the SLEIGH study. There are several key take-home points from the analysis of the SLE study. The LF model found strong evidence for an association between SLE status and passive smoke exposure as a child. Logic forest also consistently identified SNPs associated with SLE, including rs58408589, rs67898294, rs7190807, rs4632147, rs11770589, and rs11761199 (in the *IRF5*, *ITGAM*, and *ITGAX* genes). Finally, although passive smoke exposure as a child was clearly identified as a main effect (i.e., an independent risk factor), there was also evidence to suggest that it may also be involved in weak to moderate interactions with SNPs on the *ITGAM* gene (Table 2).

There are alternative statistical methods that one might consider for evaluating potential gene×gene or gene×environment interactions for SLE. For example, logistic regression is a traditional approach that could be used for such analyses. However, in order to evaluate the association between SLE and all potential two-way interactions involving the 153 predictors in our data set, one would need to examine $\binom{153}{2} = 11,628$ logistic regression models; potential three-way interactions would be even more cumbersome, as there would be almost 600,000 of them. Nonparametric decision tree methods are easily interpretable and have flexibility to identify interactions among predictors [52,53]. However, decision tree models may be unstable, in that small changes in the data can result in very different models [17,52,54,55]. Ensemble models, a collection of decision trees developed using bootstrap samples or weighted samples of a dataset improve model stability and prediction accuracy compared to single tree approaches [17,22,55–58]. Random forest (RF) and Logic Forest (LF) are ensemble extensions of two decision tree methods [17,22]. Both methods also provide a quantitative measure of the relative importance of predictors used in the model. However, LF has an additional advantage over RF in that it also has a quantitative importance measure for interactions found in the forest, rather than just individual predictors, making it ideal for identifying potential gene×gene and gene×environment interactions in SLE development.

Our findings from the SLE study are not the first to demonstrate that certain SNPs may interact with environmental exposures, such as smoking, in a way that increases the risk of developing SLE. In a Japanese cohort, investigators found significant evidence of increased risk of SLE associated with smoking, highest among those with polymorphisms in the *NAT2* gene influencing metabolic enzymes involved in reactive oxygen species production [40]. They identified a possible gene×environment interaction, where smokers with the slow acetylator genotype of *NAT2* were found to have a higher risk of SLE (Odds Ratio = 6.44, 95% CI = 3.07–13.52) when compared to non-smokers with the rapid acetylator genotype of *NAT2*. Our study was the first to find passive smoke exposure as a child (childhood exposure to secondhand smoke) to be a significant risk factor for SLE. The main effect of childhood smoke exposure and the interactions between several SNPs on the *ITGAM* gene were also significant in univariate logistic regression models of SLE status. Additionally, two SNPs on the *ITGAX* gene and two SNPs on the *IRF5* gene were also identified by the LF model, though only three of the four SNPs were also significant in subsequent logistic regression models. Logic Forest does not assume linearity in the logit link between predictors and outcome as logistic regression does, which may explain the discrepancies in significance of rs11770589 on the *ITGAX* gene.

Given the exploratory nature of these analyses and the limited sample size of our study population, replication would greatly improve the credibility of the associations identified in this study. Unfortunately, there are no large scale genetic studies of SLE (or of any related autoimmune disorder) in African Americans. Furthermore, the population selected for this study (Gullah African Americans) was chosen for their documented high genetic homogeneity [42,45] and a replication cohort of genetically similar individuals does not exist. Thus, the associations reported would need to be validated in a future study. Additional potential limitations of this study include recall bias and reliance on self-report to ascertain the individuals' smoking and exposure status. These findings should be considered as part of the "discovery" or "hypothesis generating" process of understanding whether and how smoke exposure may interact with certain genes and should not be construed as definitive proof. A detailed understanding of the mechanisms underlying SLE pathogenesis will continue to require large databases of study subjects, with well-characterized environmental exposures and genetic information. Machine learning algorithms, such as Logic Forest, will inevitably be required to help sort through the ever expanding combination of potential risk factors for disease.

5. Conclusions

This study illustrates the utility of a novel approach to identify interactions between genetic and environmental risk factors for disease. The complexity of many human diseases, which likely result from interactions between genetic and environmental factors, emphasizes the importance of evaluating such interactions when examining disease etiology. The challenge for such studies is the number of possible interactions in data with even a modest number of individual predictors. For example, in the SLE study presented here, there are $2^{153} - 1 = 5.7 \times 10^{45}$ possible interactions. The approach presented here combines candidate gene selection and a machine learning method for identification and quantification of the relative importance of interactions from among all possible interactions in determining disease state, followed by confirmation of the association between those predictors/interactions with disease outcome. Applying this approach to a study examining genetic and environmental factors in SLE identified childhood exposure to secondhand smoke (PSC) as an independent effect and interactions between PSC and SNPs on ITGAM, providing additional evidence that SLE is a disease with a complex etiology and is the first study to find childhood exposure to secondhand smoke to be a significant risk factor for SLE.

Supplementary Materials: The following are available online at http://www.mdpi.com/2073-4425/9/10/496/s1, Figure S1: Receiver operating characteristic (ROC) curve for LF model of SLE status including the recessive effect of the minor allele for all SNPS, gender, passive smoke exposure as a child and as an adult, and smoking status, Table S1: Genotype frequencies for each of the SNPs discussed in the Results, Discussion, and Conclusions.

Author Contributions: D.L.K. and G.S.G. maintain the SLEIGH cohort from which study subjects were collected, D.L.K. and P.S.R. collected genotyping information on study subjects, all authors (B.J.W., P.S.R., P.J.N., J.M.H., V.R., G.S.G., G.H., and D.L.K.) conceived of the study design and subject selection, P.S.R., B.J.W., and J.M.H. analyzed the data, B.J.W., P.S.R., P.J.N., and J.M.H. wrote the manuscript. All authors provided critical review of manuscript drafts and approved of the final version.

Funding: This work was funded, in part, by grants from the National Institutes of Health (National Institute of Arthritis and Musculoskeletal and Skin Diseases Grant No. P30-AR072582, P60-AR062755, K01-AR067280, R21-AR067459, and K24-AR068406, National Center for Advancing Translational Sciences Grant No. UL1-TR001450, and National Institute of General Medical Sciences Grant No. U54-GM104941). G.H. is grateful for support from the Medical University of South Carolina College of Medicine start-up funds and the National Institute of General Medical Sciences Grant No. R01-GM122078 and the National Cancer Institute Grant No. R21-CA209848.

Acknowledgments: We would like to thank all of the study participants for their time and commitment to the study.

Conflicts of Interest: The authors declare no conflict of interest. The founding sponsors had no role in the design of the study; in the collection, analyses, or interpretation of data; in the writing of the manuscript, and in the decision to publish the results.

References

1. Carlborg, O.; Haley, C.S. Epistasis: Too often neglected in complex trait studies? *Nat. Rev. Genet.* **2004**, *5*, 618–625. [CrossRef] [PubMed]

2. Thornton-Wells, T.A.; Moore, J.H.; Haines, J.L. Genetics, statistics and human disease: Analytical retooling for complexity. *Trends Genet.* **2004**, *20*, 640–647. [CrossRef] [PubMed]

3. Alvarez-Castro, J.M.; Carlborg, O. A unified model for functional and statistical epistasis and its application in quantitative trait Loci analysis. *Genetics* **2007**, *176*, 1151–1167. [CrossRef] [PubMed]

4. Hunter, D.J.; Kraft, P.; Jacobs, K.B.; Cox, D.G.; Yeager, M.; Hankinson, S.E.; Wacholder, S.; Wang, Z.; Welch, R.; Hutchinson, A.; et al. A genome-wide association study identifies alleles in FGFR2 associated with risk of sporadic postmenopausal breast cancer. *Nat. Genet.* **2007**, *39*, 870–874. [CrossRef] [PubMed]

5. Kotti, S.; Bickeboller, H.; Clerget-Darpoux, F. Strategy for detecting susceptibility genes with weak or no marginal effects. *Hum. Hered.* **2007**, *63*, 85–92. [CrossRef] [PubMed]

6. Dempfle, A.; Scherag, A.; Hein, R.; Beckmann, L.; Chang-Claude, J.; Schäfer, H. Gene–environment interactions for complex traits: Definitions, methodological requirements and challenges. *Eur. J. Hum. Genet.* **2008**, *16*, 1164. [CrossRef] [PubMed]

7. Ramos, R.G.; Olden, K. Gene-environment interactions in the development of complex disease phenotypes. *Int. J. Environ. Res. Public Health* **2008**, *5*, 4–11. [CrossRef] [PubMed]

8. Gilbert-Diamond, D.; Moore, J.H. Analysis of gene-gene interactions. *Curr. Protoc. Hum. Genet.* **2011**, 1–14. [CrossRef]

9. Wei, W.H.; Hemani, G.; Haley, C.S. Detecting epistasis in human complex traits. *Nat. Rev. Genet.* **2014**, *15*, 722. [CrossRef] [PubMed]

10. Cordell, H.J. Detecting gene–gene interactions that underlie human diseases. *Nat. Rev. Genet.* **2009**, *10*, 392. [CrossRef] [PubMed]

11. Khoury, M.J.; Flanders, W.D. Nontraditional epidemiologic approaches in the analysis of gene environment interaction: Case-control studies with no controls! *Am. J. Epidemiol.* **1996**, *144*, 207–213. [CrossRef] [PubMed]

12. Schmidt, S.; Schaid, D.J. Potential misinterpretation of the case-only study to assess gene-environment interaction. *Am. J. Epidemiol.* **1999**, *150*, 878–885. [CrossRef] [PubMed]

13. Yang, Q.; Khoury, M.J.; Sun, F.; Flanders, W.D. Case-only design to measure gene-gene interaction. *Epidemiology* **1999**, *10*, 167–170. [CrossRef] [PubMed]

14. Albert, P.S.; Ratnasinghe, D.; Tangrea, J.; Wacholder, S. Limitations of the case-only design for identifying gene-environment interactions. *Am. J. Epidemiol.* **2001**, *154*, 687–693. [CrossRef] [PubMed]

15. VanderWeele, T.J.; Hernandez-Diaz, S.H.M. Case-only gene-environment interaction studies: When does association imply mechanistic interaction? *Genet. Epidemiol.* **2010**, *34*, 327–334. [CrossRef] [PubMed]

16. Gatto, N.M.; Campbell, U.B.; Rundle, A.G.; Ahsan, H. Further development of the case-only design for assessing gene–environment interaction: Evaluation of and adjustment for bias. *Int. J. Epidemiol.* **2004**, *33*, 1014–1024. [CrossRef] [PubMed]

17. Breiman, L. Random Forests. *Mach. Learn.* **2001**, *45*, 5–32. [CrossRef]

18. Doniger, S.; Hofmann, T.; Yeh, J. Predicting CNS permeability of drug molecules: Comparison of neural network and support vector machine algorithms. *J. Comput. Biol.* **2002**, *9*, 849–864. [CrossRef] [PubMed]

19. Hahn, L.W.; Ritchie, M.D.; Moore, J.H. Multifactor dimensionality reduction software for detecting gene-gene and gene-environment interactions. *Bioinformatics* **2003**, *19*, 376–382. [CrossRef] [PubMed]

20. Moore, J.H.; Ritchie, M.D. The challenges of whole-genome approaches to common diseases. *J. Am. Med. Assoc.* **2004**, *291*, 1642–1643. [CrossRef] [PubMed]

21. Hastie, T.; Tibshirani, R.; Friedman, J. *The Elements of Statistical Learning: Data Mining, Inference, and Prediction*, 2nd ed.; Springer Series in Statistics: New York, NY, USA, 2009.

22. Wolf, B.J.; Hill, E.G.; Slate, E.H. Logic Forest: An ensemble classifier for discovering logical combinations of binary markers. *Bioinformatics* **2010**, *26*, 2183–2189. [CrossRef] [PubMed]

23. Wright, M.N.; Ziegler, A.; König, I.R. Do little interactions get lost in dark random forests? *BMC Bioinform.* **2016**, *17*, 145. [CrossRef] [PubMed]

24. Opala, J. *The Gullah: Rice, Slavery and the Sierra Leone-American Connection*; US Information Service (Fort Sumter National Monument): Sullivans Island, SC, USA, 1987.

25. Parra, E.J.; Marcini, A.; Akey, J.; Martinson, J.; Batzer, M.A.; Cooper, R.; Forrester, T.; Allison, D.B.; Deka, R.; Ferrell, R.E.; et al. Estimating African American admixture proportions by use of population-specific alleles. *Am. J. Hum. Genet.* **1998**, *63*, 1839–1851. [CrossRef] [PubMed]

26. Parra, E.J.; Kittles, R.A.; Argyropoulos, G.; Pfaff, C.; Hiester, K.; Bonilla, C.; Sylvester, N.; Parrish-Gause, D.; Garvey, W.; Jin, L.; et al. Ancestral proportions and admixture dynamics in geographically defined African Americans living in South Carolina. *Am. J. Phys. Anthropol.* **2001**, *114*, 18–29. [CrossRef]

27. McLean, D.C., Jr.; Spruill, I.; Gevao, S.; Morrison, E.Y.; Bernard, O.S.; Argyropoulos, G.; Garvey, W.T. Three novel mtDNA restriction site polymorphisms allow exploration of population affinities of African Americans. *Hum. Biol.* **2003**, *75*, 147–161. [CrossRef] [PubMed]

28. Jackson, B.A.; Wilson, J.L.; Kirbah, S.; Sidney, S.S.; Rosenberger, J.; Bassie, L.; Alie, J.A.; McLean, D.C.; Garvey, W.T.; Ely, B. Mitochondrial DNA genetic diversity among four ethnic groups in Sierra Leone. *Am. J. Phys. Anthropol.* **2005**, *128*, 156–163. [CrossRef] [PubMed]

29. McLean, D.C.; Spruill, I.; Argyropoulos, G.; Page, G.P.; Shriver, M.D.; Garvey, W.T. Mitochondrial DNA (mtDNA) haplotypes reveal maternal population genetic affinities of Sea Island Gullah-speaking African Americans. *Am. J. Phys. Anthropol.* **2005**, *127*, 427–438. [CrossRef] [PubMed]

30. Block, S. A brief history of twins. *Lupus* **2006**, *15*, 61–64. [CrossRef] [PubMed]

31. Deafen, D.; Escalante, A.; Weinrib, L.; Horwitz, D.; Bachman, B.; Roy-Burman, P.; Walker, A.; Mack, T.M. A revised estimate of twin concordance in systemic lupus erythematosus. *Arthritis Rheumatol.* **1992**, *35*, 311–318. [CrossRef]

32. Alarcón-Segovia, D.; Alarcón-Riquelme, M.E.; Cardiel, M.H.; Caeiro, F.; Massardo, L.; Villa, A.R.; Pons-Estel, B.A. Familial aggregation of systemic lupus erythematosus, rheumatoid arthritis, and other autoimmune diseases in 1,177 lupus patients from the GLADEL cohort. *Arthritis Rheumatol.* **2005**, *52*, 1138–1147. [CrossRef] [PubMed]

33. Deng, Y.; Tsao, B.P. Genetic susceptibility to systemic lupus erythematosus in the genomic era. *Nat. Rev. Rheumatol.* **2010**, *6*, 683. [CrossRef] [PubMed]

34. Guerra, S.G.; Vyse, T.J.; Graham, D.S.C. The genetics of lupus: A functional perspective. *Arthritis Res. Ther.* **2012**, *14*, 211. [CrossRef] [PubMed]

35. Vaughn, S.E.; Kottyan, L.C.; Munroe, M.E.; Harley, J.B. Genetic susceptibility to lupus: The biological basis of genetic risk found in B cell signaling pathways. *J. Leukoc. Biol.* **2012**, *92*, 577–591. [CrossRef] [PubMed]

36. Zhao, J.; Wu, H.; Khosravi, M.; Cui, H.; Qian, X.; Kelly, J.A.; Kaufman, K.M.; Langefeld, C.D.; Williams, A.H.; Comeau, M.E.; et al. Association of genetic variants in complement factor H and factor H-related genes with systemic lupus erythematosus susceptibility. *PLoS Genet.* **2011**, *7*, e1002079. [CrossRef] [PubMed]

37. Kim, K.; Cho, S.K.; Sestak, A.; Namjou, B.; Kang, C.; Bae, S.C. Interferon-gamma gene polymorphisms associated with susceptibility to systemic lupus erythematosus. *Ann. Rheum. Dis.* **2010**, *69*, 1247–1250. [CrossRef] [PubMed]

38. Jacob, C.O.; Zhu, J.; Armstrong, D.L.; Yan, M.; Han, J.; Zhou, X.J.; Thomas, J.A.; Reiff, A.; Myones, B.L.; Ojwang, J.O.; et al. Identification of *IRAK1* as a risk gene with critical role in the pathogenesis of systemic lupus erythematosus. *Proc. Natl. Acad. Sci. USA* **2009**, *106*, 6256–6261. [CrossRef] [PubMed]

39. Zandman-Goddard, G.; Solomon, M.; Rosman, Z.; Peeva, E.; Shoenfeld, Y. Environment and lupus-related diseases. *Lupus* **2012**, *21*, 241–250. [CrossRef] [PubMed]

40. Kiyohara, C.; Washio, M.; Horiuchi, T.; Tada, Y.; Asami, T.; Ide, S.; Takahashi, H.; Kobashi, G.; Kyushu Sapporo SLE (KYSS) Study Group. Cigarette smoking, *N*-acetyltransferase 2 polymorphisms and systemic lupus erythematosus in a Japanese population. *Lupus* **2009**, *18*, 630–638. [CrossRef] [PubMed]

41. Gilkeson, G.; James, J.; Kamen, D.; Knackstedt, T.; Maggi, D.; Meyer, A.; Ruth, N. The United States to Africa lupus prevalence gradient revisited. *Lupus* **2011**, *20*, 1095–1103. [CrossRef] [PubMed]

42. Kamen, D.L.; Barron, M.; Parker, T.M.; Shaftman, S.R.; Bruner, G.R.; Aberle, T.; James, J.A.; Scofield, R.H.; Harley, J.B.; Gilkeson, G.S. Autoantibody prevalence and lupus characteristics in a unique African American population. *Arthritis Rheumatol.* **2008**, *58*, 1237–1247. [CrossRef] [PubMed]

43. Hochberg, M.C. Updating the American College of Rheumatology revised criteria for the classification of systemic lupus erythematosus. *Arthritis Rheum.* **1997**, *40*, 1725–1725. [CrossRef] [PubMed]

44. Spruill, I.J.; Leite, R.S.; Fernandes, J.K.; Kamen, D.L.; Ford, M.E.; Jenkins, C.; Hunt, K.J.; Andrews, J.O. Successes, challenges and lessons learned: Community-engaged research with South Carolina's Gullah population. *Gatew. Int. J. Community Res. Engagem.* **2013**, *6*. [CrossRef] [PubMed]

45. Langefeld, C.D.; Ainsworth, H.C.; Graham, D.S.C.; Kelly, J.A.; Comeau, M.E.; Marion, M.C.; Howard, T.D.; Ramos, P.S.; Croker, J.A.; Morris, D.L.; et al. Transancestral mapping and genetic load in systemic lupus erythematosus. *Nat. Commun.* **2017**, *8*, 16021. [CrossRef] [PubMed]

46. Karlson, E.W.; Chang, S.C.; Cui, J.; Chibnik, L.B.; Fraser, P.A.; De Vivo, I.; Costenbader, K.H. Gene–environment interaction between HLA-DRB1 shared epitope and heavy cigarette smoking in predicting incident rheumatoid arthritis. *Ann. Rheum. Dis.* **2010**, *69*, 54–60. [CrossRef] [PubMed]

47. Costenbader, K.H.; Chang, S.C.; De Vivo, I.; Plenge, R.; Karlson, E.W. Genetic polymorphisms in PTPN22, PADI-4, and CTLA-4 and risk for rheumatoid arthritis in two longitudinal cohort studies: Evidence of gene-environment interactions with heavy cigarette smoking. *Arthritis Res. Ther.* **2008**, *10*, R52. [CrossRef] [PubMed]

48. Keenan, B.T.; Chibnik, L.B.; Cui, J.; Ding, B.; Padyukov, L.; Kallberg, H.; Bengtsson, C.; Klareskog, L.; Alfredsson, L.; Karlson, E.W. Effect of interactions of glutathione S-transferase T1, M1, and P1 and HMOX1 gene promoter polymorphisms with heavy smoking on the risk of rheumatoid arthritis. *Arthritis Rheumatol.* **2010**, *62*, 3196–3210. [CrossRef] [PubMed]

49. Davis, A.P.; Grondin, C.J.; Lennon-Hopkins, K.; Saraceni-Richards, C.; Sciaky, D.; King, B.L.; Wiegers, T.C.; Mattingly, C.J. The comparative toxicogenomics database: Update 2017. *Nucleic Acids Res.* **2017**, *45*, D972–D978. [CrossRef] [PubMed]

50. R Development Core Team. *R: A Language and Environment for Statistical Computing*; R Foundation for Statistical Computing: Vienna, Austria, 2016. Available online: http://www.R-project.org (accessed on 15 October 2016).

51. Wolf, B. *LogicForest: Logic Forest*; R package Version 2.1.0; R Foundation for Statistical Computing: Vienna, Austria, 2014.

52. Breiman, L.; Friedman, J.; Olshen, R.; Stone, C. *Classification and Regression Tees*; Chapman & Hall/CRC: Boca Raton, FL, USA, 1984.

53. Ruczinski, I.; Kooperberg, C.; LeBlanc, M. Logic regression. *J. Comput. Graph. Stat.* **2003**, *12*, 475–511. [CrossRef]

54. Svetnik, V.; Liaw, A.; Tong, C.; Culberson, J.C.; Sheridan, R.P.; Feuston, B.P. Random forest: A classification and regression tool for compound classification and QSAR modeling. *J. Chem. Inf. Comput. Sci.* **2003**, *43*, 1947–1958. [CrossRef] [PubMed]

55. Schwender, H.; Ickstadt, K. Identification of SNP interactions using logic regression. *Biostatistics* **2008**, *9*, 187–198. [CrossRef] [PubMed]

56. Breiman, L. *Bagging Predictors*; Technical Report 421; Department of Statistics, University of California at Berkley: Berkeley, CA, USA, 1994; pp. 1–19.

57. Dietterich, T.G. An experimental comparison of three methods for constructing ensembles of decision trees: Bagging, boosting, and randomization. *Mach. Learn.* **2000**, *40*, 139–157. [CrossRef]

58. Friedman, J. Greedy function approximation: A gradient boosting machine. *Ann. Stat.* **2001**, *29*, 1189–1202. [CrossRef]

 *genes*

Article

MODIMA, a Method for Multivariate Omnibus Distance Mediation Analysis, Allows for Integration of Multivariate Exposure–Mediator–Response Relationships

Bashir Hamidi [1,2], Kristin Wallace [3] and Alexander V. Alekseyenko [1,2,3,4,5,*]

[1] Program for Human Microbiome Research, Medical University of South Carolina, 135 Cannon Street MSC 200, Charleston, SC 29425, USA

[2] Biomedical Informatics Center, Medical University of South Carolina, 135 Cannon Street MSC 200, Charleston, SC 29425, USA

[3] Department of Public Health Science, Medical University of South Carolina, 135 Cannon Street MSC 200, Charleston, SC 29425, USA

[4] Department of Oral Health Sciences, Medical University of South Carolina, 135 Cannon Street MSC 200, Charleston, SC 29425, USA

[5] Department of Healthcare Leadership and Management, Medical University of South Carolina, 135 Cannon Street MSC 200, Charleston, SC 29425, USA

* Correspondence: alekseye@musc.edu

Received: 10 June 2019; Accepted: 8 July 2019; Published: 11 July 2019

Abstract: Many important exposure–response relationships, such as diet and weight, can be influenced by intermediates, such as the gut microbiome. Understanding the role of these intermediates, the mediators, is important in refining cause–effect theories and discovering additional medical interventions (e.g., probiotics, prebiotics). Mediation analysis has been at the heart of behavioral health research, rapidly gaining popularity with the biomedical sciences in the last decade. A specific analytic challenge is being able to incorporate an entire 'omics assay as a mediator. To address this challenge, we propose a hypothesis testing framework for multivariate omnibus distance mediation analysis (MODIMA). We use the power of energy statistics, such as partial distance correlation, to allow for analysis of multivariate exposure–mediator–response triples. Our simulation results demonstrate the favorable statistical properties of our approach relative to the available alternatives. Finally, we demonstrate the application of the proposed methods in two previously published microbiome datasets. Our framework adds a new tool to the toolbox of approaches to the integration of 'omics big data.

Keywords: multivariate analysis; multivariate causal mediation; distance correlation; direct effect; indirect effect; causal inference

---

1. Introduction

Natural biological phenomena are often explained using statistical methods by means of isolating the individual contexts of the phenomenon itself by establishing associations. For example, obesity, among other factors, maybe related to changes in nutrition or stress. Although these explanations fail to present a full account of the original observed phenomenon or capture the entirety of such complex dynamics, they aid in our understanding of the cause–effect relationships, especially when a plausible causal directionality can be established (e.g., increase in calorie consumption is plausibly causal to weight gain, and not the other way around). The next level of complexity is afforded by incorporating additional mechanisms arising from one or more other intermediate factors. In order to properly understand the mechanisms involved, we must understand the extent to which the exposure of interest (calorie intake) directly affects an outcome (weight gain) and the extent to which the exposure indirectly affects the outcome through intermediate factors (e.g., gut microbiome) [1–4]. Mediation analysis is at the heart of many human behavior studies and is quickly gaining traction in the biomedical research arena. With the explosion growth of the 'omics, we see the development of new analysis and tools that provide access to the integration of new knowledge and their applications as mediators of treatment–effect relationships.

Microbiome research has advanced significantly in the last decade with the rise of computational power, next-generation sequencing, and data analytics [5,6]. Naturally arising have been translational investigations assessing the interplay of human–host microbial communities with various health and diseases states [7–9], yet a notable challenge remains of understanding the extent to which and mechanisms by which such interactions take place. Three notable interactions comprise this dynamic relationship: first, the association between the environment and the host; second, the association between the microbiome and host health or disease; and third, the association between the environment and the microbiome. Because of this complexity, most available observational and experimental study designs are unable to properly assess direct causal roles of the microbiome, and, in many cases, alternative interpretations are plausible. We have seen a growing volume of evidence linking microbiome and human disease such as that of obesity, inflammatory bowel disease, and colorectal cancer [10,11]. Similarly, we have seen the relationship between environmental factors and the microbiome [12]. Now, we believe it is important to assess how outside environmental factors or host genetic characteristics affect the microbiome and, together with changes of microbiome composition, influence human health and disease. Accordingly, there is an urgent need for statistical methods that establish and isolate the mediation role of microbes in these complex dynamics.

Formal approaches to the assessment of mediation effects are primarily based on the work by Baron and Kenny [13] using the product of coefficients. The single mediator model (SMM) describes the relationship between exposure (X), response (Y), and a mediator (M), each of which are univariate random variables. SMM posits that the relationship between those can be described in terms of linear regressions, that capture the effect of the exposure on the response:

$$Y = i_1 + \gamma X + \varepsilon_1, \tag{1}$$

the effect of the exposure on the mediator:

$$M = i_3 + \alpha X + \varepsilon_3, \tag{2}$$

and the effect of both on the response:

$$Y = i_2 + \gamma' X + \beta M + \varepsilon_2. \tag{3}$$

The downside of the conceptual simplicity of the linear regression-based framework is the lack of a convenient test that could allow for the evaluation of the hypotheses about the presence of mediation without the need to estimate the regression coefficients. To this end, Boca et al. [14] have provided a mediation testing framework that casts the regression equations in terms of correlations and partial correlations. They further propose a multiple testing framework for the evaluation of the hypotheses related to the presence of multiple mediators. However, the omnibus mediation hypothesis still lacks an acceptable simple solution. Furthermore, no current approach allows for multivariate exposures and responses.

Microbiome analytics must take into account the multivariate nature of such data, and thus often use distance-based approaches. In these cases, power and type I error characteristics are often directly related to the chosen distance metric. Some proposed methods such as the work of Zhao et al. [15] utilize multiple distance and dissimilarity metrics in a regression-based association testing framework.

As another example, Tang et al. [6] assessed true association by using multiple distances simultaneously and by allowing the flexible adjustment of confounders through computing residuals by regression of the covariates on confounders. As we have seen in published reviews, limitations exist to proposed methods upon the application to 'omics data [16]. One such case is microbiome data, which are high-dimensional, under-sampled, compositional, and over-dispersed; nonetheless, we would like to be able to explain their role as a mediator just like a univariate mediator would. Within this article, we present a framework for multivariate distance mediation analysis that is suitable to such data.

In this article, we present a framework for testing multivariate distance mediation to allow for multivariate exposures, responses, and mediators. We build our test on the mediation approach published by Boca et al. [14] and extend it to high-dimensional data via distance-based methodologies. We present simulation results on the robustness and sensitivity of the proposed methods and further make comparisons with other proposed approaches, such as permutation-based testing by Boca et al. [14] and sample-wise distance matrices by Zhang et al. [17]. Lastly, we analyze two real datasets to demonstrate the power of the proposed methods and their application to high-dimensional microbiome data.

2. Materials and Methods

2.1. Availability and Implementation

Supplementary materials include reference implementation of the methods, simulation studies, and application examples and are freely available at https://github.com/alekseyenko/MODIMA.

2.2. Testing for Mediation

The testing framework developed by Boca et al. [14] expresses the relationships captured in the SMM linear regressions in terms of Pearson correlations. Thus, for a significant effect of the exposure on the response to exist, the correlation between the two has to be non-zero, $\rho(X, Y) > 0$. Furthermore, if the relationship is in fact mediated by M, both the correlation between exposure and the mediator and the conditional correlation of the mediator and the response, given the exposure, should be non-zero [13], $\rho(X, M) > 0$ and $\rho(r_{M|X}, r_{Y|X}) > 0$, respectively. Here, $r_{M|X}$ and $r_{Y|X}$ denote the residuals of the conditional correlation on regression of X on M and X on Y, respectively. These observations give rise to the following test statistic:

$$S(X, M, Y) = \rho(X, M)\, \rho(r_{M|X}, r_{Y|X}), \tag{4}$$

which is capable of capturing the presence of mediation in a hypothesis testing framework. Boca et al. [14] evaluate the significance of this test statistic using permutation testing.

2.3. Motivation for Using Energy Statistics, dCor and pdCor

Székely and Rizzo introduced a series of non-parametric tests of covariance and correlation based on energy statistics, the theoretical understanding that observations are governed by a statistical potential "energy" which is zero if and only if the underlying statistical null hypothesis is true [18]. In this context, assessments and relationships of objects are made by first calculating corresponding distances of objects and all hypothesis testing and inferences are made based on these initial distances. This allows us to compare objects against each other using their relative distance and without any knowledge about their size or other properties. In this publication, we make use of distance correlation, dCor [19], and partial distance correlation, pdCor [20], which are available in R package energy [21].

The dCor test of multivariate independence, based on the corresponding sample distance covariance dCov, has unique properties of measuring dependence. The definitions of these parallel those of the classical Pearson product moment correlation ρ with the major difference being that the centered product moment transformation is applied to the distance matrices rather than data vectors.

This test of independence can be easily applied in arbitrary dimensions—not necessarily equal—and without assumptions such as normality in the product–moment correlation counterpart. The dCov and dCor have been shown to be more powerful than the parametric counterparts, especially for nonlinear dependence structures. The practicality of applying a test and measure of dependence in high dimensions that is not only easy to apply, but also easy and intuitive to interpret, is invaluable.

The Pearson partial correlation which measures the partial correlation in vectors x and y, controlling for z, is described with the following partial correlation coefficient:

$$r(x, y; z) = \frac{r(x, y) - r(x, z)r(y, z)}{\sqrt{1 - r(x, z)^2} \sqrt{1 - r(x, y)^2}},$$ (5)

where $r(x, y)$ is the Pearson sample correlation and x, y, *and* z are one-dimensional data vectors. As an extension of the Pearson partial correlation and, in much the same way, Székely and Rizzo [20] introduced partial distance correlation pdCor:

$$pdCor(X, Y; Z) = \frac{R^*(X, Y) - R^*(X, Z)R^*(Y, Z)}{\sqrt{1 - R^*(X, Z)^2} \sqrt{1 - R^*(X, Y)^2}},$$ (6)

where $R^*(X, Y)$, $R^*(X, Z)$, *and* $R^*(Y, Z)$ denote the bias corrected distance correlation. We suggest a review of Szeékely and Rizzo, 2007 [22], 2013 [23], and 2014 [20] for a theoretical basis and deeper understanding of the methods.

2.4. Multivariate Omnibus Distance Mediation Analysis Statistic

In modeling relationships between multivariate variables, we must be able to express the relationship between those in terms similar to the Pearson correlations and partial correlations. To do so, we use distance correlation and partial distance correlation statistics that are capable of capturing relationships between vector-valued random variables. These statistics naturally flow from the definition of Pearson correlation by allowing a distance metric (such as Euclidean distance, or specialized distances for microbiome data) to serve as a sufficient statistic for the dependence relationship within each random vector. Using these, the multivariate omnibus distance mediation analysis (MODIMA) test statistic is as follows:

$$S_d(d_X(X), d_M(M), d_Y(Y)) = dCor(d_X(X), \ d_M(M))pdCor(d_X(Y), d_M(M)|d_Y(X)),$$ (7)

where $d(.)$ are appropriate pairwise distance matrices computed from the potentially multivariate observations of exposure, X, mediator, M, and response, Y.

For a more intuitive understanding of the MODIMA method, consider the illustration in Figure 1. Suppose our data consists of n observations for p_x exposure, p_m mediator, and p_y response variables. The test statistic is obtained by first calculating the $n \times n$ distance matrices from just the exposure, $d_X(X)$, just the mediator, $d_M(M)$, and just the response, $d_Y(Y)$, variables. Note that the distance (or dissimilarity) metric can potentially be different for each of these, as appropriate given the nature of these variables. The distance matrices are then used to compute the distance correlation between the exposure and mediator and the partial correlation between the mediator and the response, given the exposure. These two quantities are then multiplied together to obtain the test statistic in Equation (7).

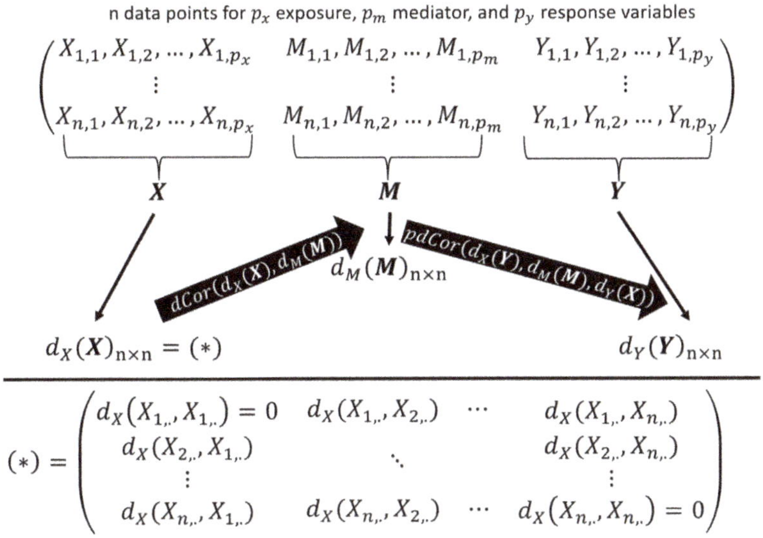

n data points for p_x exposure, p_m mediator, and p_y response variables

$$\begin{pmatrix} X_{1,1}, X_{1,2}, \ldots, X_{1,p_x} & M_{1,1}, M_{1,2}, \ldots, M_{1,p_m} & Y_{1,1}, Y_{1,2}, \ldots, Y_{1,p_y} \\ \vdots & \vdots & \vdots \\ X_{n,1}, X_{n,2}, \ldots, X_{n,p_x} & M_{n,1}, M_{n,2}, \ldots, M_{n,p_m} & Y_{n,1}, Y_{n,2}, \ldots, Y_{n,p_y} \end{pmatrix}$$

$X \qquad\qquad M \qquad\qquad Y$

$dCor(d_X(X), d_M(M)) \qquad pdCor(d_X(Y), d_M(M)), d_Y(X))$

$d_M(M)_{n \times n}$

$d_X(X)_{n \times n} = (*) \qquad\qquad d_Y(Y)_{n \times n}$

$$(*) = \begin{pmatrix} d_X(X_{1,.}, X_{1,.}) = 0 & d_X(X_{1,.}, X_{2,.}) & \cdots & d_X(X_{1,.}, X_{n,.}) \\ d_X(X_{2,.}, X_{1,.}) & & & d_X(X_{2,.}, X_{n,.}) \\ \vdots & & \ddots & \vdots \\ d_X(X_{n,.}, X_{1,.}) & d_X(X_{n,.}, X_{2,.}) & \cdots & d_X(X_{n,.}, X_{n,.}) = 0 \end{pmatrix}$$

Figure 1. Visual description of the multivariate omnibus distance mediation analysis (MODIMA) test statistic. First, the $n \times n$ distance matrices are calculated from just the exposure, $d_X(X)$, just the mediator, $d_M(M)$, and just the response, $d_Y(Y)$, variables. Using these pairwise matrices, distance correlation (dCor) of the exposure–mediator and partial distance correlation (pdCor) of mediator–response are calculated using R package energy [21]. Product of dCor and pdCor results in MODIMA test statistic.

2.5. MODIMA Permutation Testing

The permutation testing approach for the MODIMA method follows that of Boca et al. [14]. In short, to obtain the empirical distribution of the MODIMA test statistic S_d under the null hypothesis, either the relationship between the exposure and mediator, or the conditional relationship between the response and the mediator has to be scrambled. Thus, if the magnitude of the first is smaller than that of the second, we permute the rows and columns of the $d_X(X)$ matrix and re-compute the test statistic $S_d(i)$. Conversely, a permutation of the response distance matrix $d_X(Y)$ is performed to re-compute the test statistic if its partial correlation with the mediator is greater. The p-value of the observed S_d is obtained as the frequency with which the permuted statistic exceeds the observed in q permutations, $P = \frac{1}{q} \sum_{i=1}^{q} \mathbf{1}(S_d \leq S_d(i))$. Permutation testing is generally a powerful way to simulate from the null distribution; however, it is often hard to compute estimates of extremely small p-values. Although not implemented in this version of MODIMA, solutions exist to estimate small p-values based on fitting extreme value distributions to the permuted test statistics (e.g., application of Pareto distribution to permutation testing [24]). Reference R language implementation of the MODIMA test is available at https://github.com/Alekseyenko/MODIMA.

2.6. Empirical Evaluation Simulation

Single mediator. To assess statistical properties of the proposed omnibus method and compare it to existing methods, we simulated data where exposure, X, mediator, M, and response, Y, were normally distributed and followed the linear model formalism of the single mediator model (Figure 2). In this case, we varied the parameters α, β, *and* γ as follows: $\alpha = \{0, 0.25, 0.5, 0.75, 1\}$, $\beta = \{0, 0.25, 0.5, 0.75, 1\}$, and $\gamma = \{0, 0.1, 0.25, 0.5\}$. Under each combination of parameters, we simulated datasets with a varying number of observations, $n = \{20, 50, 100, 150, 200\}$. To ensure unit variance, the standard deviations of X, M, and Y were fixed at $SD_{X, M, Y} = 1$. Euclidean distance was used to compute the distance matrices. For each combination of parameters, we generated 1000 datasets for a total of 250,000 datasets. Each dataset was analyzed using our reference implementation of MODIMA as well as previous methods proposed by Sampson [14] and Chen [17].

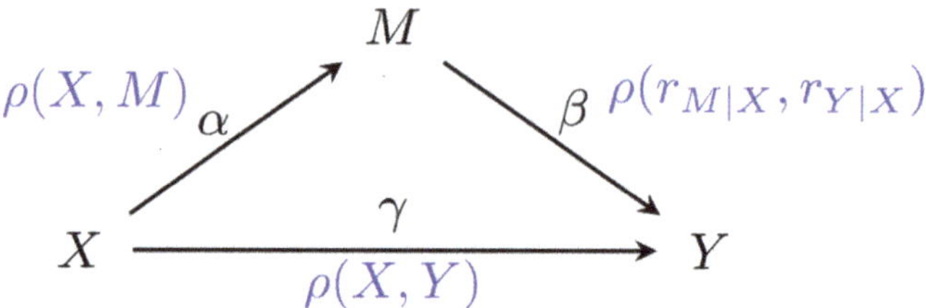

Figure 2. Linear model parameterization of the single mediator model showing exposure, *X*, mediator, *M*, and response, *Y*, and the appropriate correlation and conditional correlation coefficients.

Multiple mediators. In the presence of multiple mediators, we simulated data where exposure, *X*, and response, *Y*, were normally distributed and parameters α, β, and γ were set just as described in the single mediator case. Here, the standard deviation of *X*, $SD_X = 1$ and that of *Y*, $SD_Y = 0.01$ and Euclidean distances were computed on each. *M* was generated using a mixture of two datasets (saliva and tonsils) from the National Institutes of Health (NIH) Human Microbiome Project [25] sourced through R package HMP [26]. Saliva and tonsil datasets contain abundance data consisting of 21 taxa on 24 subjects. A proportion of used data from each dataset was determined as a function of α parameter and sample size. The abundance data were used to compute the parameter of the Dirichlet-multinomial distribution [27] modeling over-dispersion and used to generate random Dirichlet-multinomial samples for each iteration of the simulation. Additionally, rooted trees with 21 tips were generated using R package ape [28] and used for the computation of the tree-based weighted UniFrac [29] distance of the mediator. Non-tree-based distance methods of Jensen–Shannon divergence (JSD) [30] and Bray were computed for the mediators using R package phyloseq [31] and vegan [32], respectively. Parameters α, β, *and* γ were set in a similar fashion to the single mediator simulation and sample sizes were set to $n = \{20, 50, 100, 150\}$.

Details of the simulation are available as a knitted R Markdown file at https://github.com/Alekseyenko/MODIMA.

3. Results

3.1. Empirical Evaluation of MODIMA

The simulation results comprise the sample size-dependent type I error rates and power as a fraction of rejected null hypotheses at a significance threshold of 0.05 for each test (Figures 3 and 4). In the case of the single mediator, when the null hypothesis is true (Figure 3A), the association of the exposure with the mediator ($\alpha = 0$) or the effect of the mediator on the response ($\beta = 0$) are absent. A test properly controlling type I error rate is expected to have a fraction of rejections equal to the nominal error rate (0.05, in this case). In the cases of $\beta = 0$, as α and γ are increased, we observe inflation of this type I error rate for MODIMA; however, Sampson [14] and Chen [17] methods often display overly conservative type I error rates. This effect has been previously described [20] and is related to the fact that zero partial distance correlation does not correspond to conditional independence. We review this point further in the Discussion section. Our proposed method is able to demonstrate equal or better power (Figure 3B), often increasing power with the increase of mediating effect. Within a few selected parameters, all three methods performed equally. Notably, the MedTest method often shows the least power and does not perform well when the association of the mediator with the exposure is much higher than the association of the mediator with the response (Figure 3B,

$\alpha = 1$ column). In fact, that appears to be the most challenging condition for all approaches to make the necessary rejections. Our approach maintains the best performance in that instance.

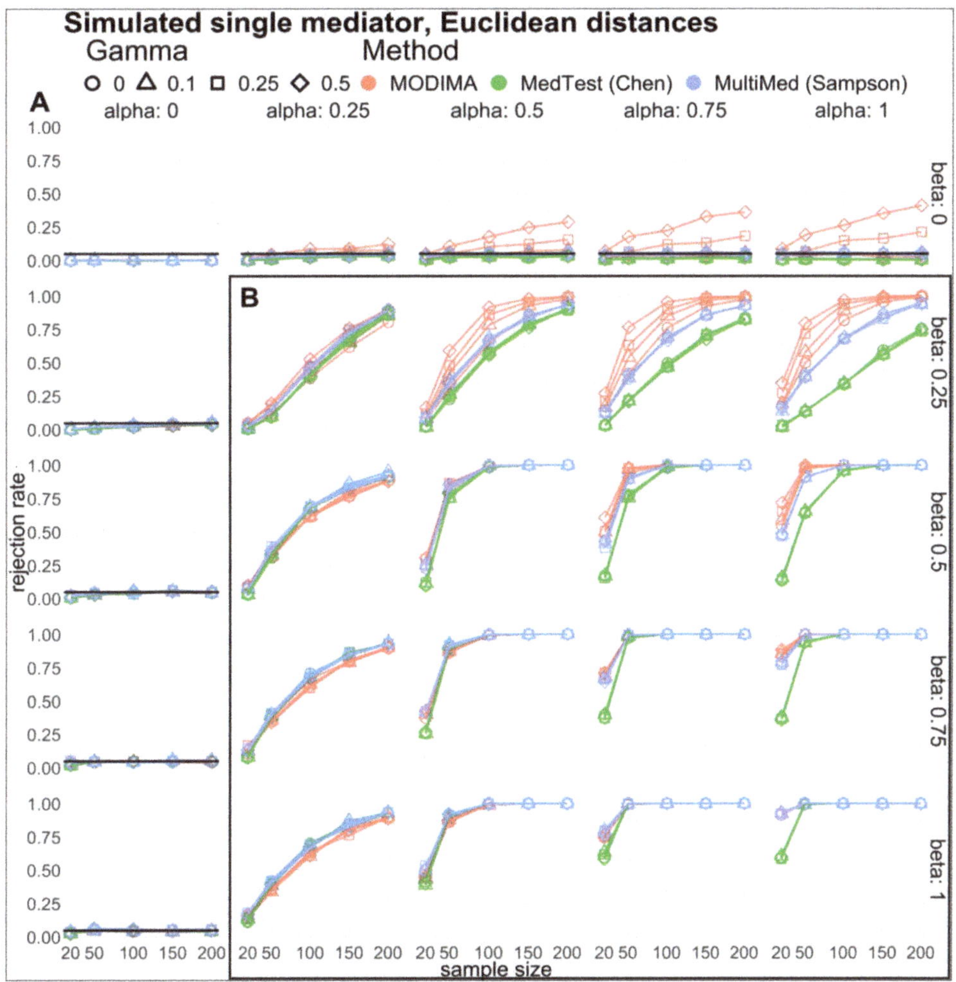

Figure 3. Simulated single mediator results. (**A**) Type I error and (**B**) power are estimated by simulation at varying α, β, γ, and sample size. Comparison between the approach proposed here and other methods proposed by Sampson (MultiMed) [14] and Chen (MedTest) [17] is portrayed using red, blue, and green, respectively. Point shapes portray the various degrees of $X - Y$ relationship, γ. Horizontal black lines in (**A**) represent 0.05 type I error rate commonly used.

In the case of the many mediators, we generated mixtures of microbiome data to be used as mediators. We computed distance metrics of Bray, Jensen–Shannon divergence (JSD), and UniFrac and present the latter here in Figure 4. The former two can be found as part of our supplementary data. When the null hypothesis is true (Figure 4A), both MODIMA and MedTest are able to maintain rejection rates. The behavior of MODIMA to inflate the rejection rates under $\beta = 0$ observed in the single mediator case is no longer present in the multiple mediator simulation case. Power curves (Figure 4B) show MODIMA excelling under certain scenarios, whereas MedTest displays better power in others. Overall, MODIMA excels under scenarios where the **X-Y** relationship noted by γ is smaller or, in other words, relatively small to no direct relationship between the exposure and outcome. Under the smaller sample sizes, we often see MODIMA performing slightly better.

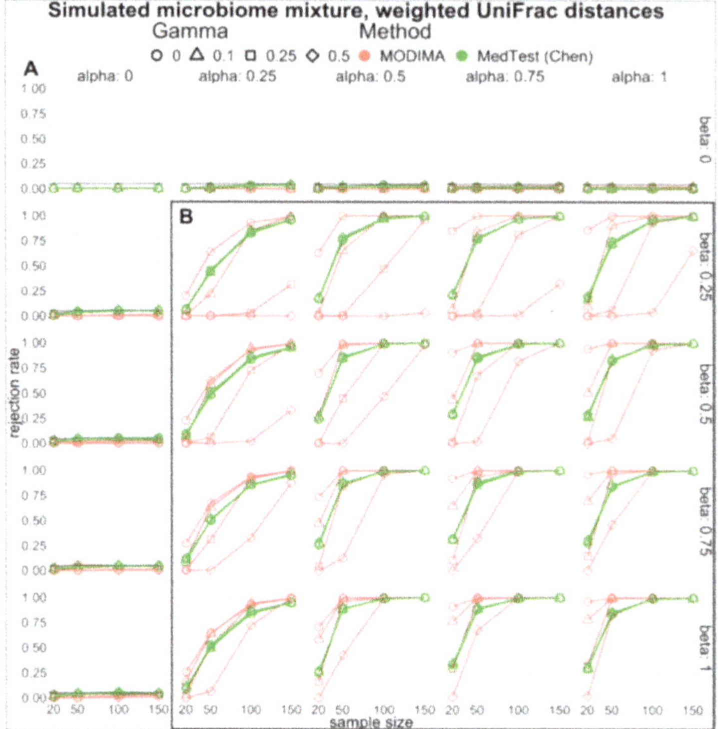

Figure 4. Simulated microbiome mixture results. (**A**) Type I error and (**B**) power are estimated by simulation at varying α, β, γ, and sample size. Shown here are the weighted UniFrac [29] distance metrics computed on microbiome matrices. Comparison between the approach proposed here and Chen (MedTest) [17] is portrayed using red and green, respectively. Point shapes portray the various degrees of $X - Y$ relationship, γ. Horizontal black lines in (**A**) represent 0.05 type I error rate commonly used. Simulated microbiome mixture results using other distance measures are provided in Figures S1 and S2, Supplementary Materials.

We next demonstrate the application of MODIMA in two empirical examples.

3.2. Application Example 1: Microbiome-Mediated Responses to Subtherapeutic Antibiotic Treatment Influencing Body Fat

Antibiotics have undoubtedly provided remarkable public health benefits in the last century. During that same time span, we see a marked increase in antibiotics use across many populations [33]. Furthermore, we see the largest use of antibacterial agents within the animal farming industry, often exclusively used in low doses to stimulate weight gain in farm animals [34]. There is growing concern about the effects from the long-term use of antibiotics and antibacterial agents [35,36]. Here, we build on the evidence on phenotypic and microbial responses to early-life subtherapeutic antibiotic treatment using murine models expanding on findings presented by Cho et al. [37].

In each experiment, each study group (control or antibiotic(s)) was composed of ten mice. The mice were allowed ad libitum access to food and water and fed standard laboratory chow. Beginning on day 28 of life, mice were given water or water containing one of the following antibiotic regimens: penicillin

VK, vancomycin, penicillin VK plus vancomycin, and chlortetracycline, each at doses equivalent to 1 μg antibiotic per g body weight. On a weekly basis, mice were weighed three times, food intake measured, and fecal pellets collected. Dual energy X-ray absorptiometry (DEXA) and a 7 Tesla MRI system were used to collect animal fat composition, lean body mass, percent body fat, and bone mineral density. The IDEAL Dixon method based on chemical shift properties was used to separate MRI images into fat and lean tissue [38]. Weight values were calculated from MRI-determined fat percent to validate scale weight. The microbiome composition was established by sequencing of the v3 region of the 16S rRNA gene using 454-FLX Titanium chemistry (Roche, Bradford, CT, USA). Preprocessing was performed using QIIME pipeline [39] at a 97% similarity threshold.

A total of 96 samples (50 cecal and 46 fecal) across 50 animals were used for all downstream analysis, resulting in 6547 unique taxa. Four antibiotic regimens of penicillin $n = 10$, vancomycin $n = 10$, penicillin plus vancomycin $n = 10$, and chlortetracycline $n = 10$ were used for comparison with controls $n = 10$. For each subject, cecal and fecal microbiome data were available.

Jensen–Shannon divergence (JSD) distances were computed for microbiota (mediator) and Euclidean distances for antibiotic use (exposure) as well as percent fat (outcome). The mediating relationship of the combined cecal and fecal microbial composition between antibiotic intake and percent fat resulted in a MODIMA statistic of 0.002 ($p = 0.99$). Assessing the relationship of antibiotic use and percent fat, we see a bias-corrected distance correlation (bcdCor) estimate of 0.173. Likewise, we see dCor estimates of microbiota (cecal and fecal combined) to antibiotic use and percent fat to be 0.113 and 0.000. Partial distance correlation (pdCor) between the relationship of antibiotic use, percent fat, and microbiota (cecal and fecal JSD) is calculated to be 0.021 ($p = 0.46$).

Fecal and cecal samples were also analyzed separately. We observe MODIMA statistics of 0.008 ($p = 0.81$) and 0.007 ($p = 0.72$) for fecal (Figure 5) and cecal (Figure S3, Supplementary Materials) samples, respectively. Using the provided estimates, we see that variable pairs do show mild distance correlation for both fecal and cecal (Additional Files 1 at https://github.com/alekseyenko/MODIMA). Partial distance correlation computations remain negligible, 0.026 ($p = 0.57$) and 0.018 ($p = 0.59$) for fecal and cecal samples, respectively, and these total effects can be seen in Figure 5C,D and Figure S3C,D (Supplementary Materials at https://github.com/alekseyenko/MODIMA)).

Further antibiotic-specific comparisons are made between individual antibiotic therapies and control. Most notable changes are seen when cecal and fecal are assessed individually with a specific antibiotic treatment. An assessment of cecal and fecal microbiome mediation of the penicillin versus control exposure results in MODIMA statistics of 0.019 ($p = 0.28$) and 0.034 ($p = 0.09$), respectively. Partial distance correlations are observed to be 0.023 and 0.110 for cecal and fecal. For fecal specifically, although mild correlation is present, and when assessed for mediation using MODIMA, effects are not detectable. Comparison of chlortetracycline and control using samples from fecal microbiome revealed the largest and only statistically significant mediation, with MODIMA statistic of 0.141 ($p = 0.016$).

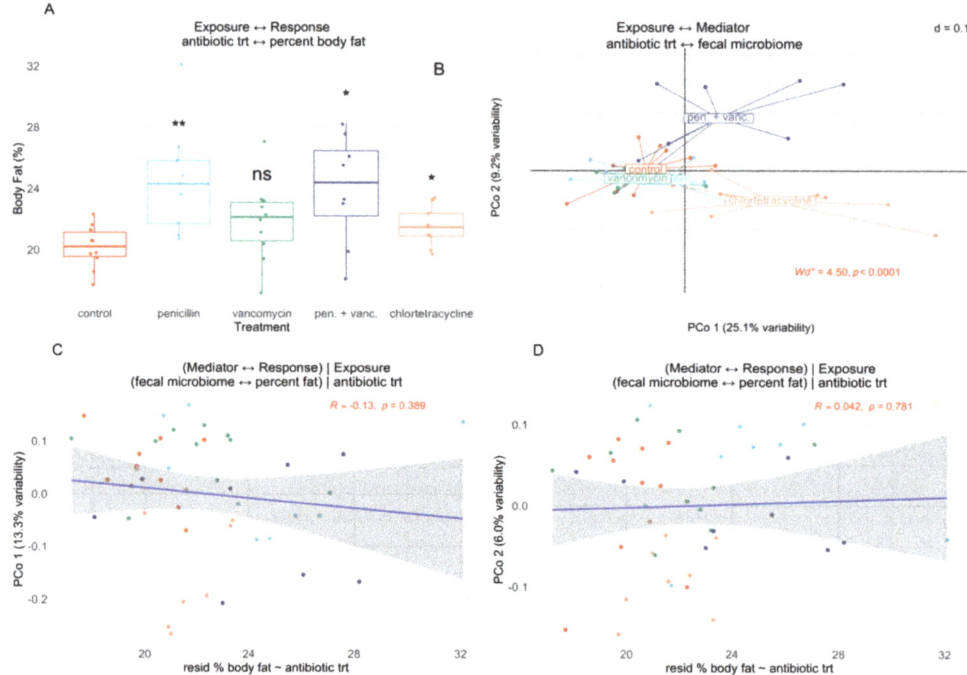

Figure 5. Empirical example 1, fecal microbiome-mediated responses to antibiotic treatment. (**A**) shows association of antibiotic treatment and percent bodyfat; (**B**) shows percent body fat was significantly increased in all antibiotic groups with the exception of vancomycin; *p*-values are noted by symbols ******, *****, and ns corresponding to $p < 0.01$, < 0.05, and not significant, respectively; association of antibiotic treatment with microbiota PCo axes 1 and 2, as well as Welch statistic and *p*-value suitable for analysis of microbiome data, W_d^* [40]; (**C**) and (**D**) show, respectively, the lack of association between microbiota and body fat using PCo axes 1 and 2, removing any effect of antibiotic treatment.

Data and analysis for this application are available at https://github.com/Alekseyenko/MODIMA.

3.3. Application Example 2: Microbiome-Mediated Responses to Dietary Fiber Intake Influencing Body Mass Index (BMI)

A growing body of evidence suggests that diet influences the compositional diversity of gut bacteria [8]. We also see an association between changes in gut bacteria diversity and human health, such as obesity [41]. Through a study of diet and 16S ribosomal DNA (rDNA) fecal samples, Wu et al. (2011) reported that long-term diet was strongly associated with enterotype clustering [42]; here, we briefly describe their methods. Healthy human subjects ($n = 98$) were enrolled in a cross-sectional study where long-term diet information was collected using self-reported questionnaires assessing usual dietary composition over the preceding year. Diet information was subsequently converted to a list of 214 nutrient categories and their corresponding intake amounts. Stool samples were collected, frozen immediately ($-80\ °C$), processed using MoBio PowerSoil kits, amplified V1–V2 region primers targeting bacterial 16S genes, and sequenced using 454/Roche. Sequences were denoised using QIIME pipeline [39] following default settings. Other demographic information including body mass index (BMI) was collected upon enrollment. Weighted and unweighted UniFrac distances for microbial communities were calculated and used for downstream analyses.

Wu et al. [42] previously reported a strong inverse association between body fat intake and microbial taxa (Spearman $\rho = -0.68$, $p < 0.0001$). These same microbial taxa were observed to be associated with BMI (PERMANOVA, unweighted $p = 0.001$, weighted $p = 0.145$). Here, we assessed the influence of dietary fiber intake on BMI mediated by microbiota. The correlation of percent fiber intake (exposure) and BMI (response) is small yet significant (Figure 6A), and fiber intake and microbiota (mediator) is small and approaching statistical significance (Figure 6B). Mediator–response relationships (total effects) are modeled in Figure 6C,D using first and second principal coordinates and residuals of exposure–response. Zhang et al. [17] applied their omnibus mediation method MedTest using multiple distance metrics to assess this mediation and showed a permutation-based p-value of 0.0309. Furthermore, mediation by individual taxa at different ranks was assessed. Zhang et al. [17] observed three ranks to be significant: family Lachnospiraceae ($p = 0.0129$), genus *Lachnospira* of the family Lachnospiraceae ($p = 0.0430$), and family Ruminococcaceae ($p = 0.0468$). We applied our proposed distance mediation analysis methods to this dataset and present our findings.

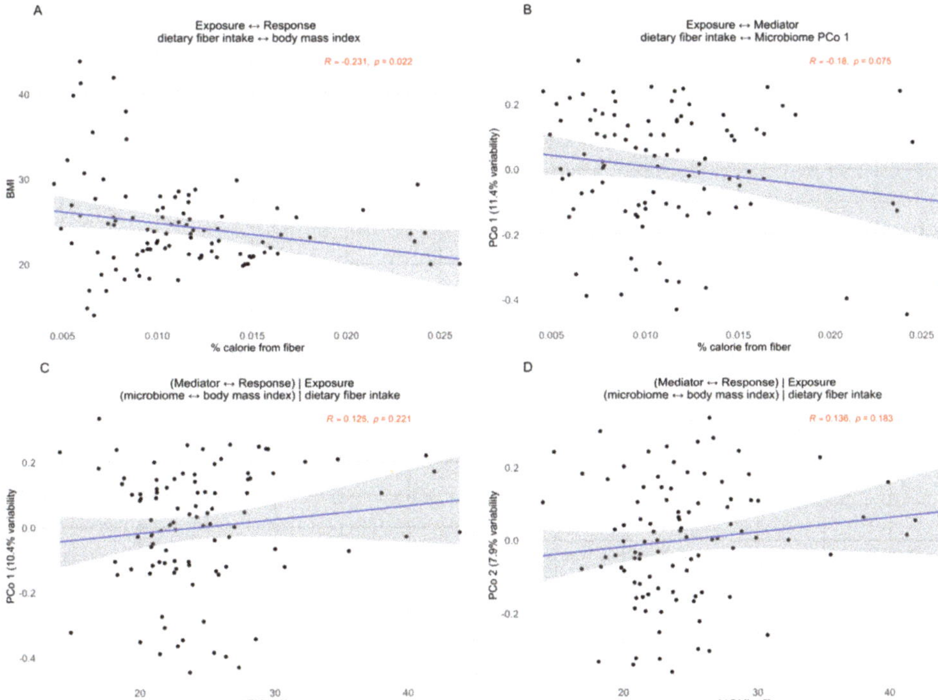

Figure 6. Empirical example 2, microbiome-mediated responses of body mass index (BMI) to dietary fiber intake. (**A**) shows association of fiber intake with outcome of BMI; (**B**) shows association of fiber intake with microbiota PCo axis 1, PCo axis 2 results can be seen in Figure S4, Supplementary Materials; (**C** and **D**) demonstrate, respectively, the lack of association between microbiota and BMI using PCo axes 1 and 2, removing any effect of dietary fiber intake. Pearson correlation coefficients and p-values are shown in red.

Distance metrics of Jensen–Shannon divergence (JSD), Bray–Curtis, Jaccard, and UniFrac (unweighted, weighted, and generalized) were computed using distance function in R package phyloseq (version 1.26.1) [31], vegdist function in R package vegan (version 2.5.4) [32], and GUniFrac function in R package GUniFrac (version 1.1) [43], respectively, following similar methodologies as presented by Zhang et al. [17]. Distance correlation (dCor) estimates between pairs of exposure, mediator, and response showed no evidence of correlation using any of the distance metrics (data not presented here but are available as Additional Files 2 at https://github.com/Alekseyenko/MODIMA). In a likewise fashion, the partial distance correlation estimate between fiber intake, microbiome, and BMI was not indicative of any correlation (estimate using JSD applied to mediator = 0.029, $p = 0.04$). MODIMA was applied using various distance metrics. Table 1 summarizes the results for MODIMA p-values for each distance metric as well as Bonferroni-adjusted test. As shown, we see that only the Jaccard metric shows a significant p-value at α of 0.05, with Bray–Curtis and unweighted UniFrac distances approaching the significance threshold. We further observe MODIMA resulting in lower p-values than MedTest in single-distance mediation tests with the exception of Jaccard.

Table 1. MODIMA and MedTest p-values for various distance metrics.

	Jensen–Shannon	Bray–Curtis	Jaccard	UniFrac	WUniFrac	GUniFrac	Bonferroni
MODIMA p-value	0.1074	0.0974	0.0321	0.0706	0.4645	0.2543	0.1926
MedTest [17] p-value	0.5423	0.5568	0.0082	0.0901	0.7859	0.5768	0.0492

To further assess potentially mediating taxa within this dataset, we sliced the released phylogenetic rooted tree using library phytools (version 0.6.60) [44], beginning at root (slice 0.01) to the height of 1 at 0.01 increments. Each of these slices resulted in subtree clades that were used for comparison testing. Each arbitrary slice resulted in clades that held taxa unique to them. We saw that, aside from a handful of the clades within certain slices, no mediation was observed with MedTest. This suggests that, if present, the mediation effect of the microbiota on the fiber intake and BMI relationship is likely small enough to be undetectable within the given sample size.

Data and analysis for this application are available at https://github.com/Alekseyenko/MODIMA.

4. Discussion

In this article, we developed a framework for multivariate omnibus distance mediation analysis (MODIMA). Although the proposed methods have wide applications to various data types, we specifically showed their robustness in high-dimensional settings by applying them to novel and previously published microbiome data. In simulations, we showed that our method to detect mediation under various scenarios is more powerful than previously published work. Simulations showed that MODIMA holds empirical type I error rates at the desired nominal significance level under the multiple mediator case.

Clearly, any analysis based on distances should not blindly and thoughtlessly pick the metric to be used. Ideally, the structure of the data and the analyst's intuition about the problem should guide the selection of an appropriate measure. A universally best distance measure may not exist for all problems and different data may result in different best performing distances, in terms of sensitivity and specificity. For example, although very popular and used in many studies, weighted UniFrac distance failed to result in a rejection in application example 2. This does not imply that this distance is universally bad. Its performance in combination with that of unweighted UniFrac is possibly telling us that beyond the phylogenetic signal the relative abundances are not informative of the mediation that may or may not exist in the data. A limitation of our method relative to MedTest is that MODIMA works on a specific distance metric rather than pooling analyses from multiple metrics. A future improvement to the method should incorporate the ability to use multiple distances. In the interim, a test-and-adjust approach can be used with MODIMA with multiple distances.

With regard to empirical power as the exposure–mediator and mediator–outcome effects are increased, MODIMA displayed increasing empirical power characteristics relative to other methods. We further see that there was increased power as sample size increased from 20 to 200, a typical expectation. It should be noted that most 'omics investigations operate on the lower end of that sample size spectrum; therefore, the ability to correctly detect differences under small sample scenarios is important.

In both of our empirical examples, although the mediating effect of the microbiota was plausible, we failed to detect the mediation. This is unsurprising for datasets of such small size and relatively small total effect of the exposure on the response (e.g., dCor = 14–17% for antibiotics–percent body fat and dCor = 9% for fiber–BMI). The microbiome in dietary fiber effect on the body mass index example demonstrates a discrepancy observed between our method and an alternative approach. Through additional analyses, we demonstrate that either the effect is again too small to be detected in a dataset of this size, or that it may be absent altogether.

Our application of the distance correlation and partial distance correlation metrics to the problem of modeling distance mediation illustrates somewhat unintuitive notions relating dependence and correlation in the context of causal analyses. First, the absence of partial correlation does not automatically imply the absence of partial dependence. The equivalence of partial correlation and conditional dependence is only true for a multivariate normal family of distributions. Furthermore, distance correlation methods demonstrate non-zero partial correlation (even asymptotically) in certain scenarios with conditionally independent univariate normal variables [20]. Specifically, as is relevant to the mediation analysis, consider X is a standard normal random variable and M and Y are each an independent linear combination of X and another standard normal variate. In that case, pdCor(M, Y | X) > 0, indicating the presence of partial distance correlation. This suggests that the notion of conditional dependence captured by partial distance correlation is different from that intuitively expected. Székely and Rizzo [20] suggest that partial distance correlation implies that there exists a pair of U-equivalent random variables that are in fact conditionally independent. The extent to which a lack of correspondence between conditional independence and zero partial distance correlation is a problem with multivariate data is unknown at the moment. However, it is easy to see via simulation that adding additional mediators uncorrelated to the exposure will decrease the population values of pdCor, which in finite sample size results in fewer rejections and thus less inflated type I errors. This is demonstrated in our multiple mediator simulations, where adding a moderate number of mediators results in better type I error control in simulation under the null hypothesis (Figure 4A).

Another potential pitfall of multivariate mediation analysis pertains to the interpretations of significant mediation results. Consider, for example, a scenario where X is a true cause for both Y and M1, while independent of M2, which is a true cause of Y. Although no univariate mediation relationships exist under this scenario, multivariately M = (M1, M2) is not conditionally independent of Y, given X and multivariate (distance) mediation does exist. This suggests that, whenever multivariate distance mediation is established, further interpretations of this relationship must be treated with caution in order not to attribute this relationship to any individual univariate marginals of the X, M, Y triple, but to treat this relationship as existing in the joint distribution.

Omnibus mediation analysis with 'omics-sized mediators is the first step towards enabling top-down approaches in genomic data. As opposed to the more widely used methods that integrate the univariate signals of individual measurements of microbes, gene expression, or genetic variants, the top-down approach starts with the collective effect of those and prunes the individual measurements down to a small set of most important ones. The significance of this approach is that top-down thinking allows for capturing effects, such as epistasis and otherwise complexly intertwined relationships. We envision that future versions of the omnibus mediation approach of this paper and of alternative approaches will allow to assign importance to components in addition to assessing the overall effect of the entire collection.

Supplementary Materials: Supplementary materials include source code for methods, simulation studies, and application examples and are freely available at https://github.com/Alekseyenko/MODIMA.

Author Contributions: A.V.A. conceived the method, derived the test statistic, developed the reference implementation in R statistical programming language, co-wrote the manuscript, and performed data analysis; B.H. implemented the code, co-wrote the manuscript, and performed data analysis; K.W. contributed to the discussion about the applications and epidemiological significance of the presented approach; all authors reviewed and approved the manuscript.

Funding: A.V.A., B.H., and K.W. were supported by the NIH/NLM R01 LM12517. A.V.A. and B.H. were supported by the NIH/NCATS R21 TR002513. A.V.A. and K.W. were supported by the Medical University of South Carolina College of Medicine Enhancing Team Science Award. A.V.A. was supported by the NIH/NCI U54 CA210962. The project described was supported by the NIH/NCATS UL1 TR001450.

Conflicts of Interest: The authors declare no conflict of interest.

References

1. VanderWeele, T. *Explanation in Causal Inference: Methods for Mediation and Interaction*; Oxford University Press: New York, NY, USA, 2015.
2. Cox, L.M.; Cho, I.; Young, S.A.; Anderson, W.H.; Waters, B.J.; Hung, S.C.; Gao, Z.; Mahana, D.; Bihan, M.; Alekseyenko, A.V.; et al. The nonfermentable dietary fiber hydroxypropyl methylcellulose modulates intestinal microbiota. *FASEB J.* **2013**, *27*, 692–702. [CrossRef] [PubMed]
3. Cox, L.M.; Yamanishi, S.; Sohn, J.; Alekseyenko, A.V.; Leung, J.M.; Cho, I.; Kim, S.G.; Li, H.; Gao, Z.; Mahana, D.; et al. Altering the intestinal microbiota during a critical developmental window has lasting metabolic consequences. *Cell* **2014**, *158*, 705–721. [CrossRef] [PubMed]
4. Nobel, Y.R.; Cox, L.M.; Kirigin, F.F.; Bokulich, N.A.; Yamanishi, S.; Teitler, I.; Chung, J.; Sohn, J.; Barber, C.M.; Goldfarb, D.S.; et al. Metabolic and metagenomic outcomes from early-life pulsed antibiotic treatment. *Nat. Commun.* **2015**, *6*, 7486. [CrossRef] [PubMed]
5. Callahan, B.J.; Sankaran, K.; Fukuyama, J.A.; McMurdie, P.J.; Holmes, S.P. Bioconductor workflow for microbiome data analysis: From raw reads to community analyses. *F1000Research* **2016**, *5*, 1492. [CrossRef] [PubMed]
6. Tang, Z.-Z.; Chen, G.; Alekseyenko, A.V. Permanova-s: Association test for microbial community composition that accommodates confounders and multiple distances. *Bioinformatics* **2016**, *32*, 2618–2625. [CrossRef]
7. Wang, Z.; Klipfell, E.; Bennett, B.J.; Koeth, R.; Levison, B.S.; DuGar, B.; Feldstein, A.E.; Britt, E.B.; Fu, X.; Chung, Y.-M.; et al. Gut flora metabolism of phosphatidylcholine promotes cardiovascular disease. *Nature* **2011**, *472*, 57–63. [CrossRef] [PubMed]
8. Arumugam, M.; Raes, J.; Pelletier, E.; Le Paslier, D.; Yamada, T.; Mende, D.R.; Fernandes, G.R.; Tap, J.; Bruls, T.; Batto, J.M.; et al. Enterotypes of the human gut microbiome. *Nature* **2011**, *473*, 174–180. [CrossRef]
9. Alekseyenko, A.V.; Perez-Perez, G.I.; De Souza, A.; Strober, B.; Gao, Z.; Bihan, M.; Li, K.; Methe, B.A.; Blaser, M.J. Community differentiation of the cutaneous microbiota in psoriasis. *Microbiome* **2013**, *1*, 31. [CrossRef]
10. Virgin, H.W.; Todd, J.A. Metagenomics and personalized medicine. *Cell* **2011**, *147*, 44–56. [CrossRef]
11. Wallace, K.; Lewin, D.N.; Sun, S.; Spiceland, C.M.; Rockey, D.C.; Alekseyenko, A.V.; Wu, J.D.; Baron, J.A.; Alberg, A.J.; Hill, E.G. Tumor-infiltrating lymphocytes and colorectal cancer survival in African American and Caucasian patients. *Cancer Epidemiol. Biomar. Prev.* **2018**, *27*, 755–761. [CrossRef]
12. Qasem, W.; Azad, M.B.; Hossain, Z.; Azad, E.; Jorgensen, S.; Castillo San Juan, S.; Cai, C.; Khafipour, E.; Beta, T.; Roberts, L.J., 2nd. Assessment of complementary feeding of canadian infants: Effects on microbiome & oxidative stress, a randomized controlled trial. *BMC Pediatr.* **2017**, *17*, 54.
13. Baron, R.M.; Kenny, D.A. The moderator-mediator variable distinction in social psychological research: Conceptual, strategic, and statistical considerations. *J. Pers. Soc. Psychol.* **1986**, *51*, 1173–1182. [CrossRef] [PubMed]
14. Boca, S.M.; Sinha, R.; Cross, A.J.; Moore, S.C.; Sampson, J.N. Testing multiple biological mediators simultaneously. *Bioinformatics* **2014**, *30*, 214–220. [CrossRef] [PubMed]
15. Zhao, N.; Chen, J.; Carroll, I.I.; Ringel-Kulka, T.; Epstein, M.P.; Zhou, H.; Zhou, J.J.; Ringel, Y.; Hongzhe, L.; Wu, M.C. Testing in microbiome-profiling studies with MiRKAT, the microbiome regression-based kernel association test. *Am. J. Hum. Genet.* **2015**, *96*, 797–807. [CrossRef] [PubMed]
16. Xia, Y.; Sun, J. Hypothesis testing and statistical analysis of microbiome. *Genes Dis.* **2017**, *4*, 138–148. [CrossRef] [PubMed]
17. Zhang, J.; Wei, Z.; Chen, J. A distance-based approach for testing the mediation effect of the human microbiome. *Bioinformatics* **2018**, *34*, 1875–1883. [CrossRef] [PubMed]

18. Székely, G.J.; Rizzo, M.L. Energy statistics: A class of statistics based on distances. *J. Stat. Plan. Inference* **2013**, *143*, 1249–1272. [CrossRef]

19. Székely, G.J.; Rizzo, M.L. Brownian distance covariance. *Ann. Appl. Stat.* **2009**, *3*, 1236–1265. [CrossRef]

20. Székely, G.J.; Rizzo, M.L. Partial distance correlation with methods for dissimilarities. *Ann. Stat.* **2014**, *42*, 2382–2412. [CrossRef]

21. Székely, G.J.; Rizzo, M.L. *Energy: E-Statistics: Multivariate inference via the energy of data*, 2018; R package version 1.7-5.

22. Székely, G.J.; Rizzo, M.L.; Bakirov, N.K. Measuring and testing dependence by correlation of distances. *Ann. Stat.* **2007**, *35*, 2769–2794. [CrossRef]

23. Székely, G.J.; Rizzo, M.L. The distance correlation t-test of independence in high dimension. *J. Multivar. Anal.* **2013**, *117*, 193–213. [CrossRef]

24. Knijnenburg, T.A.; Wessels, L.F.A.; Reinders, M.J.T.; Shmulevich, I. Fewer permutations, more accurate *P*-values. *Bioinformatics* **2009**, *25*, i161–i168. [CrossRef] [PubMed]

25. Peterson, J.; Garges, S.; Giovanni, M.; McInnes, P.; Wang, L.; Schloss, J.A.; Bonazzi, V.; McEwen, J.E.; Wetterstrand, K.A.; Deal, C.; et al. The NIH human microbiome project. *Genome Res.* **2009**, *19*, 2317–2323. [PubMed]

26. La Rosa, P.S.; Deych, E.; Carter, S.; Shands, B.; Yang, D.; Shannon, W.D. *Hmp: Hypothesis testing and power calculations for comparing metagenomic samples from hmp*, 2018; R package version 1.6.

27. Tvedebrink, T. Overdispersion in allelic counts and theta-correction in forensic genetics. *Theor. Popul. Biol.* **2010**, *78*, 200–210. [CrossRef] [PubMed]

28. Paradis, E.; Schliep, K. *Ape 5.0: An environment for modern phylogenetics and evolutionary analyses in r*, 2018; R package version 5.2. Bioinformatics.

29. Lozupone, C.A.; Hamady, M.; Kelley, S.T.; Knight, R. Quantitative and qualitative β diversity measures lead to different insights into factors that structure microbial communities. *Appl. Environ. Microbiol.* **2007**, *73*, 1576–1585. [CrossRef] [PubMed]

30. Fuglede, B.; Topsoe, F. In Proceedings of the Jensen-shannon divergence and hilbert space embedding, International Symposium on Information Theory, ISIT27. Chicago, IL, USA, 27 June–2 July 2004; p. 31.

31. McMurdie, P.J.; Holmes, S. Phyloseq: An R package for reproducible interactive analysis and graphics of microbiome census data. *PLoS ONE* **2013**, *8*, e61217. [CrossRef]

32. Oksanen, J.; Blanchet, F.G.; Friendly, M.; Kindt, R.; Legendre, P.; McGlinn, D.; Minchin, P.R.; O'Hara, R.B.; Simpson, G.L.; Solymos, P.; et al. *Vegan: Community ecology package*, 2018; R package version 2.5-4.

33. Fairlie, T.; Shapiro, D.J.; Hersh, A.L.; Hicks, L.A. National trends in visit rates and antibiotic prescribing for adults with acute sinusitis. *Arch. Intern. Med.* **2012**, *172*, 1513–1514. [CrossRef]

34. Butaye, P.; Devriese, L.A.; Haesebrouck, F. Antimicrobial growth promoters used in animal feed: Effects of less well known antibiotics on gram-positive bacteria. *Clin. Microbiol. Rev.* **2003**, *16*, 175–188. [CrossRef]

35. Blaser, M.J.; Falkow, S. What are the consequences of the disappearing human microbiota? *Nat. Rev. Genet.* **2009**, *7*, 887–894. [CrossRef]

36. Dethlefsen, L.; Relman, D.A. Incomplete recovery and individualized responses of the human distal gut microbiota to repeated antibiotic perturbation. *Proc. Natl. Acad. Sci. USA* **2011**, *108*, 4554–4561. [CrossRef]

37. Cho, I.; Yamanishi, S.; Cox, L.; Methé, B.A.; Zavadil, J.; Li, K.; Gao, Z.; Mahana, D.; Raju, K.; Teitler, I.; et al. Antibiotics in early life alter the murine colonic microbiome and adiposity. *Nature* **2012**, *488*, 621–626. [CrossRef] [PubMed]

38. Reeder, S.B.; Pineda, A.R.; Wen, Z.; Shimakawa, A.; Yu, H.; Brittain, J.H.; Gold, G.E.; Beaulieu, C.H.; Pelc, N.J. Iterative decomposition of water and fat with echo asymmetry and least-squares estimation (IDEAL): Application with fast spin-echo imaging. *Magn. Reson. Med.* **2005**, *54*, 636–644. [CrossRef] [PubMed]

39. Caporaso, J.G.; Kuczynski, J.; Stombaugh, J.; Bittinger, K.; Bushman, F.D.; Costello, E.K.; Fierer, N.; Peña, A.G.; Goodrich, J.K.; Gordon, J.I.; et al. QIIME allows analysis of high-throughput community sequencing data. *Nat. Methods* **2010**, *7*, 335–336. [CrossRef] [PubMed]

40. Hamidi, B.; Wallace, K.; Vasu, C.; Alekseyenko, A.V. Wd*-test: Robust distance-based multivariate analysis of variance. *Microbiome* **2019**, *7*, 51. [CrossRef] [PubMed]

41. Murphy, E.A.; Velazquez, K.T.; Herbert, K.M. Influence of high-fat diet on gut microbiota: A driving force for chronic disease risk. *Curr. Opin. Clin. Nutr. Metab. Care* **2015**, *18*, 515–520. [CrossRef] [PubMed]

42. Wu, G.D.; Chen, J.; Hoffmann, C.; Bittinger, K.; Chen, Y.-Y.; Keilbaugh, S.A.; Bewtra, M.; Knights, D.; Walters, W.A.; Knight, R.; et al. Linking long-term dietary patterns with gut microbial enterotypes. *Science* **2011**, *334*, 105–108. [CrossRef] [PubMed]
43. Chen, J. *Gunifrac: Generalized unifrac distances*, 2018; R package version 1.1.
44. Revell, L.J. Phytools: An R package for phylogenetic comparative biology (and other things). *Methods Ecol. Evol.* **2012**, *3*, 217–223. [CrossRef]

Article

A Computational Method for Classifying Different Human Tissues with Quantitatively Tissue-Specific Expressed Genes

JiaRui Li [1,†], Lei Chen [2,3,†], Yu-Hang Zhang [4], XiangYin Kong [4,*], Tao Huang [4,*] and Yu-Dong Cai [1,*]

1   School of Life Sciences, Shanghai University, Shanghai 200444, China; jiaruili@shu.edu.cn
2   College of Information Engineering, Shanghai Maritime University, Shanghai 201306, China; chen_lei1@163.com
3   Shanghai Key Laboratory of PMMP, East China Normal University, Shanghai 200241, China
4   Institute of Health Sciences, Shanghai Institutes for Biological Sciences, Chinese Academy of Sciences, Shanghai 200031, China; zhangyh825@163.com
*   Correspondence: xykong@sibs.ac.cn (X.K.); tohuangtao@126.com (T.H.); cai_yud@126.com (Y.-D.C.); Tel.: +86-021-6613-6132 (Y.-D.C.)
†   These authors contributed to work equally.

Received: 3 August 2018; Accepted: 4 September 2018; Published: 7 September 2018

Abstract: Tissue-specific gene expression has long been recognized as a crucial key for understanding tissue development and function. Efforts have been made in the past decade to identify tissue-specific expression profiles, such as the Human Proteome Atlas and FANTOM5. However, these studies mainly focused on "qualitatively tissue-specific expressed genes" which are highly enriched in one or a group of tissues but paid less attention to "quantitatively tissue-specific expressed genes", which are expressed in all or most tissues but with differential expression levels. In this study, we applied machine learning algorithms to build a computational method for identifying "quantitatively tissue-specific expressed genes" capable of distinguishing 25 human tissues from their expression patterns. Our results uncovered the expression of 432 genes as optimal features for tissue classification, which were obtained with a Matthews Correlation Coefficient (MCC) of more than 0.99 yielded by a support vector machine (SVM). This constructed model was superior to the SVM model using tissue enriched genes and yielded MCC of 0.985 on an independent test dataset, indicating its good generalization ability. These 432 genes were proven to be widely expressed in multiple tissues and a literature review of the top 23 genes found that most of them support their discriminating powers. As a complement to previous studies, our discovery of these quantitatively tissue-specific genes provides insights into the detailed understanding of tissue development and function.

Keywords: tissue-specific expressed genes; transcriptome; tissue classification; support vector machine; feature selection

---

1. Introduction

A biological tissue is an ensemble of similar cells residing in the same location and performing specific biological functions in multicellular organisms. As the bridge between single cells and functional organs, tissues are elementary units with both phenotypical and functional contributions to biological identity [1]. All biological functions are regulated and manipulated directly or indirectly by proteins, which can be further attributed to gene expression patterns measured by messenger RNA (mRNA) expression [2]. Therefore, different tissues and cell types could have their own unique expression patterns and a full picture of how genes are expressed in different tissues will help to unveil the molecular mechanisms involved in tissue development and function.

Ten years ago, two milestones for identifying tissue-specific gene expression were conducted and completed right after the human genome project, which built tissue-specific gene expression profiles at the protein and RNA levels [3,4]. At the protein level, the protein distribution in human tissues was explored using 718 antibodies corresponding to 650 human protein-coding genes in the Human Protein Atlas [3]. While at the RNA level, the FANTOM consortium initiated the creation of a gene atlas by integrating mouse and human expression data from multiple tissues using microarrays [4] and built the widely used BioGPS portal, which has expression data from numerous resources [5]. These projects have been extended and updated by incorporating more genes and taking advantage of advanced technologies, such as next-generation sequencing. For example, the number of genes included in the Human Protein Atlas project has been increased to 10,118, which is more than half of the protein-coding genes in humans [6]. Meanwhile, the application of RNA-Seq technologies has created an RNA-Seq Atlas in different human tissues, providing a more comprehensive and unbiased view of gene expression compared with the initial efforts using microarrays [7]. A recent study presented a map of the human tissue proteome by integrating multiple-omics approaches, including RNA-Seq and tissue microarray-based immunohistochemistry, which detected more than 90% of putative protein-coding genes and found that 2355 genes are significantly enriched in a single tissue, 3478 are enhanced in a single tissue and 1109 are enriched in a group of tissues [8]. These tissue-specific genes not only help understand human biology but can also be applied in medical research, such as pharmaceutical drug development and biomarker discovery in the field of translational medicine. However, these studies have mainly focused on "qualitatively tissue-specific expressed genes", which are enriched in a single or subgroup of tissues. The genes expressed in all or almost all tissues could also have divergent expression patterns among different cell types, which we termed as "quantitatively tissue-specific expressed genes". Although these quantitatively tissue-specific expressed genes have less expression enrichment compared with qualitatively expressed genes, they might also play important roles in tissue function and development.

In this study, we took advantage of recently published transcriptomic data in multiple tissues from the Genotype-Tissue Expression (GTEx) project [9] and present a new computational method that integrates machine learning algorithms to identify genes that are widely expressed in the human body but with different expression signatures across 25 human tissues and are capable of distinguishing different tissue types. According to our results, the 25 tissues have been taken into account and subtyped by 432 key genes using a prediction engine based on a support vector machine (SVM) [10,11] with an Matthews Correlation Coefficient (MCC) value more than 0.99, revealing the detailed expression characteristics of different tissue subtypes. The constructed classification model also had good generalization ability because it yielded MCC of 0.985 on an independent test dataset. In addition, the superiority of this model was proved by comparing it to the SVM model using tissue enriched genes. A detailed analysis was also performed on the 23 most important genes among 432 key genes. As a complement to previous studies, our results demonstrate the ability to classify tissues through a series of quantitatively tissue-specific expressed genes, suggesting that these genes could also play important roles in tissue development and function.

2. Materials and Methods

2.1. Dataset

The expression profiles from different tissue samples obtained by RNA-Seq were downloaded from GTEx V6p (http://gtexportal.org/home/datasets) [9]. Tissues with sample sizes smaller than 80 were excluded, resulting in a total of 8436 samples from 25 tissues. These samples comprised a training dataset. For an easy description, each tissue was denoted as T_i ($I = 1, 2, \ldots, 25$). For the computational analysis, we extracted expression levels of 18,365 genes that are expressed (i.e., expression level was not zero) in at least one of 8436 samples from the file "GTEx_Analysis_v6p_RNA-seq_RNA-SeQCv1.1.8_gene_rpkm.gct"; that is, each sample was represented by 18,365 features.

Besides, an independent test dataset was constructed using the new samples added in GTEx v7 after GTEx v6p. This dataset included 3367 samples from the same 25 tissues. These 25 tissues and their sample sizes in training and independent test datasets are listed in Table 1.

Table 1. The 25 tissue samples.

Tag	Tissue	Number of Samples		Tag	Tissue	Number of Samples	
		Training Dataset	Test Dataset			Training Dataset	Test Dataset
T_1	Adipose tissue	577	237	T_2	Adrenal gland	145	50
T_3	Blood	511	54	T_4	Blood vessel	689	242
T_5	Brain	1259	455	T_6	Breast	214	84
T_7	Colon	345	169	T_8	Esophagus	686	348
T_9	Heart	412	201	T_{10}	Liver	119	58
T_{11}	Lung	320	123	T_{12}	Muscle	430	155
T_{13}	Nerve	304	122	T_{14}	Ovary	97	39
T_{15}	Pancreas	171	82	T_{16}	Pituitary	103	82
T_{17}	Prostate	106	48	T_{18}	Skin	890	342
T_{19}	Small intestine	88	52	T_{20}	Spleen	104	60
T_{21}	Stomach	192	75	T_{22}	Testis	172	91
T_{23}	Thyroid	323	139	T_{24}	Uterus	83	32
T_{25}	Vagina	96	27	Total	-	8436	3367

2.2. Feature Ranking and Selection

The purpose of this study is to extract a group of genes that can contribute to tissue classification based on gene expression levels. To do that, the minimum redundancy maximum relevance (mRMR) method [12] proposed by Peng et al. was employed to analyze all the features and yield a feature list, named the mRMR feature list. This feature selection method has been applied to tackle various biological problems [13–25]. In the method, the discriminated power of a feature f is reflected by the relevance between it and a target class c, which is measured from their mutual information (*MI*). The *MI* can be evaluated according to the following equation (x and y represent two variables):

$$\iint p(x,y) log \frac{p(x,y)}{p(x)p(y)} dxdy \tag{1}$$

where $p(x)$ and $p(y)$ are the marginal probabilistic density of x and y and $p(x, y)$ is their joint probabilistic density. Additionally, the redundancy between two features f_1 and f_2 is also evaluated by their *MI*. To produce the mRMR feature list, the mRMR method repeatedly selects a non-selected feature that has a maximum relevance to the target and a minimum redundancy to the already selected features. The mRMR feature list ranks all the features according to their selection order. For the formulation, this list was denoted as follows:

$$F = [f_1, f_2, \ldots f_N] \tag{2}$$

where N is the total number of features (N = 18,365 in this study). It is clear that the top features in this list are more important for identifying tissue types.

The mRMR method only provides a feature list, in which important features receive high ranks. However, we still do not know which features should be selected for classification. In view of this, the incremental feature selection (IFS) method was adopted in this study to select a group of features that can effectively classify the human tissues. In this method, a series of feature sets, denoted as F_1, $F_2, \ldots, F_i, \ldots, F_N$, were constructed according to the mRMR feature list F, where the subscript i in F_i represents the top i features in F comprising the feature set F_i, that is,

$$F_i = [f_1, f_2, \ldots f_i] \tag{3}$$

Then, a classification model would be built on each feature set trained by a machine learning algorithm (e.g., SVM). Accordingly, the classification model yielding the best performance can be found and termed the optimal classification model. Features used in this optimal classification model comprise the optimal feature set. In this study, there were 18,365 features in total, meaning we could construct 18,365 classification models. However, due to our limited computational power, this would be time-consuming. Thus, we only examined feature sets that contained 4–500 features to construct the classification models and selected the model yielding the highest Matthews Correlation Coefficient, a powerful measurement to evaluate the performance of different models. For convenience, the obtained model was called optimal classification model.

2.3. Classification Algorithm

In the IFS method, a machine learning algorithm was used to build a classification model based on each feature set. Here, we selected one of the most classic machine learning algorithms, SVM [10,11]. The principle of SVM is to find a balance between the learning error and the minimum statistical risk. To date, several types of SVMs have been proposed to address different kinds of problems. In this study, we adopted an SVM that is trained by sequential minimum optimization (SMO) [26], which always breaks the quadratic programming (QP) problem into a series of the smallest possible sub-QP problems. Then, these sub-QP problems are solved analytically, thereby avoiding matrix storage and using numeric QP optimization steps. To quickly implement this type of SVM, a tool named "SMO" was employed in Weka (https://www.cs.waikato.ac.nz/ml/weka/) [27] and it was executed using its default settings, where the kernel was set as polynomial function and the tolerance parameter was set to 0.001.

2.4. Measurements

The prediction abilities of the constructed classification models were evaluated by a 10-fold cross-validation (10-CV) [16,28–30], which yields similar results to the stricter test called the Jackknife cross-validation (J-CV) [31,32] but saves significant computational resources.

In this study, 25 tissues were considered. Thus, based on the predicted results derived from the SVM on different feature sets and evaluated by 10-CV, we can compute the prediction accuracy for j-th tissue T_j, which was defined as follows:

$$ACC_j = \frac{n_j}{N_j}, \, j = 1, 2, \ldots, 25 \tag{4}$$

where n_j represents the number of correctly predicted samples in T_j and N_j represents the total number of samples in T_j. We can further compute the overall accuracy (*TACC*) with the following formula:

$$TACC = \frac{\sum_{j=1}^{25} n_j}{\sum_{j=1}^{25} N_j} \tag{5}$$

We can see from Table 1 that some tissues were large (e.g., brain), while some tissues had much fewer samples (e.g., uterus). The overall accuracy may strongly rely on the prediction accuracy of the tissues with large sizes. Thus, it is not proper to evaluate the performance of models using *TACC*. In binary classification, *MCC* [33] is deemed as a balanced measurement even if the class sizes are of great differences. Later, the multi-class version—called *MCC* in multiclass—was developed by Gorodkin [34], which inherits the merits of original *MCC*. Here, we give a brief description of the *MCC* in multiclass. A more detailed description can be found in Gorodkin's study [34].

Given a classification problem involving n samples, denoted by $s_1, s_2, \ldots, s_n$ and N classes, represented by 1, 2, ... , N. The true classes of all the samples can be formulated by the matrix $Y = (y_{ij})_{n \times N}$, where $y_{ij} = 1$ if s_i belongs to class j; otherwise, it is set to 0. For the predicted classes of all the samples, the classes can be used to a define another matrix, $X = \left(x_{ij}\right)_{n \times N}$, which can be defined as follows: $x_{ij} = 1$, if s_i is predicted to be in class j, otherwise, $x_{ij} = 0$. Then, the *MCC* in the multiclass is defined as follows:

$$MCC = \frac{cov(X,Y)}{\sqrt{cov(X,X)cov(Y,Y)}} \tag{6}$$

where $cov(X,Y)$ is the covariance function of X and Y, which can be computed with the following formula:

$$cov(X,Y) = \frac{1}{N}\sum_{k=1}^{N}cov(x_k, y_k) = \frac{1}{N}\sum_{k=1}^{N}\sum_{i=1}^{n}(x_{ik} - \overline{x_k})(y_{ik} - \overline{y_k}) \tag{7}$$

where x_k and y_k denote the k-th column of X and Y, respectively and $\overline{x_k}$ and $\overline{y_k}$ denote the mean values of the numbers in x_k and y_k, respectively.

Like the original *MCC* value proposed by Matthews [33], the range of the *MCC* in the multiclass is between -1 and 1, where the higher the *MCC* value obtained, the better a classifier is (1 means the given classifier yields a perfect classification, 0 indicates a classification is no better than a random prediction and -1 represents a total misclassification). In this study, we used the *MCC* values in the multiclass as the key measurements and only called *MCC* in the following text for convenience.

3. Results

3.1. Results of Feature Ranking

To avoid the unbalanced data sizes among the different tissues, we collected the transcriptome data from 8436 samples originating from 25 tissues in the GTEx project [9]; tissues with sample sizes smaller than 80 were excluded from this study. To perform a comprehensive analysis, all 18,365 genes expressed in the 8436 samples, regardless of tissue specificity or expression, were used to construct the features. All the features were rigorously analyzed with the widely used mRMR method, inducing the mRMR feature list (Supplementary Material S1).

3.2. Results of Feature Selection

The obtained mRMR feature list was used with the IFS method to discover the optimal feature set for the SVM. However, because of our limited computational power, we only tried feature sets F_4, F_5, ..., F_{500}. For each feature set, an SVM-based classification model was built, in which each sample was represented by the features in the set. Then, 10-CV was adopted to evaluate the performance of each classification model; the predicted results included the prediction accuracies of the 25 tissues, *TACC* and *MCC* (Supplementary Material S2). To clearly show the relationship between the number of used features and the performance of the corresponding classification model, the IFS curve was plotted and is shown in Figure 1 with the feature number as the X-axis and the *MCC* value as the Y-axis. The highest *MCC* value (0.994) reached almost 1.00 and was obtained when the top 432 features in the mRMR feature list were used, suggesting that these 432 features could be optimal features for the SVM for distinguishing the 25 tissues. Accordingly, the optimal classification model using 432 optimal features and SVM can be built. The detailed performance of this model, including accuracy on each tissue and *TACC*, is shown in Figure 2, from which we can see that 17 of 25 tissues received a perfect classification ($ACC_j = 1$) and all accuracies are higher than 0.930, suggesting that the performance of the optimal classification model is quite stable on different tissues.

To examine whether these 432 genes are widely expressed or tissue-specific, we searched their annotations in the Human Protein Atlas [8]. Among the 432 genes, fourteen genes were annotated as tissue-enriched, tissue-enhanced or group-enriched, which suggests that these fourteen genes might be tissue-specific or group-specific (Supplementary Material S3). However, further examination revealed that only one gene, *ASAP2*, is annotated as tissue-enriched (in testis) in the GTEx data and that this gene shows no enriched or enhanced expression in the FANTOM5 database [35], another expression atlas for RNA levels. Both the GTEx and FANTOM5 annotations were obtained through the Human Protein Atlas database to keep the criteria consistent. Therefore, these 432 genes showed the ability to distinguish different tissues that are widely expressed across multiple tissues.

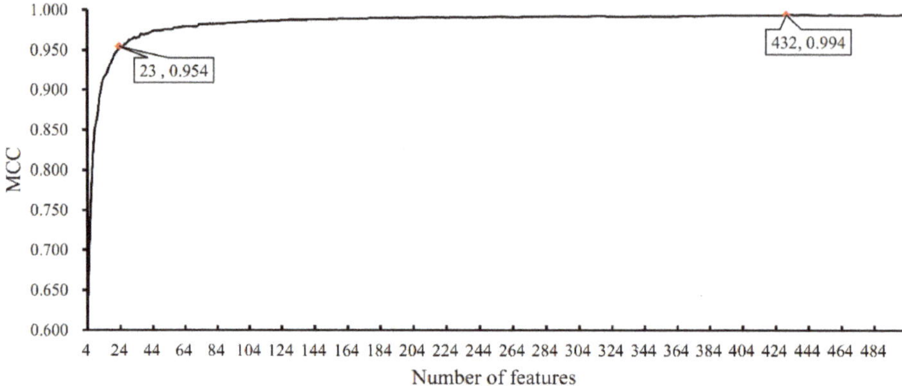

Figure 1. The incremental feature selection (IFS) curve illustrating the performance of the classification models using different numbers of features. Red diamonds represent the performance when the top 23 genes and 432 features were used for building the classification models.

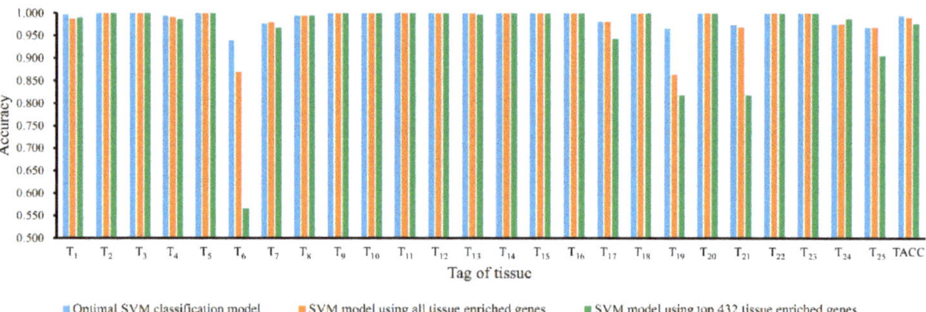

Figure 2. The performance of the optimal support vector machine (SVM) classification model, SVM model using all tissue enriched genes and SVM model using top 432 tissue enriched genes, including accuracy on each tissue and overall accuracy. The optimal SVM classification model gave better performance.

3.3. Comparison of SVM Model with Tissue Enriched Genes

As mentioned in Section 3.2, an SVM classification model was built to classify samples into 25 tissues, which yielded the *MCC* of 0.994. To indicate its effectiveness, we mapped the tissue enriched proteins retrieved from Human Protein Atlas onto the filtered GTEx expression dataset which excluded the non-expressed genes, resulting in 1981 genes (see Supplementary Material S4). The expression levels of these genes were used to represent each sample in the training dataset. Then, the SVM was executed on this data with its performance evaluated by 10-CV. The predicted results were counted as *MCC* of 0.990, which was lower than that obtained by the optimal SVM classification model. For detailed comparison, the detailed performance of these two models, including accuracy on each tissue and *TACC*, is illustrated in Figure 2. It can be observed that the optimal SVM classification model gave better or equal performance on almost all tissues (expect tissue T_7, "Colon") than SVM model using tissue enriched genes.

Besides, considering that the above two models using different number of genes, to give a fairer comparison, we also evaluated the importance of 1981 tissue enriched genes via mRMR method and extracted the top 432 genes to construct an SVM classification model and evaluate it on the training dataset, yielding *MCC* of 0.976, which was much lower than that obtained by the optimal SVM classification model. The detailed performance on each tissue and overall accuracy is shown in Figure 2, from which we can see that the optimal SVM classification model gave higher or equal accuracies on almost all tissues expect one tissue (T_{24}, "Uterus").

All these suggested that important genes for classification of samples into different tissues can be extracted via advanced computational methods used in this study and they can be adopted to build a better classification model.

3.4. Performance of the Optimal SVM Classification Model on Test Dataset

We constructed an optimal SVM classification model in Section 3.2 via mRMR and IFS methods. To test its generalization, we performed this model on the independent test dataset. The obtained *TACC* and *MCC* were 0.986 and 0.985, respectively, indicating that the prediction ability of the constructed model was quite strong. We also calculated the *MCC* of tissue enriched genes on the independent test dataset and it was 0.972, smaller than the *MCC* (0.985) obtained by the optimal SVM classification model.

3.5. Comparison of SVM Model with t-Test Genes

Beside the machine learning algorithms used in present study, there were other ways to identify the quantitatively tissue-specific expressed genes. Another more straightforward way was to identify the significantly highly expressed genes in each tissue by performing t test between one tissue and all the other tissues and then combine the highly expressed genes in each tissue to get the final quantitatively tissue-specific expressed genes. To make the t test result comparable with our result, we selected the top 19 significantly highly expressed genes from 25 tissues and obtained 422 unique *t*-test genes (see Supplementary Material S5), which had similar number of genes with the 432 optimal genes. We evaluated the performance of the 422 *t*-test genes on independent test dataset and its *MCC* was 0.984, slightly smaller than the *MCC* obtained by 432 optimal genes, 0.985. It was difficult to tell whether the increase of *MCC* from 0.984 to 0.985 was statistically significant, especially when the *MCC* was already so high and there was little space for improvement. But we strongly believe that the quantitative gene expression signatures identified either by *t*-test based method or mRMR and IFS based method, are better than traditional tissue specific gene lists that only consider the expression or not rather than the expression level differences. The results on the high-quality dataset GTEx had supported it. Although the GTEx project has ended, the Enhancing GTEx (eGTEx) project [36] carries on. We will work closely with the eGTEx Consortium to test the constructed model on larger new samples and update the model correspondingly.

4. Discussion

As we have analyzed above, RNA-Seq has been reported to be an effective classification tool for identifying cell types. Based on the transcriptome datasets from different human tissues presented in a recent study, we developed a new computational method and successfully identified 432 quantitatively tissue-specific expressed genes capable of classifying twenty-five human tissues with high accuracy (*MCC* > 0.99). To further examine the reliability of our results, we selected the top 23 genes in the mRMR feature list for detailed analyses (Table 2). The performance of the classification model built on these genes displayed an *MCC* value > 0.95 (Figure 1, Supplementary Material S2), suggesting the significant role of these genes among the 432 genes in the classification. As shown in Table 2, we searched the 23 genes in Human Protein Atlas [8] and Expression Atlas [37]. Based on Human

Protein Atlas, 16 genes were "Expressed in all"; six genes were "Mixed"; only one gene was "Tissue enhanced (thyroid gland)". Based on Expression Atlas, all genes were expressed in "Multiple tissues". Both databases suggested these genes were expressed in multiple tissues. But our method can find their expression level difference rather than whether they were expressed. For clearly displaying the expression level of these 23 genes across 25 tissues, we gave a box plot for each gene, which is illustrated in Supplementary Figure S1. It can be observed that for almost all genes, their expression levels on different tissues are quite different, indicating that they can be important biomarkers.

The first gene, *ARAF*, is a potential proto-oncogene that can be clustered into the *RAF* subfamily of Ser/Thr protein kinases and contributes to the regulation of cell proliferation and tissue development [38]. For its tissue-specific expression pattern, this gene has been reported to be steadily expressed in multiple tissues, including most of tissue/cell subtypes incorporated in this study, suggesting a wide expression signature [39]. More importantly, it has also been confirmed to have a unique expression pattern in the skin; thus, though this gene is expressed widely in multiple tissues, the expression levels could be different from one another [39].

Table 2. The top 23 genes selected for further investigation via a literature review.

Rank	Gene	Description	The Human Protein Atlas [8]	Expression Atlas of EMBL-EBI [37]
1	*ARAF*	A-Raf Proto-Oncogene, Serine/Threonine Kinase	Expressed in all	Multiple tissues
2	*ITGA3*	Integrin Subunit Alpha 3	Mixed	Multiple tissues
3	*SLAIN2*	SLAIN Motif Family Member 2	Expressed in all	Multiple tissues
4	*ZNF532*	Zinc Finger Protein 532	Mixed	Multiple tissues
5	*PPIC*	Peptidylprolyl Isomerase C	Mixed	Multiple tissues
6	*KDELR1*	KDEL Endoplasmic Reticulum Protein Retention Receptor 1	Expressed in all	Multiple tissues
7	*NBL1*	Neuroblastoma 1, DAN Family BMP Antagonist	Expressed in all	Multiple tissues
8	*PLP2*	Proteolipid Protein 2	Expressed in all	Multiple tissues
9	*STAT6*	Signal Transducer and Activator of Transcription 6	Expressed in all	Multiple tissues
10	*ARHGAP23*	Rho GTPase Activating Protein 23	Mixed	Multiple tissues
11	*LRIG3*	Leucine Rich Repeats And Immunoglobulin Like Domains 3	Tissue enhanced (thyroid gland)	Multiple tissues
12	*MANBAL*	Mannosidase Beta Like	Expressed in all	Multiple tissues
13	*PTPRA*	Protein Tyrosine Phosphatase, Receptor Type A	Expressed in all	Multiple tissues
14	*YAP1*	Yes Associated Protein 1	Mixed	Multiple tissues
15	*CLIC1*	Chloride Intracellular Channel 1	Expressed in all	Multiple tissues
16	*TMEM109*	Transmembrane Protein 109	Expressed in all	Multiple tissues
17	*MOCS2*	Molybdenum Cofactor Synthesis 2	Expressed in all	Multiple tissues
18	*PTPRF*	Protein Tyrosine Phosphatase, Receptor Type F	Mixed	Multiple tissues
19	*MYO1C*	Myosin IC	Expressed in all	Multiple tissues
20	*FAM127B*	Family with Sequence Similarity 127 Member B	Expressed in all	Multiple tissues
21	*TRIP10*	Thyroid Hormone Receptor Interactor 10	Expressed in all	Multiple tissues
22	*SERPING1*	Serpin Family G Member 1	Expressed in all	Multiple tissues
23	*TOM1L2*	Target of Myb1 Like 2 Membrane Trafficking Protein	Expressed in all	Multiple tissues

Another gene named *ITGA3* has been widely reported to function as a cell surface adhesion molecule and is involved in the malignant metastasis of certain tumor subtypes [40,41]. Therefore, it is expected that this gene has low expression in the blood, as many blood cells freely float [42], suggesting that this gene may have different expression patterns across tissues corresponding to their different cell adhesion requirements. Similarly, *SLAIN2*, a microtubule dynamics-associated regulator, has also been reported to be down-regulated in blood cells compared with other tissue subtypes in our candidates, indicating the distinctive roles of *SLAIN2* as well as microtubule dynamics in different tissues [43]. *ZNF532*, a nucleic acid binding-associated gene, has been confirmed to be involved in transcriptional regulation [44,45]. Based on recent publications, this gene has also been reported to be down-regulated in liver and whole blood cells compared with other candidate tissues, implying that this gene may be a potential marker for the identification of hepatocytes and blood cells [46].

PPIC encodes functional peptidylprolyl isomerase C (cyclophilin C), which has been confirmed to bind to the immunosuppressant cyclosporin A, implying that this gene is an immune-associated regulator [47,48]. This gene has been confirmed to have quite high expression levels in the tibial nerve but lower levels in the cortex, suggesting its differential expression levels across various cell types and indicating potential roles in tissue classification [47]. In contrast to *PPIC*, another nerve associated-gene, *NBL1*, has been confirmed to be highly expressed in the central nervous system (brain) but lowly expressed in the peripheral nervous system (such as the spinal cord) [49,50].

KDELR1, a potential endoplasmic reticulum (ER)-associated gene, has no direct evidence reporting any unique expression patterns. However, this gene has been reported to affect ion transmembrane transporter activity in the nervous system and blood cells [51,52], so it is likely that this gene may have specific expression patterns corresponding to divergent ion transmembrane transporter activity requirements in different tissues.

PLP2, which has not been reported to display unique expression patterns, has been found to increase neuronal apoptosis when down-regulated and promotes cell proliferation in leukemia when up-regulated, indicating that different expression levels are required in tissues based on their needs for cell apoptosis and proliferation [53,54].

STAT6, a transcription factor in the STAT family, has been confirmed to contribute to the proliferation and differentiation of T helper 2 cells, indicating its specific role in the immune system and certain cell types [55–57].

ARHGAP23 is an effective component of the Rho GTPase family and has been widely reported to be involved in transmembrane receptor signal transduction [58]. With regards to its detailed expression, this gene has been confirmed to be down-regulated in blood and muscles, revealing a specific expression pattern [59]. Similarly, *LRIG3* also has quite low expression levels in the blood [60]. Moreover, *LRIG3* displays specific expression patterns in the nervous system as it regulates the normal functioning of the inner ear, implying that this gene may also contribute to the identification of nerve tissues [61,62].

As a gene encoding a functional component of the Hippo signaling pathway involved in development, growth, repair and homeostasis, *YAP1* has low expression in mature blood cells [63,64]. This gene, as well as another three genes *MANBAL, PTRPA* and *CLIC1* has also been reported to be differentially expressed across different human tissues [9].

Another membrane-associated gene, *TMEM109*, has also been predicted to be distinctively expressed in different tissues. Generally, this gene has been widely reported to mediate cellular responses to DNA damage, such as ultraviolet C-induced cell death [65,66]. Not present in the bone marrow or thymus, this gene may also participate in the identification of certain cell types [66].

MOCS2, a gene encoding the eukaryotic molybdoenzyme, has been widely reported to contribute to molybdenum cofactor synthesis [67]. For its expression pattern, this gene has been confirmed to be predominantly expressed in heart and skeletal muscle, thus contributing to the identification of muscle and heart cells [67–69].

PTPRF, a member of the protein tyrosine phosphatase (PTP) family, has been reported to participate in various core survival-associated biological processes, such as cell growth, differentiation and mitosis [70,71], which suggests that differential expression patterns could be found in different tissues due to divergent levels of cell growth and differentiation. For example, the expression of this gene has been reported to be down-regulated in tissues with low proliferation potential, such as the heart and brain [72,73].

Encoding a common protein in muscle, *MYO1C* could be up-regulated in tissues that contain smooth muscle, skeletal muscle or heart muscle, which would distinguish these tissues from other tissues such as blood and cartilage [74,75].

The *TRIP10* gene has been confirmed to have quite a low expression level in the blood and nervous system, implying its potential in tissue classification [76,77]. The abnormal expression of this gene usually indicates specific pathological processes such as cancer and leukemia [76].

The last two genes *SERPING1* and *TOM1L2* both have been found to be expressed in all tissues [8]. *SERPING1* encodes the plasma protease C1 inhibitor involved in regulating important physiological pathways, including complement activation, blood coagulation, fibrinolysis and the generation of kinins [78], while *TOM1L2* may regulate growth factor-induced mitogenic signaling [79]. These discovered functional roles suggest potential requirements for differential expression patterns across different tissues.

We have discussed 22 of the 23 genes (*FAM127B* has little information in terms of its function) and shown their potential for tissue classification based on previous studies. Some genes, such as *ARAF*, *PTPRF*, *ITGA3* and *SLAIN2*, are involved in crucial cell processes, including cell proliferation, cell adhesion, cell invasion and the regulation of microtubule dynamics, which have diverse roles in different tissues, indicating that these genes might have differential expression patterns across different tissues. Other genes were confirmed to have specific functions in particular tissues or cell types, suggesting their specific roles in these tissues or cell types. By comparing the GO annotations between the 432 quantitatively-specific expressed genes identified in this study and the 2355 tissue-enriched genes annotated previously by Uhlen et al. [3] through the online tool DAVID [80], we found that these two sets of genes showed diverse clustering on biological processes. The former ones are enriched in translational, transcriptional and posttranscriptional regulation processes, while the latter ones are enriched in those biological processes specific to cell-types or tissues, such as spermatogenesis, muscle filament sliding, cell differentiation and so forth (see Supplementary Material S6). This finding indicates those quantitatively tissue-specific genes possibly contribute to tissue-specific features through transcriptional, translational, posttranscriptional and posttranslational regulations.

5. Conclusions

In summary, this study identified quantitatively tissue-specific expressed genes with discriminating power for classifying different tissue types based on their expression patterns via machine-learning algorithms. As a complement to previous studies that have identified genes enriched in a single or a few tissues, our findings of genes that are commonly but also differentially expressed in multiple tissues will provide insights into a detailed understanding of tissue development and function. In addition, we provided tissue samples represented by expression profiles of 432 optimal genes in Supplementary Material S7, in which the instructions of how to use this file to build the classification model and further make predictions were also given.

Supplementary Materials: The following are available online at http://www.mdpi.com/2073-4425/9/9/449/s1, Supplementary Material S1: The mRMR feature list yielded by the mRMR method; Supplementary Material S2: The performance of SVM classification model with different features; Supplementary Material S3: Fourteen important genes that were annotated as tissue-enriched, tissue-enhanced or group-enriched; Supplementary Material S4: Tissue enriched genes retrieved from Human Protein Atlas; Supplementary Material S5: Significantly highly expressed genes in each tissue obtained by the t test between one tissue and all other tissues. Supplementary Material S6: The GO enrichment analysis on 432 quantitatively-specific expressed genes identified in this study and 2355 tissue-enriched genes annotated previously by Uhlen et al.; Supplementary Material S7: Data and instructions used to construct the optimal SVM classification model and make predictions. Supplementary Figure S1: Box plots to show the expression level of 23 top genes across different tissues.

Author Contributions: X.K., T.H. and Y.-D.C. conceived and designed the experiments; L.C. performed the experiments; J.L. analyzed the data; J.L. and Y.-H.Z. contributed reagents/materials/analysis tools; J.L. and L.C. wrote the paper.

Funding: This research was funded by the National Natural Science Foundation of China [31701151], Natural Science Foundation of Shanghai [17ZR1412500], Shanghai Sailing Program, The Youth Innovation Promotion Association of Chinese Academy of Sciences (CAS) [2016245], the fund of the key Laboratory of Stem Cell Biology of Chinese Academy of Sciences [201703], Science and Technology Commission of Shanghai Municipality (STCSM) [18dz2271000].

Conflicts of Interest: The authors declare no conflict of interest.

References

1. Singh, S.R. Stem cell niche in tissue homeostasis, aging and cancer. *Curr. Med. Chem.* **2012**, *19*, 5965–5974. [CrossRef] [PubMed]

2. Lipscombe, D.; Andrade, A. Calcium channel cavα_1 splice isoforms—Tissue specificity and drug action. *Curr. Mol. Pharmacol.* **2015**, *8*, 22–31. [CrossRef] [PubMed]

3. Uhlen, M.; Bjorling, E.; Agaton, C.; Szigyarto, C.A.; Amini, B.; Andersen, E.; Andersson, A.C.; Angelidou, P.; Asplund, A.; Asplund, C.; et al. A human protein atlas for normal and cancer tissues based on antibody proteomics. *Mol. Cell. Proteom. MCP* **2005**, *4*, 1920–1932. [CrossRef] [PubMed]

4. Su, A.I.; Wiltshire, T.; Batalov, S.; Lapp, H.; Ching, K.A.; Block, D.; Zhang, J.; Soden, R.; Hayakawa, M.; Kreiman, G.; et al. A gene atlas of the mouse and human protein-encoding transcriptomes. *Proc. Natl. Acad. Sci. USA* **2004**, *101*, 6062–6067. [CrossRef] [PubMed]

5. Wu, C.; Orozco, C.; Boyer, J.; Leglise, M.; Goodale, J.; Batalov, S.; Hodge, C.L.; Haase, J.; Janes, J.; Huss, J.W., III; et al. BioGPS: An extensible and customizable portal for querying and organizing gene annotation resources. *Genome Biol.* **2009**, *10*, R130. [CrossRef] [PubMed]

6. Uhlen, M.; Oksvold, P.; Fagerberg, L.; Lundberg, E.; Jonasson, K.; Forsberg, M.; Zwahlen, M.; Kampf, C.; Wester, K.; Hober, S.; et al. Towards a knowledge-based human protein atlas. *Nat. Biotechnol.* **2010**, *28*, 1248–1250. [CrossRef] [PubMed]

7. Krupp, M.; Marquardt, J.U.; Sahin, U.; Galle, P.R.; Castle, J.; Teufel, A. RNA-seq atlas—A reference database for gene expression profiling in normal tissue by next-generation sequencing. *Bioinformatics* **2012**, *28*, 1184–1185. [CrossRef] [PubMed]

8. Uhlen, M.; Fagerberg, L.; Hallstrom, B.M.; Lindskog, C.; Oksvold, P.; Mardinoglu, A.; Sivertsson, A.; Kampf, C.; Sjostedt, E.; Asplund, A.; et al. Tissue-based map of the human proteome. *Science* **2015**, *347*, 1260419. [CrossRef] [PubMed]

9. The GTEx Consortium; Human genomics. The genotype-tissue expression (GTEx) pilot analysis: Multitissue gene regulation in humans. *Science* **2015**, *348*, 648–660. [CrossRef] [PubMed]

10. Meyer, D.; Leisch, F.; Hornik, K. The support vector machine under test. *Neurocomputing* **2003**, *55*, 169–186. [CrossRef]

11. Corinna, C.; Vapnik, V. Support-vector networks. *Mach. Learn.* **1995**, *20*, 273–297. [CrossRef]

12. Peng, H.; Long, F.; Ding, C. Feature selection based on mutual information: Criteria of max-dependency, max-relevance, and min-redundancy. *IEEE Trans. Pattern Anal. Mach. Intell.* **2005**, *27*, 1226–1238. [CrossRef] [PubMed]

13. Li, B.Q.; Cai, Y.D.; Feng, K.Y.; Zhao, G.J. Prediction of protein cleavage site with feature selection by random forest. *PLoS ONE* **2012**, *7*, e45854. [CrossRef] [PubMed]

14. Chen, L.; Zhang, Y.-H.; Lu, G.; Huang, T.; Cai, Y.-D. Analysis of cancer-related lncRNAs using gene ontology and kegg pathways. *Artif. Intell. Med.* **2017**, *76*, 27–36. [CrossRef] [PubMed]

15. Cai, Y.; He, J.; Lu, L. Predicting sumoylation site by feature selection method. *J. Biomol. Struct. Dyn.* **2011**, *28*, 797–804. [CrossRef] [PubMed]

16. Chen, L.; Zhang, Y.-H.; Huang, G.; Pan, X.; Wang, S.; Huang, T.; Cai, Y.-D. Discriminating cirRNAs from other lncRNAs using a hierarchical extreme learning machine (H-ELM) algorithm with feature selection. *Mol. Genet. Genom.* **2018**, *293*, 137–149. [CrossRef] [PubMed]

17. Lu, J.; Wang, S.; Cai, Y.D.; Zhang, Q. Analysis and prediction of nitrated tyrosine sites with mRMR method and support vector machine algorithm. *Curr. Bioinform.* **2017**, *13*, 3–13.

18. Liu, L.; Chen, L.; Zhang, Y.H.; Wei, L.; Cheng, S.; Kong, X.; Zheng, M.; Huang, T.; Cai, Y.D. Analysis and prediction of drug-drug interaction by minimum redundancy maximum relevance and incremental feature selection. *J. Biomol. Struct. Dyn.* **2017**, *35*, 312–329. [CrossRef] [PubMed]

19. Chen, L.; Zhang, Y.H.; Huang, T.; Cai, Y.D. Gene expression profiling gut microbiota in different races of humans. *Sci. Rep.* **2016**, *6*, 23075. [CrossRef] [PubMed]

20. Ni, Q.; Chen, L. A feature and algorithm selection method for improving the prediction of protein structural classes. *Comb. Chem. High Throughput Screen.* **2017**, *20*, 612–621. [CrossRef] [PubMed]

21. Chen, L.; Zhang, Y.H.; Zheng, M.; Huang, T.; Cai, Y.D. Identification of compound-protein interactions through the analysis of gene ontology, kegg enrichment for proteins and molecular fragments of compounds. *Mol. Genet. Genom.* **2016**, *291*, 2065–2079. [CrossRef] [PubMed]

22. Wang, S.; Zhang, Y.-H.; Huang, G.; Chen, L.; Cai, Y.-D. Analysis and prediction of myristoylation sites using the mRMR method, the ifs method and an extreme learning machine algorithm. *Comb. Chem. High Throughput Screen.* **2017**, *20*, 96–106. [CrossRef] [PubMed]

23. Chen, L.; Wang, S.; Zhang, Y.-H.; Wei, L.; Xu, X.; Huang, T.; Cai, Y.-D. Prediction of nitrated tyrosine residues in protein sequences by extreme learning machine and feature selection methods. *Comb. Chem. High Throughput Screen.* **2018**, *21*, 393–402. [CrossRef] [PubMed]

24. Li, B.Q.; Zheng, L.L.; Hu, L.L.; Feng, K.Y.; Huang, G.; Chen, L. Prediction of linear B-ceel epitopes with mRMR feature selection and analysis. *Curr. Bioinform.* **2016**, *11*, 22–31. [CrossRef]

25. Chen, L.; Pan, X.; Hu, X.; Zhang, Y.-H.; Wang, S.; Huang, T.; Cai, Y.-D. Gene expression differences among different MSI statuses in colorectal cancer. *Int. J. Cancer* **2018**. [CrossRef] [PubMed]

26. Platt, J. *Sequential Minimal Optimizaton: A Fast Algorithm for Training Support Vector Machines*; Technical Report MSR-TR-98-14; Microsoft Res: Redmon, WA, USA, 1998.

27. Frank, E.; Hall, M.; Trigg, L.; Holmes, G.; Witten, I.H. Data mining in bioinformatics using weka. *Bioinformatics* **2004**, *20*, 2479–2481. [CrossRef] [PubMed]

28. Kohavi, R. A study of cross-validation and bootstrap for accuracy estimation and model selection. In Proceedings of the International Joint Conference on Artificial Intelligence, Montreal, QC, Canada, 20–25 August 1995; Lawrence Erlbaum Associates Ltd.: Mahwah, NJ, USA, 1995; pp. 1137–1145.

29. Chen, L.; Wang, S.; Zhang, Y.-H.; Li, J.; Xing, Z.-H.; Yang, J.; Huang, T.; Cai, Y.-D. Identify key sequence features to improve CRISPR sgRNA efficacy. *IEEE Access* **2017**, *5*, 26582–26590. [CrossRef]

30. Wang, D.; Li, J.-R.; Zhang, Y.-H.; Chen, L.; Huang, T.; Cai, Y.-D. Identification of differentially expressed genes between original breast cancer and xenograft using machine learning algorithms. *Genes* **2018**, *9*, 155. [CrossRef] [PubMed]

31. Chen, L.; Chu, C.; Zhang, Y.-H.; Zheng, M.-Y.; Zhu, L.; Kong, X.; Huang, T. Identification of drug-drug interactions using chemical interactions. *Curr. Bioinform.* **2017**, *12*, 526–534. [CrossRef]

32. Chen, L.; Zeng, W.M.; Cai, Y.D.; Feng, K.Y.; Chou, K.C. Predicting anatomical therapeutic chemical (ATC) classification of drugs by integrating chemical-chemical interactions and similarities. *PLoS ONE* **2012**, *7*, e35254. [CrossRef] [PubMed]

33. Matthews, B.W. Comparison of the predicted and observed secondary structure of T4 phage lysozyme. *Biochim. Biophys. Acta* **1975**, *405*, 442–451. [CrossRef]

34. Gorodkin, J. Comparing two K-category assignments by a K-category correlation coefficient. *Comput. Biol. Chem.* **2004**, *28*, 367–374. [CrossRef] [PubMed]

35. Lizio, M.; Harshbarger, J.; Shimoji, H.; Severin, J.; Kasukawa, T.; Sahin, S.; Abugessaisa, I.; Fukuda, S.; Hori, F.; Ishikawa-Kato, S.; et al. Gateways to the FANTOM5 promoter level mammalian expression atlas. *Genome Biol.* **2015**, *16*, 22. [CrossRef] [PubMed]

36. eGTEx Project; Stranger, B.E.; Brigham, L.E.; Hasz, R.; Hunter, M.; Johns, C.; Johnson, M.; Kopen, G.; Leinweber, W.F.; Lonsdale, J.T.; et al. Enhancing gtex by bridging the gaps between genotype, gene expression, and disease. *Nat. Genet.* **2017**, *49*, 1664. [CrossRef] [PubMed]

37. Papatheodorou, I.; Fonseca, N.A.; Keays, M.; Tang, Y.A.; Barrera, E.; Bazant, W.; Burke, M.; Fullgrabe, A.; Fuentes, A.M.; George, N.; et al. Expression atlas: Gene and protein expression across multiple studies and organisms. *Nucleic Acids Res.* **2018**, *46*, D246–D251. [CrossRef] [PubMed]

38. Lee, A.W. The role of atypical protein kinase C in CSF-1-dependent ERK activation and proliferation in myeloid progenitors and macrophages. *PLoS ONE* **2011**, *6*, e25580. [CrossRef] [PubMed]

39. Kang, Z.H.; Xu, F.; Zhang, Q.A.; Wu, Z.Y.; Zhang, X.J.; Xu, J.H.; Luo, Y.; Guan, M. Oncogenic mutations in extramammary Paget's disease and their clinical relevance. *Int. J. Cancer* **2013**, *132*, 824–831. [CrossRef] [PubMed]

40. Li, D.; Lu, Z.Y.; Jia, J.Y.; Zheng, Z.F.; Lin, S. Changes in microRNAs associated with podocytic adhesion damage under mechanical stress. *J. Renin-Angiotensin Aldosterone Syst.* **2013**, *14*, 97–102. [CrossRef] [PubMed]

41. Pinatel, E.M.; Orso, F.; Penna, E.; Cimino, D.; Elia, A.R.; Circosta, P.; Dentelli, P.; Brizzi, M.F.; Provero, P.; Taverna, D. miR-223 is a coordinator of breast cancer progression as revealed by bioinformatics predictions. *PLoS ONE* **2014**, *9*, e84859. [CrossRef] [PubMed]

42. O'Connell, G.C.; Treadway, M.B.; Petrone, A.B.; Tennant, C.S.; Lucke-Wold, N.; Chantler, P.D.; Barr, T.L. Peripheral blood AKAP7 expression as an early marker for lymphocyte-mediated post-stroke blood brain barrier disruption. *Sci. Rep.* **2017**, *7*, 1172. [CrossRef] [PubMed]

43. Van der Vaart, B.; Franker, M.A.M.; Kuijpers, M.; Hua, S.S.; Bouchet, B.P.; Jiang, K.; Grigoriev, I.; Hoogenraad, C.C.; Akhmanova, A. Microtubule plus-end tracking proteins SLAIN1/2 and ch-TOG promote axonal development. *J. Neurosci.* **2012**, *32*, 14722–14729. [CrossRef] [PubMed]

44. Suchy-Dicey, A.; Heckbert, S.R.; Smith, N.L.; McKnight, B.; Rotter, J.I.; Chen, Y.I.; Psaty, B.M.; Enquobahrie, D.A. Gene expression in thiazide diuretic or statin users in relation to incident type 2 diabetes. *Int. J. Mol. Epidemiol. Genet.* **2014**, *5*, 22–30. [PubMed]

45. Cowell, J.K.; Lo, K.C.; Luce, J.; Hawthorn, L. Interpreting aCGH-defined karyotypic changes in gliomas using copy number status, loss of heterozygosity and allelic ratios. *Exp. Mol. Pathol.* **2010**, *88*, 82–89. [CrossRef] [PubMed]

46. Zhou, M.; Ye, Z.; Gu, Y.; Tian, B.; Wu, B.; Li, J. Genomic analysis of drug resistant pancreatic cancer cell line by combining long non-coding RNA and mRNA expression profiling. *Int. J. Clin. Exp. Pathol.* **2015**, *8*, 38–52. [PubMed]

47. Gao, Y.F.; Zhu, T.; Mao, C.X.; Liu, Z.X.; Wang, Z.B.; Mao, X.Y.; Li, L.; Yi, J.Y.; Zhou, H.H.; Liu, Z.Q. PPIC, EMP3 and CHI3L1 are novel prognostic markers for high grade glioma. *Int. J. Mol. Sci.* **2016**, *17*, 1808. [CrossRef] [PubMed]

48. Romero-Saavedra, F.; Laverde, D.; Wobser, D.; Michaux, C.; Budin-Verneuil, A.; Bernay, B.; Benachour, A.; Hartke, A.; Huebner, J. Identification of peptidoglycan-associated proteins as vaccine candidates for enterococcal infections. *PLoS ONE* **2014**, *9*, e111880. [CrossRef] [PubMed]

49. Krizhanovsky, V.; Ben-Arie, N. A novel role for the choroid plexus in BMP-mediated inhibition of differentiation of cerebellar neural progenitors. *Mech. Dev.* **2006**, *123*, 67–75. [CrossRef] [PubMed]

50. Ohtori, S.; Yamamoto, T.; Ino, H.; Hanaoka, E.; Shinbo, J.; Ozaki, T.; Takada, N.; Nakamura, Y.; Chiba, T.; Nakagawara, A.; et al. Differential screening-selected gene aberrative in neuroblastoma protein modulates inflammatory pain in the spinal dorsal horn. *Neuroscience* **2002**, *110*, 579–586. [CrossRef]

51. Yi, C.H.; Zheng, T.Z.; Leaderer, D.; Hoffman, A.; Zhu, Y. Cancer-related transcriptional targets of the circadian gene NPAS2 identified by genome-wide ChIP-on-chip analysis. *Cancer Lett.* **2009**, *284*, 149–156. [CrossRef] [PubMed]

52. Siggs, O.M.; Popkin, D.L.; Krebs, P.; Li, X.H.; Tang, M.; Zhan, X.M.; Zeng, M.; Lin, P.; Xia, Y.; Oldstone, M.B.A.; et al. Mutation of the er retention receptor kdelr1 leads to cell-intrinsic lymphopenia and a failure to control chronic viral infection. *Proc. Natl. Acad. Sci. USA* **2015**, *112*, E5706–E5714. [CrossRef] [PubMed]

53. Zhang, L.; Wang, T.; Valle, D. Reduced PLP2 expression increases ER-stress-induced neuronal apoptosis and risk for adverse neurological outcomes after hypoxia ischemia injury. *Hum. Mol. Genet.* **2015**, *24*, 7221–7226. [CrossRef] [PubMed]

54. Zhu, H.; Miao, M.H.; Ji, X.Q.; Xue, J.; Shao, X.J. miR-664 negatively regulates PLP2 and promotes cell proliferation and invasion in T-cell acute lymphoblastic leukaemia. *Biochem. Biophys. Res. Commun.* **2015**, *459*, 340–345. [CrossRef] [PubMed]

55. Dorsey, N.J.; Chapoval, S.P.; Smith, E.P.; Skupsky, J.; Scott, D.W.; Keegan, A.D. STAT6 controls the number of regulatory T cells in vivo, thereby regulating allergic lung inflammation. *J. Immunol.* **2013**, *191*, 1517–1528. [CrossRef] [PubMed]

56. Myklebust, J.H.; Irish, J.M.; Brody, J.; Czerwinski, D.K.; Houot, R.; Kohrt, H.E.; Timmerman, J.; Said, J.; Green, M.R.; Delabie, J.; et al. High PD-1 expression and suppressed cytokine signaling distinguish T cells infiltrating follicular lymphoma tumors from peripheral T cells. *Blood* **2013**, *121*, 1367–1376. [CrossRef] [PubMed]

57. Weber, M.S.; Prod'homme, T.; Youssef, S.; Dunn, S.E.; Steinman, L.; Zamvil, S.S. Neither T-helper type 2 nor Foxp3+ regulatory T cells are necessary for therapeutic benefit of atorvastatin in treatment of central nervous system autoimmunity. *J. Neuroinflamm.* **2014**, *11*, 29. [CrossRef] [PubMed]

58. Martin-Vilchez, S.; Whitmore, L.; Asmussen, H.; Zareno, J.; Horwitz, R.; Newell-Litwa, K. RhoGTPase regulators orchestrate distinct stages of synaptic development. *PLoS ONE* **2017**, *12*, e0170464. [CrossRef] [PubMed]

59. Katoh, M.; Katoh, M. Characterization of human ARHGAP10 gene in silico. *Int. J. Oncol.* **2004**, *25*, 1201–1206. [PubMed]

60. Hellstrom, M.; Ericsson, M.; Johansson, B.; Faraz, M.; Anderson, F.; Henriksson, R.; Nilsson, S.K.; Hedman, H. Cardiac hypertrophy and decreased high-density lipoprotein cholesterol in Lrig3-deficient mice. *Am. J. Physiol. Regul. Integr. Comp. Physiol.* **2016**, *310*, R1045–R1052. [CrossRef] [PubMed]

61. Abraira, V.E.; Satoh, T.; Fekete, D.M.; Goodrich, L.V. Vertebrate Lrig3-erbb interactions occur in vitro but are unlikely to play a role in Lrig3-dependent inner ear morphogenesis. *PLoS ONE* **2010**, *5*, e8981. [CrossRef] [PubMed]

62. Abraira, V.E.; del Rio, T.; Tucker, A.F.; Slonimsky, J.; Keirnes, H.L.; Goodrich, L.V. Cross-repressive interactions between Lrig3 and netrin 1 shape the architecture of the inner ear. *Development* **2008**, *135*, 4091–4099. [CrossRef] [PubMed]

63. Jansson, L.; Larsson, J. Normal hematopoietic stem cell function in mice with enforced expression of the hippo signaling effector YAP1. *PLoS ONE* **2012**, *7*, e32013. [CrossRef] [PubMed]

64. Hoshiba, T.; Otaki, T.; Nemoto, E.; Maruyama, H.; Tanaka, M. Blood-compatible polymer for hepatocyte culture with high hepatocyte-specific functions toward bioartificial liver development. *ACS Appl. Mater. Interfaces* **2015**, *7*, 18096–18103. [CrossRef] [PubMed]

65. Loke, S.Y.; Wong, P.T.; Ong, W.Y. Global gene expression changes in the prefrontal cortex of rabbits with hypercholesterolemia and/or hypertension. *Neurochem. Int.* **2017**, *102*, 33–56. [CrossRef] [PubMed]

66. Yamashita, A.; Taniwaki, T.; Kaikoi, Y.; Yamazaki, T. Protective role of the endoplasmic reticulum protein mitsugumin23 against ultraviolet C-induced cell death. *FEBS Lett.* **2013**, *587*, 1299–1303. [CrossRef] [PubMed]

67. Reiss, J.; Hahnewald, R. Molybdenum cofactor deficiency: Mutations in *GPHN*, *MOCS1*, and *MOCS2*. *Hum. Mutat.* **2011**, *32*, 10–18. [CrossRef] [PubMed]

68. Wang, J.; Krizowski, S.; Fischer-Schrader, K.; Niks, D.; Tejero, J.; Sparacino-Watkins, C.; Wang, L.; Ragireddy, V.; Frizzell, S.; Kelley, E.E.; et al. Sulfite oxidase catalyzes single-electron transfer at molybdenum domain to reduce nitrite to nitric oxide. *Antioxid. Redox Signal.* **2015**, *23*, 283–294. [CrossRef] [PubMed]

69. Ricketts, C.D.; Bates, W.R.; Reid, S.D. The effects of acute waterborne exposure to sublethal concentrations of molybdenum on the stress response in rainbow trout, oncorhynchus mykiss. *PLoS ONE* **2015**, *10*, e0115334. [CrossRef] [PubMed]

70. Stewart, K.; Uetani, N.; Hendriks, W.; Tremblay, M.L.; Bouchard, M. Inactivation of LAR family phosphatase genes Ptprs and Ptprf causes craniofacial malformations resembling pierre-robin sequence. *Development* **2013**, *140*, 3413–3422. [CrossRef] [PubMed]

71. Unoki, M.; Shen, J.C.; Zheng, Z.M.; Harris, C.C. Novel splice variants of ing4 and their possible roles in the regulation of cell growth and motility. *J. Biol. Chem.* **2006**, *281*, 34677–34686. [CrossRef] [PubMed]

72. Silver, D.J.; Siebzehnrubl, F.A.; Schildts, M.J.; Yachnis, A.T.; Smith, G.M.; Smith, A.A.; Scheffler, B.; Reynolds, B.A.; Silver, J.; Steindler, D.A. Chondroitin sulfate proteoglycans potently inhibit invasion and serve as a central organizer of the brain tumor microenvironment. *J. Neurosci.* **2013**, *33*, 15603–15617. [CrossRef] [PubMed]

73. Park, J.; Lee, J.; Choi, C. Evaluation of drug-targetable genes by defining modes of abnormality in gene expression. *Sci. Rep.* **2015**, *5*, 13576. [CrossRef] [PubMed]

74. Desh, H.; Gray, S.L.; Horton, M.J.; Raoul, G.; Rowlerson, A.M.; Ferri, J.; Vieira, A.R.; Sciote, J.J. Molecular motor MYO1C, acetyltransferase KAT6B and osteogenetic transcription factor RUNX2 expression in human masseter muscle contributes to development of malocclusion. *Arch. Oral Biol.* **2014**, *59*, 601–607. [CrossRef] [PubMed]

75. Toyoda, T.; An, D.; Witczak, C.A.; Koh, H.J.; Hirshman, M.F.; Fujii, N.; Goodyear, L.J. *Myo1c* regulates glucose uptake in mouse skeletal muscle. *J. Biol. Chem.* **2011**, *286*, 4133–4140. [CrossRef] [PubMed]

76. Akahane, K.; Inukai, T.; Zhang, X.; Hirose, K.; Kuroda, I.; Goi, K.; Honna, H.; Kagami, K.; Nakazawa, S.; Endo, K.; et al. Resistance of t-cell acute lymphoblastic leukemia to tumor necrosis factor–related apoptosis-inducing ligand-mediated apoptosis. *Exp. Hematol.* **2010**, *38*, 885–895. [CrossRef] [PubMed]

77. Yu, R.; Mao, J.; Yang, Y.; Zhang, Y.; Tian, Y.; Zhu, J. Protective effects of calcitriol on diabetic nephropathy are mediated by down regulation of TGF-β1 and CIP4 in diabetic nephropathy rat. *Int. J. Clin. Exp. Pathol.* **2015**, *8*, 3503–3512. [PubMed]

78. Aulak, K.S.; Davis, A.E., III; Donaldson, V.H.; Harrison, R.A. Chymotrypsin inhibitory activity of normal C1-inhibitor and a P1 arg to his mutant: Evidence for the presence of overlapping reactive centers. *Protein Sci. Publ. Protein Soc.* **1993**, *2*, 727–732. [CrossRef] [PubMed]

79. Katoh, Y.; Imakagura, H.; Futatsumori, M.; Nakayama, K. Recruitment of clathrin onto endosomes by the Tom1-Tollip complex. *Biochem. Biophys. Res. Commun.* **2006**, *341*, 143–149. [CrossRef] [PubMed]
80. Huang, D.W.; Sherman, B.T.; Lempicki, R.A. Systematic and integrative analysis of large gene lists using david bioinformatics resources. *Nat. Protoc.* **2009**, *4*, 44–57. [CrossRef] [PubMed]

Article

Improving the Gene Ontology Resource to Facilitate More Informative Analysis and Interpretation of Alzheimer's Disease Data

Barbara Kramarz [1], Paola Roncaglia [2], Birgit H. M. Meldal [2], Rachael P. Huntley [1], Maria J. Martin [2], Sandra Orchard [2], Helen Parkinson [2], David Brough [3], Rina Bandopadhyay [4], Nigel M. Hooper [3] and Ruth C. Lovering [1,*]

[1] UCL Institute of Cardiovascular Science, University College London, Rayne Building, 5 University Street, London WC1E 6JF, UK; barbara.kramarz@ucl.ac.uk (B.K.); r.huntley@ucl.ac.uk (R.P.H.)

[2] European Bioinformatics Institute (EMBL-EBI), European Molecular Biology Laboratory, Wellcome Genome Campus, Hinxton, Cambridge CB10 1SD, UK; paola@ebi.ac.uk (P.R.); bmeldal@ebi.ac.uk (B.H.M.M.); martin@ebi.ac.uk (M.J.M.); orchard@ebi.ac.uk (S.O.); parkinson@ebi.ac.uk (H.P.)

[3] Division of Neuroscience and Experimental Psychology, School of Biological Sciences, Faculty of Biology, Medicine and Health, Manchester Academic Health Science Centre, University of Manchester, AV Hill Building, Oxford Road, Manchester M13 9PT, UK; david.brough@manchester.ac.uk (D.B.); nigel.hooper@manchester.ac.uk (N.M.H.)

[4] UCL Queen Square Institute of Neurology and Reta Lila Weston Institute of Neurological Studies, 1 Wakefield Street, London WC1N 1PJ, UK; rina.bandopadhyay@ucl.ac.uk

[*] Correspondence: r.lovering@ucl.ac.uk or goannotation@ucl.ac.uk; Tel.: +44-207-679-6965

Received: 31 October 2018; Accepted: 23 November 2018; Published: 29 November 2018

Abstract: The analysis and interpretation of high-throughput datasets relies on access to high-quality bioinformatics resources, as well as processing pipelines and analysis tools. Gene Ontology (GO, geneontology.org) is a major resource for gene enrichment analysis. The aim of this project, funded by the Alzheimer's Research United Kingdom (ARUK) foundation and led by the University College London (UCL) biocuration team, was to enhance the GO resource by developing new neurological GO terms, and use GO terms to annotate gene products associated with dementia. Specifically, proteins and protein complexes relevant to processes involving amyloid-beta and tau have been annotated and the resulting annotations are denoted in GO databases as 'ARUK-UCL'. Biological knowledge presented in the scientific literature was captured through the association of GO terms with dementia-relevant protein records; GO itself was revised, and new GO terms were added. This literature biocuration increased the number of Alzheimer's-relevant gene products that were being associated with neurological GO terms, such as 'amyloid-beta clearance' or 'learning or memory', as well as neuronal structures and their compartments. Of the total 2055 annotations that we contributed for the prioritised gene products, 526 have associated proteins and complexes with neurological GO terms. To ensure that these descriptive annotations could be provided for Alzheimer's-relevant gene products, over 70 new GO terms were created. Here, we describe how the improvements in ontology development and biocuration resulting from this initiative can benefit the scientific community and enhance the interpretation of dementia data.

Keywords: Alzheimer's disease; dementia; cognitive impairment; neurodegeneration; Gene Ontology; annotation; biocuration; amyloid-beta; microtubule-associated protein tau

1. Introduction

Understanding the cellular bases of Alzheimer's disease (AD) and other dementias is an essential step in helping develop improved therapies for their treatment and prevention, as well as supporting early diagnosis. Although several genes associated with monogenic AD have been identified [1], the majority of cases are likely to be due to multiple genetic, as well as environmental, risk factors [2,3]. In order to understand the cellular processes and risk factors associated with AD and other dementias, numerous transcriptomic, proteomic, and genome-wide association (GWA) studies have been conducted [2,4–6]. Analyses and the interpretation of such high-throughput datasets rely on high-quality annotations describing the biological roles of the implicated gene products, as well as appropriate methodologies being used to generate the data, the application of appropriate statistical methods, and implementation in analysis tools [4,6–8]. Gene annotation provides functional knowledge of gene products (proteins and RNAs) in a format that can be fully exploited by systems biology and genomic investigators. Thus, annotation resources bridge the gap between data collection and data analysis [4,9]. The main resources used to identify significantly enriched pathways in GWA, proteomic, and transcriptomic studies are those provided by the Gene Ontology (GO) [10], Reactome [11], KEGG [12], and molecular interaction databases [13,14]. This data is imported into independent enrichment analysis tools, such as g:Profiler [15], VLAD [16], or DAVID [17], which can provide different enrichment options and/or settings, and are updated with different frequencies, all of which will affect analyses' outcomes.

The GO resource is a biomedical ontology that describes the physiological functional attributes of gene products, across all species, in a consistent and computer-accessible manner using a controlled vocabulary that facilitates the integration of public biological data [10,18,19]. Fully defined GO terms are associated with gene products across many species, providing a computable summary of individual experiments that demonstrate functional information (Figure 1). These links between GO terms and gene products are known as 'annotations', and enable the description of gene products according to their molecular functions (e.g., 'scavenger receptor activity'), the biological processes they contribute towards (e.g., 'microtubule cytoskeleton organisation'), and their subcellular locations (or cellular components, e.g., 'extracellular region') [10].

Symbol	GO Term		Reference	Taxon	Assigned By
CDK5	GO:0031175	neuron projection development	PMID:17491008	9606 Homo sapiens	ARUK-UCL
Epha4	GO:0106030	neuron projection fasciculation	PMID:12761826	10090 Mus musculus	ARUK-UCL
ROCK1	GO:0031175	neuron projection development	PMID:27160703	9606 Homo sapiens	ARUK-UCL
ROCK1	GO:0140058	neuron projection arborization	PMID:27160703	9606 Homo sapiens	ARUK-UCL
SEMA3A	GO:0150020	basal dendrite arborization	PMID:22683681	9606 Homo sapiens	ARUK-UCL
SEMA3A	GO:0150020	basal dendrite arborization	PMID:22683681	9606 Homo sapiens	ARUK-UCL
CDK5R1	GO:0031175	neuron projection development	PMID:17491008	9606 Homo sapiens	ARUK-UCL
Map3k13	GO:0150012	positive regulation of neuron projection arborization	PMID:27511108	10090 Mus musculus	ARUK-UCL
Fzd4	GO:0150012	positive regulation of neuron projection arborization	PMID:25424568	10090 Mus musculus	ARUK-UCL
Nlgn1	GO:0031175	neuron projection development	PMID:22750515	10116 Rattus norvegicus	ARUK-UCL
Nlgn1	GO:0140058	neuron projection arborization	PMID:22750515	10116 Rattus norvegicus	ARUK-UCL
Taok2	GO:0150019	basal dendrite morphogenesis	PMID:22683681	10090 Mus musculus	ARUK-UCL
Taok2	GO:0150020	basal dendrite arborization	PMID:22683681	10090 Mus musculus	ARUK-UCL
NLGN1	GO:0031175	neuron projection development	GO_REF:0000024	9606 Homo sapiens	ARUK-UCL
NLGN1	GO:0140058	neuron projection arborization	GO_REF:0000024	9606 Homo sapiens	ARUK-UCL

Figure 1. *Cont.*

Symbol	GO Term	Reference	Taxon	Assigned By
DVL3	GO:0150012 positive regulation of neuron projection arborization	GO_REF:0000024	9606 Homo sapiens	ARUK-UCL
Taok2	GO:0150019 basal dendrite morphogenesis	PMID:22683681	10116 Rattus norvegicus	ARUK-UCL
Taok2	GO:0150020 basal dendrite arborization	PMID:22683681	10116 Rattus norvegicus	ARUK-UCL
Wnt5a	GO:0150012 positive regulation of neuron projection arborization	PMID:25424568	10116 Rattus norvegicus	ARUK-UCL
Fzd4	GO:0150012 positive regulation of neuron projection arborization	PMID:25424568	10116 Rattus norvegicus	ARUK-UCL
TAOK2	GO:0150019 basal dendrite morphogenesis	PMID:22683681	9606 Homo sapiens	ARUK-UCL
TAOK2	GO:0150020 basal dendrite arborization	PMID:22683681	9606 Homo sapiens	ARUK-UCL
FZD4	GO:0150012 positive regulation of neuron projection arborization	GO_REF:0000024	9606 Homo sapiens	ARUK-UCL
Dvl1	GO:0150012 positive regulation of neuron projection arborization	PMID:25424568	10116 Rattus norvegicus	ARUK-UCL
GRIP1	GO:0150012 positive regulation of neuron projection arborization	GO_REF:0000024	9606 Homo sapiens	ARUK-UCL

Figure 1. A selection of Gene Ontology (GO) annotations created by Alzheimer's Research United Kingdom (ARUK)—University College London (UCL), as a part of our project that focused on capturing knowledge on amyloid-beta and tau by the expert biocuration of experimental data available through the biomedical literature. These annotations describe related biological processes using GO terms, which are descendants of the 'neuron projection development' GO term. We created 13 and revised two of the 24 GO terms in this branch. We created a total of 50 new annotations using these new neurological GO terms (Results sections: Revisions and development of neuron projection GO branches; Figure 4). 'Symbol' corresponds to the HUGO Gene Nomenclature Committee (HGNC)-approved gene symbol [20] encoding the gene product being annotated; the 'GO Term' section provides the GO identifier and term name; the 'Reference' column lists the source of the data supporting the annotation; this may be the curated article or information about the electronic annotation pipeline. The 'Taxon' column describes the species of origin of the protein being annotated; and the 'Assigned By' column shows that these annotations resulted from this project (Screenshot from QuickGO—accessed: 31 October 2018).

One of the key advantages of the GO resource is that it can describe a protein's likely role in a process or its probable location in a cell, even when its actual functional roles are still under investigation. For instance, GO will capture the role of a protein kinase in a signalling cascade even when the step in this cascade at which the enzyme is acting has not yet been fully elucidated. In contrast, Reactome and KEGG provide more 'specific' information about the role of a protein within a pathway by creating networks where the details of each 'reaction' being catalysed or facilitated are provided [11,12]. Consequently, proteins whose roles have not been fully elucidated cannot be included in these databases. However, occasionally Reactome will capture 'BlackBox' events when not all of the steps in a pathway are known [11]. In summary, Reactome provides annotations for 10,800 human proteins, whereas GO provides manual annotations for 20,000 human proteins (data from QuickGO and Reactome, accessed: 20 August 2018). A common strength of all of these resources is that for any curated model organism, data can be conservatively mapped across to their human (and other species) orthologs; however, because many gene products are involved in multiple cellular processes, the comprehensive annotation of all gene products has not been achieved. Furthermore, although human and mammalian phenotype ontologies (HPO, MP) are being used to interpret next generation sequencing (NGS) data [21], understanding how multiple genes contribute to a single disease or phenotype will require resources, such as GO, that describe the cellular functions of these gene products.

Amyloid-beta fibril deposits commonly known as plaques, and neurofibrillary tangles of phosphorylated tau, have for decades been the pathological hallmarks of Alzheimer's disease [22]. For many years, researchers have been studying the formation and the effects of these protein aggregates on brain function [23–26]. However, much less attention and fewer resources have been committed towards elucidating the underlying cellular mechanisms that drive the formation of these protein aggregates, or towards understanding the roles of the various cell types involved in the maintenance of normal brain functions, including not only neurons, but also glia and the cells of the brain microvasculature [27–29]. It is currently commonly accepted within the dementia research communities that impairments in cellular processes that lead to disease onset can occur decades before the first manifestations of clinical symptoms [27]. Accordingly, the research focus has been steadily shifting from protein aggregates towards protein homeostasis and the implicated cellular processes [30]. This has coincided with the 'explosion' of 'big data' in biomedicine. GWA and biomarker expression studies have resulted in the generation of extensive datasets of gene loci linked to dementia and AD [31–35]. In order to reveal the biological processes that are impaired in dementia due to changes in expression levels and/or the regulation of these genes, high-throughput datasets can be interpreted using GO term enrichment analysis or pathway analysis. By using the contextual information available in GO, it would also be possible to identify the cell type(s) in which an enriched process most likely occurred, if gene expression had been measured across the whole brain rather than in specific cell types or individual cells [35]. Identification of the key cells that are involved is also likely to contribute to our understanding of the cellular mechanisms underlying AD. For example, Espuny-Camacho et al. performed GO gene set enrichment analysis on a novel chimeric model of AD, and observed an increased expression of genes involved in myelination, and a decreased expression of genes related to memory and cognition, synaptic transmission, and neuron projection [36]. Furthermore, two recent studies have successfully used GO to interpret AD patient data. Xu et al. [37] investigated regional protein expression in the brain of human AD patients, and showed the involvement of some of the differentially expressed genes in key neurological processes, such as 'nervous system development' (GO:0007399) and 'neuron projection development' (GO:0031175). Kunkle et al. [38] described GWA data from the largest cohort to date of late-onset AD cases, and demonstrated the involvement of the differentially expressed genes in dementia-relevant processes such as 'regulation of amyloid-beta formation' (GO:1902003), 'tau protein binding' (GO:0048156) and 'activation of immune response' (GO:0002253). Another study applied a new statistical approach that used GO with neuroimaging phenotypes and GWA data to identify gene products that are involved in processes impaired in disease [39]. A peculiarity of that work was the inclusion of a focussed GO annotation of 21 gene products, the majority of which were encoded by AD-associated Mendelian genes. The 400 manual annotations that the authors associated with these and eight other genes were included in GO annotation files prior to carrying out the analysis. Last but not least, GO can also be used in the reanalysis of previously published datasets, which is a potentially productive exercise, because the availability of new GO data impacts on the information content [40].

In order to meet the urgent requirement to improve the interpretation of and draw informative conclusions from high-throughput dementia studies, the focus of the project described here was to provide GO annotations that capture knowledge on amyloid-beta and tau by the expert biocuration of experimental data available through the biomedical literature (annotations resulting from this project are assigned by 'ARUK-UCL'). We have now completed the first phase of enhancing the GO resource by developing new neurological GO terms and capturing the roles of proteins interacting with amyloid-beta and tau using functional GO annotations. The curation of 40 amyloid-beta binding proteins (i.e., proteins and protein complexes acting as amyloid-beta receptors [41]) has provided over 2000 annotations describing the normal role of these proteins in signalling, receptor-mediated endocytosis, and phagocytosis and/or clearance of amyloid-beta. In addition, the normal roles of microtubule-associated protein tau and its interacting partners [42] are now also captured in GO annotation files, following the annotation of 33 proteins and the creation of over 1700 annotations. Here, we provide summaries of key areas of improvement, as well as an in-depth discussion of the specific challenges that arose during this work.

2. Materials and Methods

2.1. Community Involvement

Collaborations have been established between members of the Gene Ontology Consortium [43] and community experts to ensure that our biocuration efforts aligned with the needs of the neuroscience research community. Project progress and direction were discussed and, if required, revised and updated during project meetings and through regular correspondence.

2.2. Selection of Experimental Data to Annotate

Two sets of AD-relevant high-priority human proteins and protein complexes were compiled based on recent review articles. The first set included amyloid-beta species and their binding partners, which have been shown to act as receptors for the amyloid-beta monomers and/or their oligomers, and were fully based on the amyloid-beta receptors listed in the Jarosz-Griffiths el al. (2016) review [41]. The second set was a collection of proteins interacting with the microtubule-associated protein tau, based on the review article by Guo, Noble, and Hanger (2017) [42]. This review provided a vast list of tau-binding partners, of which 33 were selected for annotation after consultations with the neuroscience community. Primary research articles cited in both reviews were prioritised for annotation. The PubMed database [44] was also used to identify additional research articles that contained experimental data describing each of the human gene products prioritised for annotation. For each high priority protein/protein complex, PubMed searches were performed using the gene symbols and names approved by the HUGO Gene Nomenclature Committee (HGNC) [20], as well as their synonyms. For well-researched gene products, relevant papers were identified with secondary searches for gene symbols and names combined with the following keywords (one at a time): 'dementia', 'Alzheimer's', 'Alzheimer', 'AD', 'amyloid', 'APP' (amyloid precursor protein), 'tau', '*MAPT*' (microtubule-associated protein tau), 'microtubule', 'neurology', 'neurological', 'neurobiology', or 'neurodegeneration'. If no, or insufficient, information on the human gene product was found, orthologs, identified using the HGNC ortholog prediction tool 'HCOP', were curated [20]. Peer-reviewed articles were selected for curation based on criteria described by Denny et al. [40], i.e., (1) they contained experimental research data; (2) the curation of any given article would result in new information being added to the current GO annotation data associated with the prioritised gene products; (3) it was possible to identify the species from which the gene products and/or expression constructs were derived, crucial information enabling biocurators to assign the appropriate database identifiers to gene products. Selected research articles were annotated fully, resulting in GO annotations of not only the dementia-relevant high-priority gene products, but also any other proteins or protein complexes described in those articles.

2.3. Gene Ontology Annotation of Proteins and Protein Complexes—Manual Curation Process

Primary research articles were read by skilled GO biocurators to describe the molecular functions, biological processes, and cellular locations of the gene products and capture them using GO terms, following established GO Consortium guidelines [40,45]. In addition, often terms from other ontologies were included in the GO annotation extension field to provide additional contextual information [46]. The gene product identifiers that were used in this project included UniProt accessions [47], Complex Portal accessions [14], as well as RNAcentral accessions [48]. Specific evidence codes were included in each biocurator-generated annotation, based on the type of experimental data reported in the research articles (e.g., IPI: inferred from physical interaction, or IMP: inferred from mutant phenotype), or to infer evidence from statements made in reviews (e.g., TAS: traceable author statement [49]). GO annotations created for rodent or other mammalian gene products, based on experimental evidence, were transferred by biocurators to human orthologs using the evidence code ISS (inferred from sequence similarity), if 1-to-1 orthology could be confirmed. This biocuration process was consistently applied to describe the published experimental data for each of the human high-priority genes and/or their orthologs. The annotations contributed by this project to the GO resource are attributed to ARUK-UCL, and were captured using the European Bioinformatics Institute (EBI) GO annotation tool Protein2GO [50]. All of the ARUK-UCL annotations are included in the GO Consortium annotation files, and thus made available through various ftp sites and the GO browsers QuickGO [51–53] and AmiGO [54,55], which are updated on a weekly and monthly basis, respectively. Our GO annotations are also propagated to other major biological databases, including NCBI Gene [56], Ensembl [57], UniProt [47], RNAcentral [48], and miRBase [58]. Additionally, the GO annotation work that was completed as a part of this project contributed to the biocuration of new Complex Portal [14] entries.

2.4. Ontology Development and Integration of Resources

The ontology editor Protégé [59] was used to generate new GO terms and, if required, modify and/or update the existing terms. All of the ontology changes were integrated into the public ontology version using GitHub [60]. In the QuickGO browser [51–53], GO term entries that were created and/or updated as a result of this project include an acknowledgment for ARUK. In the AmiGO browser [54,55], the same GO terms are identified by the source 'GOC:aruk' [61].

2.5. Functional Analysis of Hippocampal Proteomic Data

To identify the over-representation of GO terms in a hippocampal protein dataset (Table S8), which was identified as differentially expressed in Alzheimer's disease patients compared to age-matched controls [37], the functional analysis tool g:GOSt available from the g:Profiler server [15] was used. The analysis was undertaken on 19 November 2018, using the annotated human proteome as the background 'population' set. The g:SCS method was used for computing multiple testing correction for the GO enrichment analysis p-values, with an experiment-wide threshold of a = 0.05. This algorithm considers the ontology structure underlying the gene sets annotated to each GO term. The analysis used two g:GOSt preloaded GO datasets: December 2016 includes the annotation file available in Ensembl 87 (build date December 2016) and the ontology file released 13 December 2016; November 2018 includes the annotation file available in Ensembl 93 (build date July 2018) and ontology file released 3 August 2018. All of the protein identifiers (Table S8) were provided by Xu et al. [37]. In addition, this dataset was also 'seeded' with the full list of priority proteins, in order to further examine the impact of this annotation project; this list is subsequently referred to as the 'priority protein-seeded hippocampal' list.

3. Results

3.1. Assignment of Database Identifiers

The gene products that were prioritised for Gene Ontology (GO) annotation were identified from two key reviews [41,42] and finalised following consultations with dementia experts. Forty-nine gene products implicated in amyloid-beta biology were prioritised for annotation; this list included 40 cellular receptors reviewed by Jarosz-Griffiths et al. [41], to which nine monomeric and oligomeric amyloid-beta species have been shown to bind. Among the tau-interacting partners, 33 were prioritised for annotation based on the review by Guo, Noble, and Hanger [42]; the priority list therefore includes 34 proteins (also counting tau itself). The focus of this project was the annotation of human gene products; therefore, all of the 84 prioritised entities (76 proteins and eight complexes) are referred to using database identifiers corresponding to human gene products (Table S1).

In order to capture amyloid-beta biology accurately, it was necessary to be able to distinguish between amyloid-beta monomers, dimers, and oligomers, because of the different cellular effects of monomeric amyloid-beta and its dimeric or oligomeric forms [62]. Consequently, a collaboration was established with Complex Portal (CP) [14,63] biocurators, which resulted in the generation of 18 new

CP entries for the different amyloid-beta monomers, dimers, and oligomers in three species (human, rat, and mouse) (Table S2).

3.2. Gene Ontology Annotation

Research articles were selected as described in the methods section and annotated following the established GO guidelines for manual biocuration [40]. Of the 226 PubMed-indexed articles that we curated to capture knowledge about our 84 prioritised gene products, 191 described roles of human gene products (Table 1, Table S3C). Twenty-five articles that had appeared in the PubMed searches and had been read were deemed not suitable for annotation; 12 of these provided no information about the species of the gene product(s) being annotated; six provided information about either disease, patients, or pharmacological agents; three provided expression data without functional information; and three were either opinion or review articles; whereas one article contained data images of insufficient quality for a biocurator to be able to review and capture the experimental information in a GO annotation(s) (Table S3D). Gene products were associated with relevant GO terms according to evidence provided in the published research articles. Rather than capturing the experimental data describing only the gene products on the priority list, all of the experimental data in each curated paper was associated with GO terms, thus increasing the number of gene products curated, which is an efficient approach that is often referred to as 'full paper curation'.

Table 1. Summary of key curation statistics. (Data from QuickGO [51–53] and the Complex Portal [14,63]—accessed: 18 October 2018—summarised in Supplementary Tables S1–S6).

ARUK-UCL GO Annotation Project: Priorities and Curation Summary	Total	
Prioritised human gene products	84	
Prioritised human amyloid-beta-relevant gene products [#]	50	
Prioritised human tau-relevant gene products [##]	34	
	All species	Human
PubMed identifiers (PMIDs) curated *	226	191
GO annotations **	3886	2770
Gene products annotated **	561	308
GO annotations associated with 84 human gene products prioritised for annotation *[#]	n/a	2055
New Complex Portal (CP) entries ***	18	6

[#] For priority amyloid-beta-relevant gene products annotations, see Table S1A; [##] For priority tau-relevant gene products annotations, see Table S1B; * For PMIDs annotated by ARUK-UCL, see Table S3C; ** For all of the ARUK-UCL annotations and details of gene products, see Table S3A; *[#] For ARUK-UCL annotations to prioritised gene products, see Table S4 and filter 'Assigned By' A → Z to group 'ARUK-UCL' annotations together; *** For details about new CP details, see Table S2.

Through our whole-paper curation approach, a total of 3886 GO annotations have been contributed for a total of 561 gene products, including proteins (UniProt [47,64] identifiers), microRNAs (RNAcentral [48] identifiers), and protein complexes (Complex Portal [14,63] identifiers) as a part of this ARUK-UCL annotation initiative. Among these, 2770 GO annotations were associated with 308 human gene products, and 2055 were associated specifically with the proteins and complexes prioritised for annotation (Table 1, Tables S1–S4); based on data from QuickGO, accessed 17 August 2018.

The annotations that we have provided ensure that the role of these gene products in dementia-relevant processes is captured; overall, this project created just over a quarter of the annotations now associated with these 84 protein and complex records. For example, several of the priority genes (such as *APP*, *APOE*, *FYN*, and *HSP90AB1*) had over 200 associated GO annotations prior to the start of this project, but few of these described neurological processes. Our work has resulted in 526 new annotations to neurological GO terms for all but one of the prioritised gene products (Figures 2 and 3, Table S5), capturing their dementia-relevant roles in neurological processes. For one, *SCARB2*, no evidence to support the association of a neurological or amyloid-related GO term was identified in the published literature.

Overall, approximately two-thirds (526 of the 834) of the neurological, amyloid-beta, or tau-relevant GO terms, which are now associated with 84 of our prioritised gene products, were created by the ARUK-UCL initiative. Previously, only 49 of these gene products were associated with neurological process, a neuronal cellular location, or a molecular function relevant to amyloid-beta or tau biology (Figures 1 and 3, Table S5). In summary, the ARUK-UCL contribution greatly increased the number of our prioritised gene products being associated with neurological process GO terms, e.g., 'amyloid-beta clearance' and 'learning or memory', as well as localising to neuronal structures, including axons, dendrites, and their compartments.

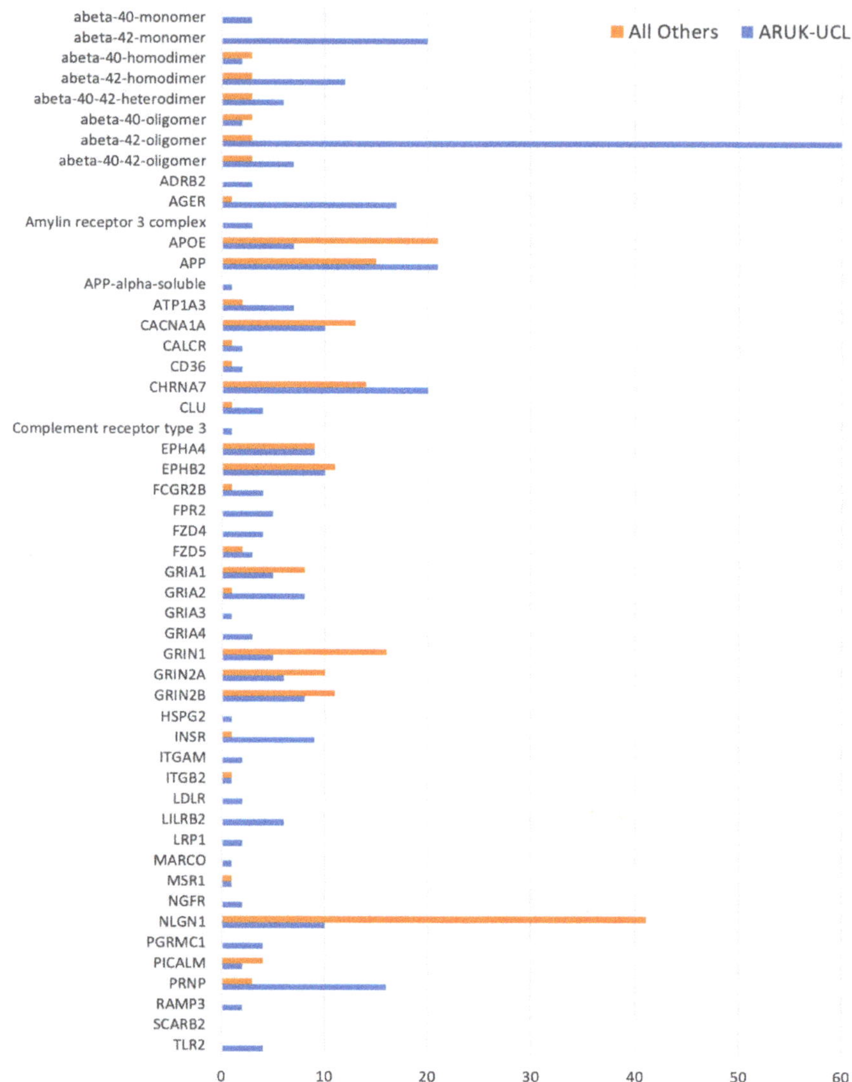

Figure 2. A histogram of all of the manually curated neurological GO annotations for the prioritised amyloid-beta-relevant human gene products, including annotations contributed by ARUK-UCL as well as by other groups. Data for this figure was derived from Table S5.

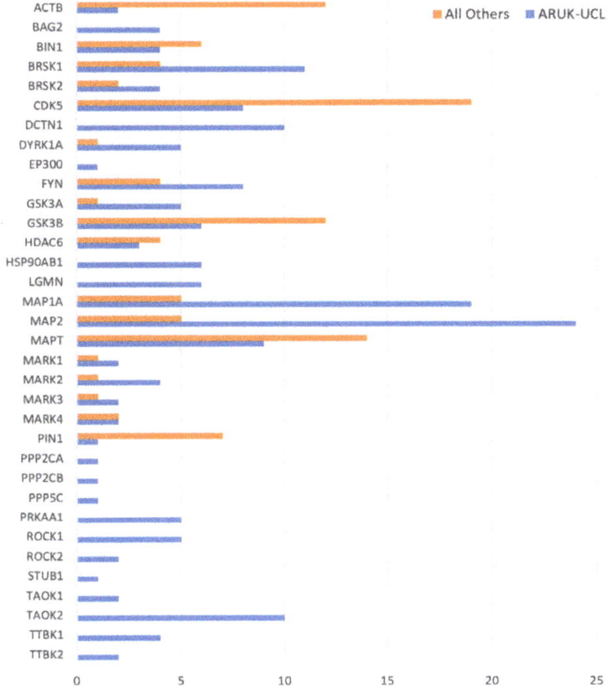

Figure 3. A histogram of all the manually curated neurological annotations for the prioritised tau-relevant human gene products, including annotations contributed by ARUK-UCL as well as by other groups. Data for this figure was derived from Table S5.

3.3. Gene Ontology Development and Revisions

3.3.1. New and Revised Gene Ontology Terms

As the literature curation progressed, new GO terms were needed to represent normal cellular functions or compartments altered in dementia. In some cases existing GO terms were also revised and updated with recent knowledge, or with synonyms that were useful for biocuration and literature mining (Table S6). For instance, 10 new GO terms have been added to the 'neuron projection morphogenesis' biological process branch, and numerous corresponding 'neuron projection' cellular component terms have been revised. In addition, new molecular function and biological process terms have been added to lipoprotein particle-related terms. Moreover, revisions have been made to 'synaptic vesicle endocytosis', and new child terms, for this GO term, have been added. Overall, 84 GO term entries were curated as part of this project; of these, 71 new GO terms were created, and 13 existing GO terms were modified (Table S6). Sixty-eight of these new, or modified, GO terms represent biological processes; 15 are cellular component terms, and one is a molecular function term. These terms were used to curate proteins and protein complexes prioritised for annotation (Table S1) due to their association with dementia. Details of these new additions and revisions can be tracked on the GitHub Gene Ontology pages [60] by filtering issues marked with 'ARUK-UCL' labels.

3.3.2. Revisions and Development of Neuron Projection Gene Ontology Branches

The revision and modification, as well as creation, of GO terms describing neuron projection-relevant processes are key contributions of this project, and were undertaken to improve the representation of dementia-relevant brain biology. Axons and dendrites, i.e., pre-synaptic and post-synaptic neuron projections, enable neuron-to-neuron signalling through synapses, whereas synaptic plasticity, which is dependent on the structural and functional integrity of synapses between neuron projections, is key for memory formation and learning [65]. Early in this project, the need to revise the 'neuron projection' GO domain was identified in order to capture full functional information that describes how the binding of amyloid-beta to receptors on pre-synapses and post-synapses or the presence of phospho-tau neurofibrillary tangles in synaptic compartments may lead to the modulation of synaptic plasticity [66,67].

Literature curation did confirm the involvement of many of the prioritised gene products (Table S1) in 'neuron projection development' and 'neuron projection organisation'. This resulted in 33 experimental, or orthology-based, GO annotations applying terms descended from 'neuron projection development' (Figures 1 and 4); all of these annotations were made to the GO terms created, or revised, by ARUK-UCL. Biocurators from the Rat Genome Database (RGD) [68] also contributed to the development of this GO branch or terms descended from 'neuron projection development' by requesting regulation terms of 'neuron projection arborisation'.

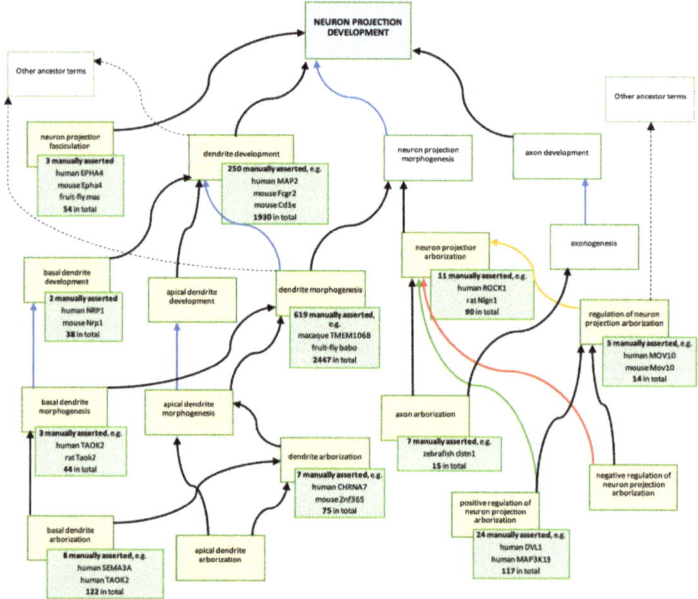

Figure 4. Descendants of biological process 'neuron projection development' GO term created or modified, as a result of this ARUK-UCL project. Ontology: The blue box represents the ancestor term 'neuron projection development' (GO:0031175). The yellow boxes represent the terms created, or modified, as a result of the ARUK-UCL project (Table S6); the majority of these are new terms; two terms were modified: 'dendrite development' (GO:0016358) and 'dendrite morphogenesis' (GO:0048813). The white boxes represent related terms in the ontology; no changes have been made to these terms. Arrows indicate the following ontology relationships: black—is_a, blue—part_of, yellow—regulates, green—positively_regulates, red—negatively_regulates. Annotations: Each green box provides information about gene products associated with the GO term in the yellow box, which it overlaps (data from QuickGO [51–53], accessed: 20 November 2018). For total numbers of annotations, provided in the green boxes, the following QuickGO filter was used: 'GO terms → Options → Use these terms as an exact match'; an additional filter 'Evidence → ECO:0000352: evidence used in manual assertion' was then applied to retrieve the numbers and examples of manually asserted annotations; these include annotations contributed by ARUK-UCL and by other groups. Examples of annotations including GO terms shown above are presented in Figure 1.

New additions and revisions of the biological process terms involving 'neuron projection' naturally led to revisions of the related cellular component terms (Figure 5). In parallel, revisions of the dendrite-related component terms were additionally prompted by RGD requests for new GO terms to describe 'primary dendrite', 'distal dendrite', and 'dendritic spine origin'. This also led to a revision of the 'dendritic tree' (GO:0097447), 'dendritic branch' (GO:0044307), 'dendrite' (GO:0030425), and 'basal dendrite' (GO:0097441) terms; specifically, of their definitions and relations to other GO terms in the ontology. Among others, several of the prioritised gene products (Table S1) have been associated with descendants of the 'neuron projection' GO term (Figure 5, green boxes).

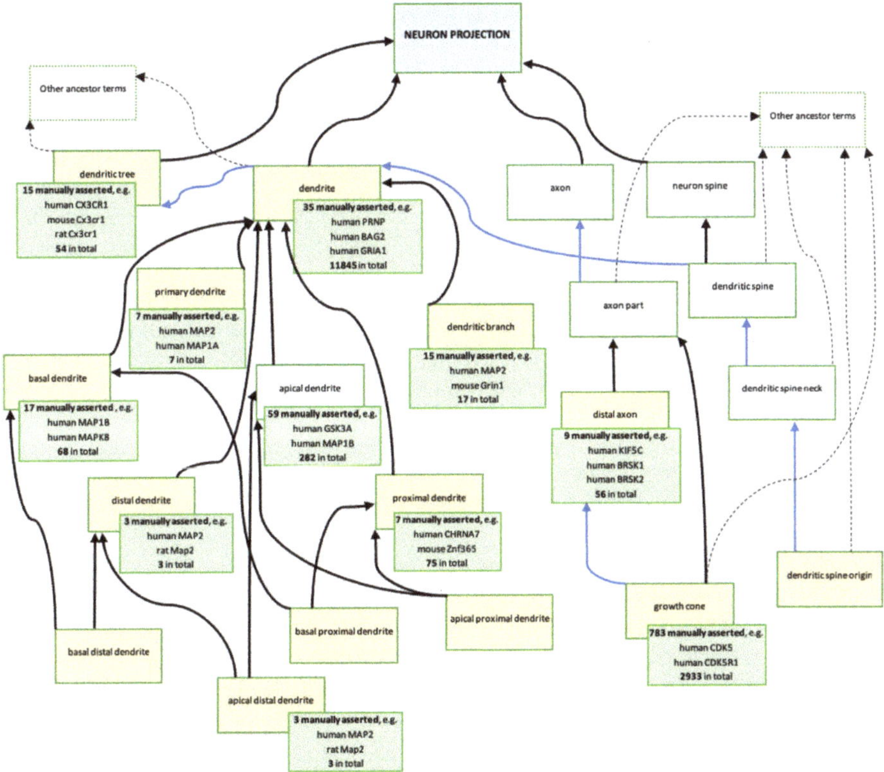

Figure 5. Descendants of cellular component 'neuron projection' GO terms created or modified, as a result of this ARUK-UCL project. Ontology: The blue box represents the ancestor term 'neuron projection' (GO:0043005). The yellow boxes represent the terms created, or modified, as a result of the ARUK-UCL project (Table S6). The majority of these are new terms; five terms were modified: 'dendritic tree' (GO:0097447), 'dendrite' (GO:0030425), 'basal dendrite' (GO:0097441), 'dendritic branch' (GO:0044307), and 'growth cone' (GO:0030426). The white boxes represent related terms in the ontology; no changes have been made to these terms. Arrows indicate the following ontology relationships: black—is_a, blue—part_of. Annotations: Each green box provides information about the gene products associated with the GO term in the yellow box, which it overlaps (data from QuickGO [51–53]—accessed: 20 November 2018). For the total numbers of annotations, which is provided in the green boxes, the following QuickGO filter was used: 'GO terms → Options → Use these terms as an exact match'; an additional filter 'Evidence → ECO:0000352: evidence used in manual assertion' was then applied to retrieve the numbers and examples of manually asserted annotations. The green boxes include examples of gene products annotated by ARUK-UCL and other groups.

A major change to the cellular component 'neuron projection' branch, which resulted from the ARUK-UCL literature curation and RGD new term requests, concerned the conflicting definitions of 'dendrite' and 'dendritic tree', which previously stated that both of these neuron projections were branched, suggesting that both terms referred to the same structure. These definitions have now been modified in order to distinguish the terms from each other. The first sentence in the 'dendrite' definition has been changed from 'a neuron projection that has a short, tapering, often branched morphology ... ' to 'a neuron projection that has a short, tapering, morphology ... '. Whereas, the 'dendritic tree' definition now includes the sentence, 'the entire complement of dendrites for a neuron, consisting of the primary dendrite and all its branches'. In addition, it was also necessary to revise the relationships between the different descendants of 'dendritic tree' as well as with its parent terms (Figure 5). The GO term 'dendritic tree' now has the is_a parent 'neuron projection', and the part_of parent 'somatodendritic compartment', and the direct is_a parent 'cell part' has been removed. In contrast, the term 'somatodendritic compartment' is no longer a direct part_of the parent of 'dendrite', which instead has the is_a parent 'plasma membrane-bound cell projection part' and the part_of parent 'dendritic tree'.

Furthermore, as a result of the ARUK-UCL literature curation, even more specific 'dendrite' terms were added, including 'apical distal dendrite' (GO:0150014), 'apical proximal dendrite' (GO:0150015), 'basal distal dendrite' (GO:0150016), and 'basal proximal dendrite' (GO:0150017). The creation of these new terms was prompted by the necessity to capture the specific neuronal localisation of the MAP2 gene product, which was prioritised for annotation of tau-relevant biology (Table S1), to 'apical distal dendrite' (Figure 6). The RGD also requested the new term 'distal axon' (GO:0150034), which led to revisions of the axonal 'neuron projection' GO branch and resulted in the modification of definition and relations of 'growth cone' (GO:0030426) (Figure 5).

Gene Product	Symbol	Qualifier	GO Term	Evidence	Reference	With / From	Taxon	Assigned By	Annotation Extension
UniProtKB:P11137	MAP2	part_of	GO:0150014 apical distal dendrite	ECO:0000250 ISS	GO_REF:0000024	UniProtKB:P15146	9606 Homo sapiens	ARUK-UCL	part_of (CL:1001571)
UniProtKB:P15146	Map2	part_of	GO:0150014 apical distal dendrite	ECO:0000314 IDA	PMID:3517689		10116 Rattus norvegicus	ARUK-UCL	part_of (CL:1001571)

Figure 6. Rat Map2 and human MAP2 GO annotations to 'apical distal dendrite' (GO:0150014) assigned by ARUK-UCL. (Screenshot from QuickGO [51–53]—accessed: 31 October 2018).

3.3.3. Concerted Effort of GO Consortium Members: Revisions of 'Synaptic Vesicle Endocytosis'

The 'synaptic vesicle endocytosis' GO term was originally defined as dependent on clathrin; however, not all synaptic vesicle endocytosis is clathrin-mediated [69,70]. Some of the GO annotations that had been made to 'synaptic vesicle endocytosis' may not have been based on clathrin-mediated mechanisms; the gene products may have been involved in an alternative mechanism of 'bulk synaptic vesicle endocytosis'. Consequently, GO biocurators agreed to review 97 experimentally evidenced annotations to 'synaptic vesicle endocytosis' (GO:0048488) and its child terms, including '(positive/negative) regulation of synaptic vesicle endocytosis' terms. At least a third of the 'synaptic vesicle endocytosis' annotations lacked experimental evidence that confirmed the involvement of clathrin in the curated data. There was unanimous agreement among the GO contributors that the definition should be made more general, in order to encompass the mechanisms of synaptic vesicle import independent of clathrin. Additionally, in order to allow for more specific annotations, we created two new child terms of 'synaptic vesicle endocytosis' (GO:0048488): 'clathrin-dependent synaptic vesicle endocytosis' (GO:0150007) and 'bulk synaptic vesicle endocytosis' (GO:0150008). These terms are distinguished by their definitions: GO:0150007: 'Clathrin-dependent endocytosis of presynaptic membrane regions comprising synaptic vesicles' membrane constituents. This is a relatively slow process occurring in the range of tens of seconds.'; and: GO:0150008: 'Endocytosis of large regions of presynaptic membrane after intense stimulation-mediated fusion of multiple synaptic vesicles. Bulk endocytosis is triggered by high loads of membrane addition through exocytosis of synaptic vesicles and elevated concentration of calcium in the presynapse'.

3.3.4. Challenges Related to Emphasis on 'Normal' Processes

The GO has been designed to capture only 'normal' processes, functions, and cellular locations. When annotating gene products relevant to a specific disease, this presents a major challenge, as it is necessary to determine what is 'normal'. For the annotation of AD-relevant gene products, this was a significant hurdle, particularly when capturing the roles of amyloid-beta species, including monomers, dimers, and oligomers (Table S2). This annotation project recognised that 'ageing' (GO:0007568) is a normal physiological process and consequently teased out data that described the association of dementia with ageing and the involvement of amyloid-beta in AD pathology [71–75]. Furthermore, we decided that it was necessary to capture the evidence that amyloid-beta (as well as tau) has normal physiological roles at the synapse [66,67], and, consequently, the normal roles of amyloid-beta complexes have been annotated as part of this project [62,66,76]. GO annotations describing age-related effects associated with amyloid-beta dimers and/or oligomers were also captured [71–75].

On the other hand, the literature review also revealed that there are some types of amyloid-beta oligomers, specifically the so-called amyloid-beta globulomers [77,78] and amylospheroids [79,80], which have only ever been observed in pathological cases; consequently, their roles have not been captured in GO.

3.4. Impact of Improved GO Annotation on Data Analysis

In order to evaluate the impact of this annotation project on high-throughput data interpretation, we undertook a functional analysis (Tables S7–S10) of a hippocampal proteomic dataset [37]. This dataset identified the hippocampal proteins that were differentially expressed in AD versus age-matched controls (Table S8), and analysed using the current (November 2018) GO annotation and ontology files preloaded in g:GOSt [15] as well as archived files (December 2016).

The previous analysis by Xu et al. [37] had identified a variety of enriched pathways in this dataset, including innate immune response, carbohydrate metabolism, and a variety of specific signalling pathways, such as those involved in the regulation of apoptosis and cell cycle, and pathways leading to neurotransmitter synthesis. Both our 2016 and 2018 g:GOSt analyses also identified these processes, as well as other neurologically-relevant terms such as synaptic signalling and axonal transport (Table S7). By grouping the enriched GO terms into broad classes (Tables S9 and S10), it was apparent that the majority of the enriched terms were associated with transport (i.e., localisation, transport, vesicle and vacuole terms), as demonstrated, for instance, by annotations to microtubule-related terms (Table S7, filter on column A by 'microtubule'), in addition to metabolism and the organisation and biogenesis of specific cellular components.

A comparison between the GO terms enriched using the 2016 versus the 2018 files (Table S11) demonstrates how the continued contribution of annotations to the GO Consortium resource and the revision of the ontology continues to impact the analysis of high-throughput datasets. The 2016 to 2018 comparison identified a difference between these analyses, with 84 terms identified only in the 2016 analysis, and 181 enriched GO terms only present in the 2018 analysis (Table S11A, column F, rows 5, 7, 14). Of particular note is the increase in the number of hippocampal proteins associated with immune system terms: using the 2016 GO files, only antigen presentation terms were identified as dysregulated in the hippocampus, while the 2018 analysis identified that 23% of dysregulated hippocampal proteins may be contributing to an immune response, with a suggestion that neutrophils are mediating this effect (Table S7, filter on column E by 'immune system process'). Similarly, the GO term 'neuron death' is only enriched when using the 2018 GO files, with 4% of hippocampal proteins associated with this term (Table S7, filter on column E by 'neuron death').

To demonstrate the impact of the ARUK-UCL project on functional analysis (Tables S7–S10), we repeated the above analyses using the same hippocampal protein list, but with the addition of all the proteins that we had prioritised for annotation (the 'priority protein-seeded hippocampal' list). However, 10 of these proteins (CLU, DCTN1, FYN, GSK3A, GSK3B, HSP90AB1, HSPG2, PIN1, PPP2CA, and ROCK2) were present in both the hippocampal protein and priority protein lists (Table S8). While the same 633 GO terms can be seen as enriched in both the 2016 and 2018 analyses, for this 'priority protein-seeded hippocampal' list, there are 356 GO terms that are only enriched using the 2018 GO files (Tables S10 and S11). As the ARUK-UCL project has used many of the terms as only enriched in the 2018 analysis to capture the role of dementia-relevant gene products, it is likely that some of these differences can be attributed to this project. Notably, amyloid-beta clearance and neuroinflammation-relevant GO terms are enriched in the 2018 analysis, but not the 2016 analysis (Table S7, filter on column E by 'amyloid-beta clearance' and 'neuroinflammation'). An investigation into the source of the 'amyloid-beta clearance' annotations associated with human proteins confirms that the ARUK-UCL project is responsible, either directly or indirectly (through the annotation of an ortholog), for all but one of these annotations (data from QuickGO—accessed: 20 November 2018). In addition, many of enriched GO terms that are identified only in the 2018 files with this 'priority protein-seeded hippocampal' list, are relevant to signalling and transport, reflecting the role of many of our priority proteins in these areas.

4. Discussion

Controlled biomedical vocabularies, or ontologies, have been used to semantically capture and describe AD-specific knowledge to enhance its sharing and exchange, as well as to aid collaborative funding and research efforts [81–84]. The use of ontologies has also been suggested to help with the prioritisation of genes for neuroimaging studies [39], as well as with the diagnosis of cognitive impairments [85]. With one exception [6], these vocabularies have been intended to capture and describe the domain of knowledge specific to disease, or phenotype, and not normal physiological processes.

Here, we used Gene Ontology (GO), a controlled biological vocabulary, to describe the normal roles of gene products implicated in AD and dementia. The advantage of GO is that it is a well-established and regularly maintained resource, which is commonly used for enrichment analyses across biomedical fields [40,86–88] and even in clinical practice [89]. In addition, during curation, we focus on a specific biological area, such as amyloid-beta receptors, and we annotate whole research articles using GO to capture the roles of all the gene products described. Thus, through our project, we enhanced GO more broadly, which not only benefits the Alzheimer's disease researchers, but also the biomedical research communities overall.

4.1. Cellular Events Underlying Dementia

Amyloid-beta plaques and phospho-tau neurofibrillary tangles (NFTs) are the pathological hallmarks of dementias, and historically, they were believed to be the underlying causes of the dementia associated with AD [22,90–92]. It is now understood that in terms of amyloid-beta, AD pathology is exacerbated by its oligomers (amyloid-beta intermediates formed from the monomeric peptides prior to sedimentation into plaques) rather than the plaques, and the oligomers' neurotoxicity depends on their concentrations and/or the presence of specific forms, such as globulomers or amylospheroids [71–75,77–80]. Whereas, phospho-tau has been demonstrated to contribute to dementia pathology due to its prion-like seeding activity [93–96], which allows the aberrantly phosphorylated tau aggregates to spread to as-yet unaffected brain regions, thus advancing the clinical symptoms. Yet, in recent years, scientific research has generated evidence, which demonstrates that the formation of these neurotoxic protein aggregates results from prior impairments in underlying cellular processes [27]. When these cellular brain processes are malfunctioning, this eventually leads to disruptions in protein homeostasis and the clinical manifestations of dementia [30]. Consequently, the plaques and/or NFTs that are characteristic of the brain pathology associated with advanced-stage dementia are currently believed to be the by-products of compensatory processes, which become activated in the brain at the cellular level to address the imbalances underlying the disease [30,97].

Therefore, until disease progression and its mechanisms are fully elucidated and understood, it should not be assumed that the fibrils of amyloid-beta and/or phospho-tau cause dementia, or specifically AD. Studies on post-mortem brain tissue sections of healthy controls revealed vast deposits of amyloid-beta in the brain despite a lack of symptoms [98]. Despite findings from animal models and cell culture studies, it is currently understood that the onset of dementia pathology occurs decades before the first symptoms manifest themselves [27], supporting the hypotheses that protein homeostasis mechanisms and other cellular functions begin to deteriorate long before amyloid-beta deposition and NFT formation become apparent [30,97]. Furthermore, research focus has been shifting to other cell types in addition to neurons, specifically glia and the cells of the brain vasculature, as evidence suggests that impaired communication between these cell types and neurons also promotes disease progression [27].

This very tight link of amyloid-beta and phospho-tau with dementia pathology and AD has posed major challenges for this GO annotation project. The purpose of GO is to capture the 'normal' physiological roles of gene products, not their roles in disease pathology. Although the impact of amyloid-beta and phospho-tau can be captured using phenotype ontologies, a combined disease and GO database that applies GO terms in the context of a pathological environment is currently missing from the resources that are available to the biomedical research communities. The question of what is 'normal', especially in the context of amyloid-beta and a lack of clear boundaries between ageing-related biological aspects and disease processes, still exists [98]. Therefore, the first goal of this project was to establish whether any roles of amyloid-beta, and, more importantly, its multimeric forms, could be captured using GO.

A literature search revealed that in normal physiological conditions, amyloid-beta levels are regulated by neuronal activity [99–101], and that amyloid-beta in turn modulates synaptic plasticity in neural circuits [66,102]. Amyloid-beta has also been shown to positively affect neuronal growth [62,76]. In addition, scientific findings have demonstrated that the effects of amyloid-beta are age-related [71–75]. 'Ageing' (GO:0007568) is a normal physiological process involving the 'loss of functions such as resistance to disease, homeostasis, and fertility, as well as wear and tear', i.e., aspects that are directly translatable to impairments in cellular processes underlying dementia. Based on this published evidence, we curated 18 new Complex Portal (CP) entries for physiologically occurring human, mouse, and rat amyloid-beta dimers and oligomers (Table S2), and we annotated their roles using GO. Our annotations will enable users to extract a list of the proteins known to interact with amyloid-beta. Furthermore, the detection of dysregulated amyloid-beta-interacting proteins within '-omics' datasets will be facilitated.

In addition to amyloid-beta monomers, dimers, and oligomers as well as tau, 74 gene products were prioritised for GO annotation, following dementia experts' advice, and these included proteins and protein complexes interacting with either amyloid-beta or tau (Table S1). As we take a process-focussed approach to curation, we annotate research articles fully and capture the roles of all of the gene products that were investigated and described in any given research paper. Consequently, the total number of gene products annotated in the context of amyloid-beta and tau biology was 561, of which 308 were human, i.e., far more gene products were annotated than those in the prioritised list. The majority of the annotated human gene products were proteins, but some protein complexes (including six types of amyloid-beta complexes and two of their interacting partners prioritised for annotation) and two microRNAs were also annotated as a part of this project (Table S3A). These 561 gene products have been associated with 3886 GO terms, based on published evidence; of these, 308 human gene products have been associated with 2770 GO terms (Table 1). Hence, by contributing these new ARUK-UCL annotations, we have greatly expanded the representation of dementia-relevant knowledge in GO, incorporating new GO annotations for four (gene products of *CLU*, *PICALM*, *APOE*, and *BIN1*) of the 21 gene products previously annotated as a part of the AD-focussed initiative at the University of Toronto [39]. Furthermore, prior to this project, there were no GO terms associated with either the complement receptor type 3 complex (CP-1826), or the amylin receptor 3 complex (CP-3187), which are both products of genes prioritised for annotation (Table S1). As a result of this project, there are currently 14 and 21 annotations, respectively, for these two complexes in the GO database (data from QuickGO, accessed: 30 October 2018).

Collectively, the 3886 annotations describing the normal physiological roles of the prioritised gene products interacting with either amyloid-beta or tau, as well as other proteins, complexes, and microRNAs implicated in processes impaired in AD, have greatly expanded the representation of dementia-relevant neurological processes, functions, and cellular compartments in GO. Over the past two years, increasing numbers of annotations have been included in the GO Consortium annotation files, and we have demonstrated that this improves the interpretation of AD proteomic data (Tables S7–S10). Our focus on dementia-relevant processes has contributed to this, for example, by creating 'regulation of neuron death' annotations to some of the prioritised proteins. By capturing published knowledge about these processes, we are providing the AD and dementia research communities with a resource that is potentially useful for diagnostic purposes and/or disease prevention. The annotation of dementia-relevant gene products involved in these early cellular processes makes knowledge available for analyses of clinical or experimental datasets related to cognition, which could help delineate the biological pathways that can be targeted for diagnostic or preventive purposes decades prior to the onset of this debilitating disorder.

4.2. Representing Neurobiology Using Gene Ontology

The dementia-focussed ARUK-UCL project described here and the University of Toronto GO annotation project [39] are not the only GO initiatives intended to improve the representation of the neurobiological knowledge domain in GO. Our team was previously funded to capture the knowledge relevant to Parkinson's disease (annotations assigned by ParkinsonsUK-UCL) [40,103,104], and we have also participated in a large multi-center collaborative effort to curate published information about the synapse (annotations assigned by SynGO and SynGO-UCL) [105]. Furthermore, our interactions with other members of the GO Consortium, including biocurators of model organism resources, have been very important for the success of this project, especially in the context of GO development and revisions.

Overall, the project has resulted in the addition of 71 new GO terms to the ontology (Table S6; data from AmiGO2, accessed: 30 May 2018). These GO terms were created for the annotation of research findings involving the prioritised gene products, but many of them have also been used for manual annotation by other curation teams, as well as included in automatic annotation approaches [64]. For instance, the ARUK-UCL-contributed 'neuron projection arborisation' (GO:0140058) GO term has been used in a total of 90 annotations, of which 11 are manual annotations, although only three of these annotations resulted from this project (data from QuickGO, accessed: 11 October 2018). This shows that other members of the GO Consortium are also using GO terms resulting from this project for the annotation of their biological domains. This is just one example of how work resulting from one focussed GO annotation initiative aids the work of other curation teams, and thus benefits wider communities of researchers who use GO in their data analyses.

In addition to contributing new GO terms, work on this project also led to the revision of thirteen previously existing GO terms, as well as the logical relations between them (Table S6). A number of these revisions were prompted because new terms were required to annotate gene products prioritised in this project. However, the revisions were improved by additional new term requests and suggestions from biocurators at the Rat Genome Database (RGD). GO branches that were revised and enhanced through these concerted efforts organised the knowledge about descendants of the biological process 'neuron projection development' (GO:0031175) GO terms (Figure 4) and the cellular component 'neuron projection' (GO:0043005) GO terms (Figure 5). These improvements to the ontology are crucial for the accurate representation of the dementia-relevant neurobiological domain using GO. In fact, this project has resulted in 57 annotations to 'neuron projection development', or one of its descendant terms, and 249 annotations to 'neuron projection', or one of its descendants (data from QuickGO, accessed: 11 October 2018).

A close collaboration is of even greater importance when ontology revisions need to be coordinated with revisions of existing GO annotations. This was demonstrated when it became apparent that updates to the annotations associated with the GO term 'synaptic vesicle endocytosis' were required. 'Synaptic vesicle endocytosis' was originally defined as a clathrin-mediated process; however, FlyBase [106] biocurators identified that not all of the endocytic processes occurring in the synapse depend on clathrin. Prior to this, there had been a total of 97 experimentally evidenced annotations to 'synaptic vesicle endocytosis' assigned by 10 different curation groups. Thus, in order to reach a GO Consortium-wide consensus about whether to broaden the definition of this term to include endocytic mechanisms not requiring the involvement of clathrin, or whether to update the GO term name to 'clathrin-dependent synaptic vesicle endocytosis', it was crucial to first review the existing annotations to determine whether the experimental support did indeed provide evidence for clathrin mediating the endocytosis. As a result of this combined effort, and following consultations with synapse biology experts, a decision was made to broaden the definition of the existing 'synaptic vesicle endocytosis' GO term, and to add two more descriptive child GO terms to capture information about the specific types of synaptic endocytosis. Biocurators then updated the annotations using these new terms if there was sufficient experimental data to support this.

Overall, the new GO terms resulting from this project have broadened the representation of the neuroscience domain in GO, making a greater number of terms available to biocurators for the GO annotation of neurological concepts. In addition, ontology revisions helped to ensure the highest possible quality of this resource by reviewing existing annotations and ontology entries describing neurobiological concepts, and by ensuring that they are accurately defined and appropriately related to each other in the GO term hierarchy.

5. Conclusions

In order to understand the cellular processes underlying Alzheimer's disease and other dementias, numerous transcriptomic, proteomic, and genome-wide association (GWA) studies have been conducted [2,4–6]. The analyses and interpretation of results from such high-throughput analyses greatly rely on functional annotation data provided by resources such as GO [10], KEGG [12], or Reactome [11], as well as on appropriate experimental design, the analysis tools used, and the settings applied. Among these, analyses performed using the latter two resources can yield informative results only if the dataset being analysed contains gene products with known functions in biological pathways. In contrast, the GO resource captures the cellular roles and locations of a higher number of gene products, even when the molecular functions of gene products are unknown. Therefore, GO is suitable for analyses of datasets that are likely to contain gene products whose specific function and biological role have not yet been fully investigated. Yet, prior to this project, the applicability of GO for the analysis of dementia-relevant neurological datasets was limited, because there had been no previous efforts to comprehensively annotate gene products with roles in this biological domain, with the exception of a project undertaken by the University of Toronto [39] that focussed only on a small group of AD risk genes.

So far, our commitment to improving the GO resource for dementia-relevant research has focussed on amyloid-beta, the microtubule-associated protein tau and its interacting partners, as described here. Our ongoing and future efforts aim to capture the biology of neuroinflammatory processes, e.g., the activation of glial cells in response to an inflammatory stimulus [107], the functional information about proteins involved in these processes, as well as knowledge about the microRNAs that regulate the expression of these proteins. As an initial focus, we have been annotating the roles of proteins involved in the biology of glial cells, primarily microglia; the lists of proteins that we have prioritised for annotation are available on our website [108]. Additionally, we have begun the GO annotation of microRNAs involved in the regulation of the expression of these proteins, primarily focussing on capturing experimentally evidenced microRNA–target interaction.

Through our systematic and focussed development of GO terms relevant to dementia and revision of previously existing neurological GO terms, as well as by detailed manual GO annotation of gene products implicated in dementia, we have enhanced the suitability of the GO resource for analysis of neurological '-omics' datasets. These ongoing and future contributions to the GO resource will help provide insights into the molecular bases of dementia, thus supporting the development of treatments and of tools for early diagnosis.

All of the GO terms and annotations are freely available and can be downloaded from QuickGO [51–53] and AmiGO [54,55]. Information on protein complexes can be found at the Complex Portal [63,109]. We encourage scientists to become involved in the GO annotation of their own papers and/or gene products of interest. In order to contribute to GO, please contact UCL Functional Gene Annotation [108], GOA (https://www.ebi.ac.uk/GOA/contactus) [110] or the GO Consortium (http://geneontology.org/page/contributing-go) [111].

Supplementary Materials: The following are available online at http://www.mdpi.com/2073-4425/9/12/593/s1, Table S1: Interacting partners of amyloid-beta(A) and microtubule-associated protein tau (B) prioritised for GO annotation, Table S2: Amyloid-beta monomers and/or oligomers, Table S3: All ARUK-UCL GO annotations, Table S4: All manually-curated GO annotations for the prioritised gene products (Table S1), contributed by ARUK-UCL, as well as by other groups (Data from QuickGO - accessed: 31 October 2018; filtered by database identifiers from Table S1 and evidence used in manual assertion (ECO:0000352), Table S5: Manually curated neurological GO term annotations for the prioritised gene products, contributed by ARUK-UCL, as well as by other groups (Data from QuickGO, accessed: 31 October 2018), Table S6: GO development and revisions, Table S7: Summary of four separate g:GOSt analyses of AD hippocampal datasets. Table S8: AD hippocampal protein list, Table S9: Alzheimer's disease (AD)-relevant filter terms included in Table S7, Table S10: Number of enriched GO terms within each grouping term included in Table S7, Table S11: Summary of the number of GO terms significantly enriched in the different analyses in Table S7.

Author Contributions: Conceptualization, B.K., P.R., R.B., N.M.H. and R.C.L.; Data curation, B.K., P.R., B.H.M.M. and R.P.H.; Formal analysis, B.K.; Funding acquisition, M.J.M., H.P., D.B., R.B., N.M.H. and R.C.L.; Project administration, R.C.L.; Supervision, S.O., H.P. and R.C.L.; Writing—original draft, B.K., P.R. and R.C.L.; Writing—review & editing, B.K., P.R., B.H.M.M., H.P., D.B., R.B., N.M.H. and R.C.L.

Funding: The University College London functional annotation team is supported by ARUK-NSG2016-13, ARUK-NAS2017A-1 and the National Institute for Health Research University College London Hospitals Biomedical Research Centre. The GO resource is supported by grant from the National Human Genome Research Institute [grant U41 HG002273 to P.D. Thomas, P.W. Sternberg, S.E. Lewis, J.M. Cherry, J.A. Blake].

Acknowledgments: We are grateful to Selina Wray for her advice in the prioritisation of genes for annotation. We also thank the many biocurators, editors and other members of the GO Consortium who have contributed to the annotation of dementia-relevant genes and the development of the Gene Ontology, and the g:Profiler team. With particular thanks to Pascale Gaudet (Swiss Institute of Bioinformatics), David P. Hill (The Jackson Laboratory), Stan Laulederkind (Rat Genome Database), David Osumi-Sutherland (EMBL-EBI), Helen Attrill (FlyBase), Pim van Nierop (SynGO and Vrije Universiteit Amsterdam) and Kimberly M. Van Auken (Caltech) for their help with ontology development, and also identification of ontology domains in need of revision. We are also grateful to our expert advisors: John Hardy and Paul Whiting (UCL), Elizabeth Wu and Kelly Dakin (Alzforum), Jo Knight (University of Lancaster) and Tracy Hussell (University of Manchester), who have been involved in other ARUK discussions.

Conflicts of Interest: The authors declare no conflict of interest. The funders had no role in the design of the study; in the collection, analyses, or interpretation of data; in the writing of the manuscript, or in the decision to publish the results.

References

1. Van Cauwenberghe, C.; Van Broeckhoven, C.; Sleegers, K. The genetic landscape of Alzheimer disease: Clinical implications and perspectives. *Genet. Med.* **2016**, *18*, 421–430. [CrossRef] [PubMed]

2. Sassi, C.; Nalls, M.A.; Ridge, P.G.; Gibbs, J.R.; Ding, J.; Lupton, M.K.; Troakes, C.; Lunnon, K.; Al-Sarraj, S.; Brown, K.S.; et al. ABCA7 p.G215S as potential protective factor for Alzheimer's disease. *Neurobiol. Aging* **2016**, *46*, 235.e1–235.e9. [CrossRef] [PubMed]

3. Barnes, D.E.; Yaffe, K. The projected effect of risk factor reduction on Alzheimer's disease prevalence. *Lancet Neurol.* **2011**, *10*, 819–828. [CrossRef]

4. Cooper-Knock, J.; Kirby, J.; Ferraiuolo, L.; Heath, P.R.; Rattray, M.; Shaw, P.J. Gene expression profiling in human neurodegenerative disease. *Nat. Rev. Neurol.* **2012**, *8*, 518–530. [CrossRef] [PubMed]

5. Guerreiro, R.; Wojtas, A.; Bras, J.; Carrasquillo, M.; Rogaeva, E.; Majounie, E.; Cruchaga, C.; Sassi, C.; Kauwe, J.S.; Younkin, S.; et al. TREM2 variants in Alzheimer's disease. *N. Engl. J. Med.* **2013**, *368*, 117–127. [CrossRef] [PubMed]

6. Kang, M.G.; Byun, K.; Kim, J.H.; Park, N.H.; Heinsen, H.; Ravid, R.; Steinbusch, H.W.; Lee, B.; Park, Y.M. Proteogenomics of the human hippocampus: The road ahead. *Biochim. Biophys. Acta* **2015**, *1854*, 788–797. [CrossRef] [PubMed]

7. Guio-Vega, G.P.; Forero, D.A. Functional genomics of candidate genes derived from genome-wide association studies for five common neurological diseases. *Int. J. Neurosci.* **2017**, *127*, 118–123. [CrossRef]

8. Ebbert, M.T.; Ridge, P.G.; Kauwe, J.S. Bridging the gap between statistical and biological epistasis in Alzheimer's disease. *Biomed. Res. Int.* **2015**, *2015*, 870123. [CrossRef]

9. Cambiaghi, A.; Ferrario, M.; Masseroli, M. Analysis of metabolomic data: Tools, current strategies and future challenges for omics data integration. *Brief Bioinform.* **2017**, *18*, 498–510. [CrossRef]

10. Ashburner, M.; Ball, C.A.; Blake, J.A.; Botstein, D.; Butler, H.; Cherry, J.M.; Davis, A.P.; Dolinski, K.; Dwight, S.S.; Eppig, J.T.; et al. Gene ontology: Tool for the unification of biology. The Gene Ontology Consortium. *Nat. Genet.* **2000**, *25*, 25–29. [CrossRef]

11. Fabregat, A.; Sidiropoulos, K.; Garapati, P.; Gillespie, M.; Hausmann, K.; Haw, R.; Jassal, B.; Jupe, S.; Korninger, F.; McKay, S.; et al. The Reactome pathway knowledgebase. *Nucleic Acids Res.* **2016**, *44*, D481–D487. [CrossRef] [PubMed]

12. Kanehisa, M.; Sato, Y.; Kawashima, M.; Furumichi, M.; Tanabe, M. KEGG as a reference resource for gene and protein annotation. *Nucleic Acids Res.* **2016**, *44*, D457–D462. [CrossRef] [PubMed]

13. Orchard, S.; Ammari, M.; Aranda, B.; Breuza, L.; Briganti, L.; Broackes-Carter, F.; Campbell, N.H.; Chavali, G.; Chen, C.; del-Toro, N.; et al. The MIntAct project—IntAct as a common curation platform for 11 molecular interaction databases. *Nucleic Acids Res.* **2014**, *42*, D358–D363. [CrossRef] [PubMed]

14. Meldal, B.H.M.; Bye, A.J.H.; Gajdos, L.; Hammerova, Z.; Horackova, A.; Melicher, F.; Perfetto, L.; Pokorny, D.; Lopez, M.R.; Turkova, A.; et al. Complex Portal 2018: Extended content and enhanced visualization tools for macromolecular complexes. *Nucleic Acids Res.* **2018**. [CrossRef] [PubMed]

15. Reimand, J.; Arak, T.; Adler, P.; Kolberg, L.; Reisberg, S.; Peterson, H.; Vilo, J. g:Profiler—A web server for functional interpretation of gene lists (2016 update). *Nucleic Acids Res.* **2016**. [CrossRef] [PubMed]

16. Richardson, J.E.; Bult, C.J. Visual annotation display (VLAD): A tool for finding functional themes in lists of genes. *Mamm. Genome* **2015**, *26*, 567–573. [CrossRef] [PubMed]

17. Huang, D.W.; Sherman, B.T.; Lempicki, R.A. Systematic and integrative analysis of large gene lists using DAVID Bioinformatics Resources. *Nat. Protoc.* **2009**, *4*, 44–57. [CrossRef]

18. Alam-Faruque, Y.; Huntley, R.P.; Khodiyar, V.K.; Camon, E.B.; Dimmer, E.C.; Sawford, T.; Martin, M.J.; O'Donovan, C.; Talmud, P.J.; Scambler, P.; et al. The impact of focused Gene Ontology curation of specific mammalian systems. *PLoS ONE* **2011**, *6*, e27541. [CrossRef]

19. Patel, S.; Roncaglia, P.; Lovering, R.C. Using Gene Ontology to describe the role of the neurexin-neuroligin-SHANK complex in human, mouse and rat and its relevance to autism. *BMC Bioinform.* **2015**, *16*, 186. [CrossRef]

20. Gray, K.A.; Yates, B.; Seal, R.L.; Wright, M.W.; Bruford, E.A. Genenames.org: The HGNC resources in 2015. *Nucleic Acids Res.* **2015**, *43*, D1079–D1085. [CrossRef]

21. Masino, A.J.; Dechene, E.T.; Dulik, M.C.; Wilkens, A.; Spinner, N.B.; Krantz, I.D.; Pennington, J.W.; Robinson, P.N.; White, P.S. Clinical phenotype-based gene prioritization: An initial study using semantic similarity and the human phenotype ontology. *BMC Bioinform.* **2014**, *15*, 248. [CrossRef] [PubMed]

22. Kametani, F.; Hasegawa, M. Reconsideration of amyloid hypothesis and tau hypothesis in Alzheimer's Disease. *Front. Neurosci.* **2018**, *12*, 25. [CrossRef] [PubMed]

23. Hardy, J.; Allsop, D. Amyloid deposition as the central event in the aetiology of Alzheimer's disease. *Trends Pharmacol. Sci.* **1991**, *12*, 383–388. [CrossRef]

24. Goedert, M. Tau protein and the neurofibrillary pathology of Alzheimer's disease. *Ann. N. Y. Acad. Sci.* **1996**, *777*, 121–131. [CrossRef] [PubMed]

25. Goedert, M. Tau protein and the neurofibrillary pathology of Alzheimer's disease. *Trends Neurosci.* **1993**, *16*, 460–465. [CrossRef]

26. Selkoe, D.J.; Hardy, J. The amyloid hypothesis of Alzheimer's disease at 25 years. *EMBO Mol. Med.* **2016**, *8*, 595–608. [CrossRef] [PubMed]

27. De Strooper, B.; Karran, E. The cellular phase of Alzheimer's disease. *Cell* **2016**, *164*, 603–615. [CrossRef] [PubMed]

28. Schott, J.M.; Revesz, T. Inflammation in Alzheimer's disease: Insights from immunotherapy. *Brain* **2013**, *136*, 2654–2656. [CrossRef] [PubMed]

29. Nelson, A.R.; Sweeney, M.D.; Sagare, A.P.; Zlokovic, B.V. Neurovascular dysfunction and neurodegeneration in dementia and Alzheimer's disease. *Biochim. Biophys. Acta* **2016**, *1862*, 887–900. [CrossRef] [PubMed]

30. Kundra, R.; Ciryam, P.; Morimoto, R.I.; Dobson, C.M.; Vendruscolo, M. Protein homeostasis of a metastable subproteome associated with Alzheimer's disease. *Proc. Natl. Acad. Sci. USA* **2017**, *114*, E5703–E5711. [CrossRef] [PubMed]

31. Cuyvers, E.; Sleegers, K. Genetic variations underlying Alzheimer's disease: Evidence from genome-wide association studies and beyond. *Lancet Neurol.* **2016**, *15*, 857–868. [CrossRef]

32. Shen, L.; Jia, J. An Overview of genome-wide association studies in Alzheimer's disease. *Neurosci. Bull.* **2016**, *32*, 183–190. [CrossRef] [PubMed]

33. De Matos, M.R.; Ferreira, C.; Herukka, S.K.; Soininen, H.; Janeiro, A.; Santana, I.; Baldeiras, I.; Almeida, M.R.; Lleo, A.; Dols-Icardo, O.; et al. Quantitative genetics validates previous genetic variants and identifies novel genetic players influencing Alzheimer's disease cerebrospinal fluid biomarkers. *J. Alzheimers Dis.* **2018**, *66*, 639–652. [CrossRef] [PubMed]

34. Abu-Rumeileh, S.; Mometto, N.; Bartoletti-Stella, A.; Polischi, B.; Oppi, F.; Poda, R.; Stanzani-Maserati, M.; Cortelli, P.; Liguori, R.; Capellari, S.; Parchi, P. Cerebrospinal fluid biomarkers in patients with frontotemporal dementia spectrum: A single-center study. *J. Alzheimers Dis.* **2018**, *66*, 551–563. [CrossRef] [PubMed]

35. Verheijen, J.; Sleegers, K. Understanding Alzheimer disease at the interface between genetics and transcriptomics. *Trends Genet.* **2018**, *34*, 434–447. [CrossRef] [PubMed]

36. Espuny-Camacho, I.; Arranz, A.M.; Fiers, M.; Snellinx, A.; Ando, K.; Munck, S.; Bonnefont, J.; Lambot, L.; Corthout, N.; Omodho, L.; et al. Hallmarks of Alzheimer's disease in stem-cell-derived human neurons transplanted into mouse brain. *Neuron* **2017**, *93*, 1066–1081. [CrossRef] [PubMed]

37. Xu, J.; Patassini, S.; Rustogi, N.; Riba-Garcia, I.; Hale, B.D.; Phillips, A.M.; Waldvogel, H.; Haines, R.; Bradbury, P.; Stevens, A.; et al. Regional protein expression in human Alzheimer's brain correlates with disease severity. *bioRxiv* **2018**. [CrossRef]

38. Kunkle, B.W.; Grenier-Boley, B.; Sims, R.; Bis, J.C.; Naj, A.C.; Boland, A.; Vronskaya, M.; van der Lee, S.J.; Amlie-Wolf, A.; Bellenguez, C.; et al. Meta-analysis of genetic association with diagnosed Alzheimer's disease identifies novel risk loci and implicates Abeta, Tau, immunity and lipid processing. *bioRxiv* **2018**. [CrossRef]

39. Patel, S.; Park, M. Gene prioritization for imaging genetics studies using gene ontology and a stratified false discovery rate approach. *Front. Neuroinform.* **2016**, *10*, 14. [CrossRef]

40. Denny, P.; Feuermann, M.; Hill, D.P.; Lovering, R.C.; Plun-Favreau, H.; Roncaglia, P. Exploring autophagy with Gene Ontology. *Autophagy* **2018**, *14*, 419–436. [CrossRef]

41. Jarosz-Griffiths, H.H.; Noble, E.; Rushworth, J.V.; Hooper, N.M. Amyloid-beta receptors: The good, the bad, and the prion protein. *J. Biol. Chem.* **2016**, *291*, 3174–3183. [CrossRef]

42. Guo, T.; Noble, W.; Hanger, D.P. Roles of tau protein in health and disease. *Acta Neuropathol.* **2017**, *133*, 665–704. [CrossRef]

43. The Gene Ontology Consortium. Expansion of the Gene Ontology knowledgebase and resources. *Nucleic Acids Res.* **2017**, *45*, D331–D338. [CrossRef]

44. NCBI PubMed. Available online: https://www.ncbi.nlm.nih.gov/pubmed/ (accessed on 30 May 2018).

45. Balakrishnan, R.; Harris, M.A.; Huntley, R.; Van Auken, K.; Cherry, J.M. A guide to best practices for Gene Ontology (GO) manual annotation. *Database* **2013**, *2013*, bat054. [CrossRef]

46. Huntley, R.P.; Harris, M.A.; Alam-Faruque, Y.; Blake, J.A.; Carbon, S.; Dietze, H.; Dimmer, E.C.; Foulger, R.E.; Hill, D.P.; Khodiyar, V.K.; et al. A method for increasing expressivity of Gene Ontology annotations using a compositional approach. *BMC Bioinform.* **2014**, *15*, 155. [CrossRef]

47. Pundir, S.; Martin, M.J.; O'Donovan, C.; UniProt Consortium. UniProt tools. *Curr. Protoc. Bioinform.* **2016**, *53*, 1–15.

48. The RNAcentral Constortium. RNAcentral: A hub of information for non-coding RNA sequences. *Nucleic Acids Res.* **2018**. [CrossRef]

49. Gene Ontology Consortium. Gene Ontology Evidence Code Documentation. 2016. Available online: http://www.geneontology.org/page/guide-go-evidence-codes (accessed on 10 May 2017).

50. Huntley, R.P.; Sawford, T.; Mutowo-Meullenet, P.; Shypitsyna, A.; Bonilla, C.; Martin, M.J.; O'Donovan, C. The GOA database: Gene Ontology annotation updates for 2015. *Nucleic Acids Res.* **2015**, *43*, D1057–D1063. [CrossRef]

51. Huntley, R.P.; Binns, D.; Dimmer, E.; Barrell, D.; O'Donovan, C.; Apweiler, R. QuickGO: A user tutorial for the web-based Gene Ontology browser. *Database* **2009**, *2009*, bap010. [CrossRef]

52. Binns, D.; Dimmer, E.; Huntley, R.; Barrell, D.; O'Donovan, C.; Apweiler, R. QuickGO: A web-based tool for Gene Ontology searching. *Bioinformatics* **2009**, *25*, 3045–3046. [CrossRef]

53. EMBL-EBI, QuickGO. Available online: https://www.ebi.ac.uk/QuickGO/ (accessed on 30 October 2018).

54. Carbon, S.; Ireland, A.; Mungall, C.J.; Shu, S.; Marshall, B.; Lewis, S. AmiGO: Online access to ontology and annotation data. *Bioinformatics* **2009**, *25*, 288–289. [CrossRef]

55. AmiGO 2. Available online: http://amigo.geneontology.org/amigo/landing (accessed on 30 May 2018).

56. Brown, G.R.; Hem, V.; Katz, K.S.; Ovetsky, M.; Wallin, C.; Ermolaeva, O.; Tolstoy, I.; Tatusova, T.; Pruitt, K.D.; Maglott, D.R.; Murphy, T.D. Gene: A gene-centered information resource at NCBI. *Nucleic Acids Res.* **2015**, *43*, D36–D42. [CrossRef]

57. Newman, V.; Moore, B.; Sparrow, H.; Perry, E. The Ensembl Genome Browser: Strategies for accessing eukaryotic genome data. *Methods Mol. Biol.* **2018**, *1757*, 115–139.

58. Griffiths-Jones, S.; Grocock, R.J.; van Dongen, S.; Bateman, A.; Enright, A.J. miRBase: microRNA sequences, targets and gene nomenclature. *Nucleic Acids Res.* **2006**, *34*, D140–D144. [CrossRef]

59. Kibbe, W.A.; Arze, C.; Felix, V.; Mitraka, E.; Bolton, E.; Fu, G.; Mungall, C.J.; Binder, J.X.; Malone, J. Disease Ontology 2015 update: An expanded and updated database of human diseases for linking biomedical knowledge through disease data. *Nucleic Acids Res.* **2015**, *43*, D1071–D1078. [CrossRef]

60. GitHub. Available online: https://github.com/ (accessed on 30 October 2018).

61. AmiGO 2. Available online: http://amigo.geneontology.org/amigo/landing (accessed on 30 May 2018).

62. Lopez-Toledano, M.A.; Shelanski, M.L. Neurogenic effect of beta-amyloid peptide in the development of neural stem cells. *J. Neurosci.* **2004**, *24*, 5439–5444. [CrossRef]

63. EMBL-EBI. Complex Portal. Available online: https://www.ebi.ac.uk/complexportal/home (accessed on 30 October 2018).

64. Huntley, R.P.; Sawford, T.; Martin, M.J.; O'Donovan, C. Understanding how and why the Gene Ontology and its annotations evolve: The GO within UniProt. *Gigascience* **2014**, *3*, 4. [CrossRef]

65. Citri, A.; Malenka, R.C. Synaptic plasticity: Multiple forms, functions, and mechanisms. *Neuropsychopharmacology* **2008**, *33*, 18–41. [CrossRef]

66. Puzzo, D.; Privitera, L.; Leznik, E.; Fa, M.; Staniszewski, A.; Palmeri, A.; Arancio, O. Picomolar amyloid-beta positively modulates synaptic plasticity and memory in hippocampus. *J. Neurosci.* **2008**, *28*, 14537–14545. [CrossRef]

67. Perez-Nievas, B.G.; Stein, T.D.; Tai, H.C.; Dols-Icardo, O.; Scotton, T.C.; Barroeta-Espar, I.; Fernandez-Carballo, L.; de Munain, E.L.; Perez, J.; Marquie, M.; et al. Dissecting phenotypic traits linked to human resilience to Alzheimer's pathology. *Brain* **2013**, *136*, 2510–2526. [CrossRef]

68. Shimoyama, M.; De Pons, J.; Hayman, G.T.; Laulederkind, S.J.; Liu, W.; Nigam, R.; Petri, V.; Smith, J.R.; Tutaj, M.; Wang, S.J.; et al. The Rat Genome Database 2015: Genomic, phenotypic and environmental variations and disease. *Nucleic Acids Res.* **2015**, *43*, D743–D750. [CrossRef]

69. Milosevic, I. Revisiting the role of clathrin-mediated endoytosis in synaptic vesicle recycling. *Front. Cell. Neurosci.* **2018**, *12*, 27. [CrossRef]

70. Gan, Q.; Watanabe, S. Synaptic vesicle endocytosis in different model systems. *Front. Cell. Neurosci.* **2018**, *12*, 171. [CrossRef]

71. Chen, G.; Chen, K.S.; Knox, J.; Inglis, J.; Bernard, A.; Martin, S.J.; Justice, A.; McConlogue, L.; Games, D.; Freedman, S.B.; et al. A learning deficit related to age and beta-amyloid plaques in a mouse model of Alzheimer's disease. *Nature* **2000**, *408*, 975–979. [CrossRef]

72. Janus, C.; Pearson, J.; McLaurin, J.; Mathews, P.M.; Jiang, Y.; Schmidt, S.D.; Chishti, M.A.; Horne, P.; Heslin, D.; French, J.; et al. A beta peptide immunization reduces behavioural impairment and plaques in a model of Alzheimer's disease. *Nature* **2000**, *408*, 979–982. [CrossRef]

73. Westerman, M.A.; Cooper-Blacketer, D.; Mariash, A.; Kotilinek, L.; Kawarabayashi, T.; Younkin, L.H.; Carlson, G.A.; Younkin, S.G.; Ashe, K.H. The relationship between Abeta and memory in the Tg2576 mouse model of Alzheimer's disease. *J. Neurosci.* **2002**, *22*, 1858–1867. [CrossRef]

74. Takeda, S.; Hashimoto, T.; Roe, A.D.; Hori, Y.; Spires-Jones, T.L.; Hyman, B.T. Brain interstitial oligomeric amyloid-beta increases with age and is resistant to clearance from brain in a mouse model of Alzheimer's disease. *FASEB J.* **2013**, *27*, 3239–3248. [CrossRef]

75. Shankar, G.M.; Leissring, M.A.; Adame, A.; Sun, X.; Spooner, E.; Masliah, E.; Selkoe, D.J.; Lemere, C.A.; Walsh, D.M. Biochemical and immunohistochemical analysis of an Alzheimer's disease mouse model reveals the presence of multiple cerebral Abeta assembly forms throughout life. *Neurobiol. Dis.* **2009**, *36*, 293–302. [CrossRef]

76. Yankner, B.A.; Duffy, L.K.; Kirschner, D.A. Neurotrophic and neurotoxic effects of amyloid-beta protein: Reversal by tachykinin neuropeptides. *Science* **1990**, *250*, 279–282. [CrossRef]

77. Barghorn, S.; Nimmrich, V.; Striebinger, A.; Krantz, C.; Keller, P.; Janson, B.; Bahr, M.; Schmidt, M.; Bitner, R.S.; Harlan, J.; et al. Globular amyloid-beta-peptide oligomer—A homogenous and stable neuropathological protein in Alzheimer's disease. *J. Neurochem.* **2005**, *95*, 834–847. [CrossRef]

78. Nimmrich, V.; Grimm, C.; Draguhn, A.; Barghorn, S.; Lehmann, A.; Schoemaker, H.; Hillen, H.; Gross, G.; Ebert, U.; Bruehl, C. Amyloid-β oligomers (A-β_{1-42} globulomer) suppress spontaneous synaptic activity by inhibition of P/Q-type calcium currents. *J. Neurosci.* **2008**, *28*, 788–797. [CrossRef]

79. Noguchi, A.; Matsumura, S.; Dezawa, M.; Tada, M.; Yanazawa, M.; Ito, A.; Akioka, M.; Kikuchi, S.; Sato, M.; Ideno, S.; et al. Isolation and characterization of patient-derived, toxic, high mass amyloid-β-protein (A-β) assembly from Alzheimer disease brains. *J. Biol. Chem.* **2009**, *284*, 32895–32905. [CrossRef]

80. Ohnishi, T.; Yanazawa, M.; Sasahara, T.; Kitamura, Y.; Hiroaki, H.; Fukazawa, Y.; Kii, I.; Nishiyama, T.; Kakita, A.; Takeda, H.; et al. Na, K-ATPase $\alpha 3$ is a death target of Alzheimer patient amyloid-β assembly. *Proc. Natl. Acad. Sci. USA* **2015**, *112*, E4465–E4474. [CrossRef]

81. Malhotra, A.; Younesi, E.; Gundel, M.; Muller, B.; Heneka, M.T.; Hofmann-Apitius, M. ADO: A disease ontology representing the domain knowledge specific to Alzheimer's disease. *Alzheimers Dement.* **2014**, *10*, 238–246. [CrossRef]

82. Drame, K.; Diallo, G.; Delva, F.; Dartigues, J.F.; Mouillet, E.; Salamon, R.; Mougin, F. Reuse of termino-ontological resources and text corpora for building a multilingual domain ontology: An application to Alzheimer's disease. *J. Biomed. Inform.* **2014**, *48*, 171–182. [CrossRef]

83. Refolo, L.M.; Snyder, H.; Liggins, C.; Ryan, L.; Silverberg, N.; Petanceska, S.; Carrillo, M.C. Common Alzheimer's disease research ontology: National institute on aging and Alzheimer's association collaborative project. *Alzheimers Dement.* **2012**, *8*, 372–375. [CrossRef]

84. Liggins, C.; Snyder, H.M.; Silverberg, N.; Petanceska, S.; Refolo, L.M.; Ryan, L.; Carrillo, M.C. International Alzheimer's disease research portfolio (IADRP) aims to capture global Alzheimer's disease research funding. *Alzheimers Dement.* **2014**, *10*, 405–408. [CrossRef]

85. Zhang, X.; Hu, B.; Ma, X.; Moore, P.; Chen, J. Ontology driven decision support for the diagnosis of mild cognitive impairment. *Comput. Methods Prog. Biomed.* **2014**, *113*, 781–791. [CrossRef]

86. Lovering, R.C.; Roncaglia, P.; Howe, D.G.; Laulederkind, S.J.F.; Khodiyar, V.K.; Berardini, T.Z.; Tweedie, S.; Foulger, R.E.; Osumi-Sutherland, D.; Campbell, N.H.; et al. Improving Interpretation of cardiac phenotypes and enhancing discovery with expanded knowledge in the gene ontology. *Circ. Genom Precis Med.* **2018**, *11*, e001813. [CrossRef]

87. Ferrari, R.; Grassi, M.; Salvi, E.; Borroni, B.; Palluzzi, F.; Pepe, D.; D'avila, F.; Padovani, A.; Archetti, S.; Rainero, I.; et al. A genome-wide screening and SNPs-to-genes approach to identify novel genetic risk factors associated with frontotemporal dementia. *Neurobiol. Aging* **2015**, *36*, 2904.e13–2904.e26. [CrossRef]

88. Welton, J.L.; Loveless, S.; Stone, T.; Von Ruhland, C.; Robertson, N.P.; Clayton, A. Cerebrospinal fluid extracellular vesicle enrichment for protein biomarker discovery in neurological disease; multiple sclerosis. *J. Extracell. Vesicles* **2017**, *6*, 1369805. [CrossRef]

89. Hirsch, T.; Rothoeft, T.; Teig, N.; Bauer, J.W.; Pellegrini, G.; De Rosa, L.; Scaglione, D.; Reichelt, J.; Klausegger, A.; Kneisz, D.; et al. Regeneration of the entire human epidermis using transgenic stem cells. *Nature* **2017**, *551*, 327–332. [CrossRef]

90. Ittner, L.M.; Ke, Y.D.; Delerue, F.; Bi, M.; Gladbach, A.; Van Eersel, J.; Wolfing, H.; Chieng, B.C.; Christie, M.J.; Napier, I.A.; et al. Dendritic function of tau mediates amyloid-beta toxicity in Alzheimer's disease mouse models. *Cell* **2010**, *142*, 387–397. [CrossRef]

91. Iqbal, K.; Liu, F.; Gong, C.X.; Grundke-Iqbal, I. Tau in Alzheimer disease and related tauopathies. *Curr. Alzheimer Res.* **2010**, *7*, 656–664. [CrossRef]

92. Butterfield, D.A.; Boyd-Kimball, D. Amyloid-beta-peptide(1-42) contributes to the oxidative stress and neurodegeneration found in Alzheimer disease brain. *Brain Pathol.* **2004**, *14*, 426–432. [CrossRef]

93. Mudher, A.; Colin, M.; Dujardin, S.; Medina, M.; Dewachter, I.; Alavi Naini, S.M.; Mandelkow, E.M.; Mandelkow, E.; Buee, L.; Goedert, M.; et al. What is the evidence that tau pathology spreads through prion-like propagation? *Acta Neuropathol. Commun.* **2017**, *5*, 99. [CrossRef]

94. Kaufman, S.K.; Thomas, T.L.; Del Tredici, K.; Braak, H.; Diamond, M.I. Characterization of tau prion seeding activity and strains from formaldehyde-fixed tissue. *Acta Neuropathol. Commun.* **2017**, *5*, 41. [CrossRef]

95. Kaufman, S.K.; Del Tredici, K.; Thomas, T.L.; Braak, H.; Diamond, M.I. Tau seeding activity begins in the transentorhinal/entorhinal regions and anticipates phospho-tau pathology in Alzheimer's disease and PART. *Acta Neuropathol.* **2018**, *136*, 57–67. [CrossRef]

96. DeVos, S.L.; Corjuc, B.T.; Oakley, D.H.; Nobuhara, C.K.; Bannon, R.N.; Chase, A.; Commins, C.; Gonzalez, J.A.; Dooley, P.M.; Frosch, M.P.; et al. Synaptic tau seeding precedes tau pathology in human Alzheimer's disease brain. *Front. Neurosci.* **2018**, *12*, 267. [CrossRef]

97. Hyttinen, J.M.; Amadio, M.; Viiri, J.; Pascale, A.; Salminen, A.; Kaarniranta, K. Clearance of misfolded and aggregated proteins by aggrephagy and implications for aggregation diseases. *Ageing Res. Rev.* **2014**, *18*, 16–28. [CrossRef]

98. Rodrigue, K.M.; Kennedy, K.M.; Park, D.C. Beta-amyloid deposition and the aging brain. *Neuropsychol. Rev.* **2009**, *19*, 436–450. [CrossRef]

99. Kamenetz, F.; Tomita, T.; Hsieh, H.; Seabrook, G.; Borchelt, D.; Iwatsubo, T.; Sisodia, S.; Malinow, R. APP processing and synaptic function. *Neuron* **2003**, *37*, 925–937. [CrossRef]

100. Cirrito, J.R.; Yamada, K.A.; Finn, M.B.; Sloviter, R.S.; Bales, K.R.; May, P.C.; Schoepp, D.D.; Paul, S.M.; Mennerick, S.; Holtzman, D.M. Synaptic activity regulates interstitial fluid amyloid-beta levels in vivo. *Neuron* **2005**, *48*, 913–922. [CrossRef]

101. Li, X.; Uemura, K.; Hashimoto, T.; Nasser-Ghodsi, N.; Arimon, M.; Lill, C.M.; Palazzolo, I.; Krainc, D.; Hyman, B.T.; Berezovska, O. Neuronal activity and secreted amyloid-beta lead to altered amyloid-beta precursor protein and presenilin 1 interactions. *Neurobiol. Dis.* **2013**, *50*, 127–134. [CrossRef]

102. Cao, L.; Schrank, B.R.; Rodriguez, S.; Benz, E.G.; Moulia, T.W.; Rickenbacher, G.T.; Gomez, A.C.; Levites, Y.; Edwards, S.R.; Golde, T.E.; et al. Abeta alters the connectivity of olfactory neurons in the absence of amyloid plaques in vivo. *Nat. Commun.* **2012**, *3*, 1009. [CrossRef]

103. Denny, P.; Feuermann, M.; Hill, D.P.; Roncaglia, P.; Lovering, R.C. Exploring autophagy with Gene Ontology. *F1000Research (Poster)* **2016**. Available online: https://f1000research.com/posters/5-754 (accessed on 30 October 2018). [CrossRef]

104. Foulger, R.E.; Denny, P.; Hardy, J.; Martin, M.J.; Sawford, T.; Lovering, R.C. Using the Gene Ontology to annotate key players in Parkinson's disease. *Neuroinformatics* **2016**, *14*, 297–304. [CrossRef]

105. Gene Ontology Consortium, SynGO—Synapse Biology. 2018. Available online: http://www.geneontology.org/page/syngo-synapse-biology (accessed on 30 October 2018).

106. Thurmond, J.; Goodman, J.L.; Strelets, V.B.; Attrill, H.; Gramates, L.S.; Marygold, S.J.; Matthews, B.B.; Millburn, G.; Antonazzo, G.; Trovisco, V.; et al. FlyBase 2.0: the next generation. *Nucleic Acids Res.* **2018**. [CrossRef]

107. Li, Q.; Barres, B.A. Microglia and macrophages in brain homeostasis and disease. *Nat. Rev. Immunol.* **2018**, *18*, 225–242. [CrossRef]

108. UCL Functional Gene Annotation, Neurological Gene Ontology. Available online: https://www.ucl.ac.uk/functional-gene-annotation/neurological (accessed on 30 October 2018).

109. Meldal, B.H.; Forner-Martinez, O.; Costanzo, M.C.; Dana, J.; Demeter, J.; Dumousseau, M.; Dwight, S.S.; Gaulton, A.; Licata, L.; Melidoni, A.N.; et al. The complex portal—an encyclopaedia of macromolecular complexes. *Nucleic Acids Res.* **2015**, *43*, D479–D484. [CrossRef]
110. GOA Contact Us. Available online: https://www.ebi.ac.uk/GOA/contactus (accessed on 30 October 2018).
111. Contributing to GO. Available online: http://geneontology.org/page/contributing-go (accessed on 30 October 2018).

Article

NeVOmics: An Enrichment Tool for Gene Ontology and Functional Network Analysis and Visualization of Data from OMICs Technologies

Eduardo Zúñiga-León, Ulises Carrasco-Navarro and Francisco Fierro *

Departamento de Biotecnología, Universidad Autónoma Metropolitana-Unidad Iztapalapa, Ciudad de Mexico 09340, Mexico; pgen10@hotmail.com (E.Z.-L.); ulises.c.n@gmail.com (U.C.-N.)
* Correspondence: degfff@yahoo.com or fierrof@xanum.uam.mx

Received: 26 September 2018; Accepted: 16 November 2018; Published: 23 November 2018

Abstract: The increasing number of OMICs studies demands bioinformatic tools that aid in the analysis of large sets of genes or proteins to understand their roles in the cell and establish functional networks and pathways. In the last decade, over-representation or enrichment tools have played a successful role in the functional analysis of large gene/protein lists, which is evidenced by thousands of publications citing these tools. However, in most cases the results of these analyses are long lists of biological terms associated to proteins that are difficult to digest and interpret. Here we present NeVOmics, Network-based Visualization for Omics, a functional enrichment analysis tool that identifies statistically over-represented biological terms within a given gene/protein set. This tool provides a hypergeometric distribution test to calculate significantly enriched biological terms, and facilitates analysis on cluster distribution and relationship of proteins to processes and pathways. NeVOmics is adapted to use updated information from the two main annotation databases: Gene Ontology and Kyoto Encyclopedia of Genes and Genomes (KEGG). NeVOmics compares favorably to other Gene Ontology and enrichment tools regarding coverage in the identification of biological terms. NeVOmics can also build different network-based graphical representations from the enrichment results, which makes it an integrative tool that greatly facilitates interpretation of results obtained by OMICs approaches. NeVOmics is freely accessible at https://github.com/bioinfproject/bioinfo/.

Keywords: Gene Ontology; KEGG pathways; enrichment analysis; proteomic analysis; plot visualization

---

1. Introduction

Omics technologies are revolutionizing biological research by enabling genome-scale analysis of complex biological systems and processes [1]. Functional annotation of data from these approaches is essential to reduce the huge complexity of lists with hundreds to thousands of genes/proteins to a few processes or pathways in which they are involved, which will have more explanatory power than a simple list of identifiers. Several bioinformatic tools have been developed to perform functional annotations [2,3]. Over-representation analysis (ORA) is the most popular bioinformatic methodology to obtain significant functional information (enrichment) from sets of related genes/proteins [4]. The ORA method consists of searching in biological databases (e.g., Gene Ontology, GO [5] or Kyoto Encyclopedia of Genes and Genomes, KEGG [6]), using statistical testing to find biological terms, and functional annotations that are significantly enriched in a list of genes/proteins. The aim of the enrichment analysis is finding biological annotations that are over-represented in the query gene/protein list compared to what would be expected in a reference list (usually the whole proteome) [7]. In other words, if in a set of proteins certain biological processes or pathways are significantly enriched, the proteins with such signatures are likely to play these roles in vivo. However, in most cases, the results of these analyses are very long lists of biological terms or pathways associated to genes/proteins that are difficult to digest and interpret. Most available enrichment tools do not include comprehensible graphical visualizations and present the results as simple bar or pie chart plots, which do not allow insight into the functional relationships existing between the identified genes/proteins and the enriched GO terms and pathways [8]. As an example, a protein could be involved in three or more relevant biological processes or pathways, and a bar or pie chart plot cannot provide such information.

Network analysis has become an increasingly popular tool to deal with the complexity of large datasets of all sorts. The importance of using networks lies in their ability to reveal relationships between factors, rather than seeing them as isolated entities [9]. Intersection networks are bipartite networks which, when applied to biological systems, allow detection of multifunctional proteins, i.e., genes/proteins with more than one function and involved in more than one process or pathway.

Here we present NeVOmics (Network-based Visualization for Omics), a bioinformatic tool that facilitates the functional characterization of data from OMICs technologies such as transcriptomics and proteomics. NeVOmics has been developed in programming language Python, it integrates ORA methodology and network-based visualization with R packages, allowing the generation of four different types of graphical visualization to show the enrichment results. NeVOmics applies a hypergeometric statistical test to identify significantly enriched GO terms and pathways in a list of genes or proteins. The tool supports all organisms deposited in UniProt Knowledgebase (UniProtKB) and KEGG databases, and incorporates a functionality to assign pathways to organisms with no annotated genome information available from orthologous gene pathways deposited in the KEGG database.

2. Materials and Methods

2.1. General Features

NeVOmics has been developed in programming language Python. It integrates ORA methodology to obtain significantly over-represented GO terms and pathways in a list of genes/proteins. In addition, it uses R packages to provide network-based visualization to show the enrichment results. The code is configured to use updated information from UniProt-Gene Ontology Annotation (UniProt-GOA) [10,11], UniProtKB [12], and KEGG [6] databases. NeVOmics can be used in both Linux and Windows operating systems and it provides 13 additional protein lists from diverse organisms for the user to have the optionality to test the tool.

2.2. Usage

The flowchart for data processing is depicted in Figure 1. NeVOmics documentation is available for download from GitHub (https://github.com/bioinfproject/bioinfo/). It includes NeVOmics Python script, instructions for use, system requirements, and some examples.

NeVOmics can perform three different enrichment analyses using updated databases (Figure S1). The first analysis (1) is Gene Ontology using all information stored in UniProt-GOA (Complete GO Annotation) and UniProtKB Annotations. The second analysis (2) uses all annotations stored in the KEGG database to find relevant pathways. Finally, the third analysis (3) is more flexible because it can identify KEGG pathways from protein sequences by performing identity searches, and thus independent of the availability or not of annotated genome information for a particular organism. After executing each of the previous analyses, the information is stored in specific directories.

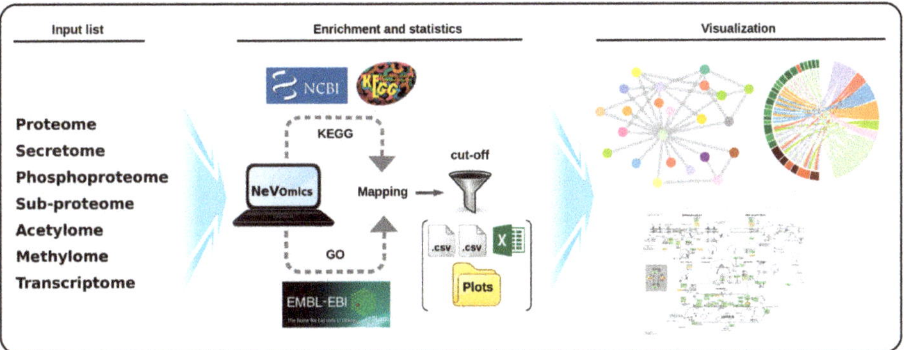

Figure 1. Schematic overview of the steps performed by NeVOmics (Network-based Visualization for Omics). The tool is composed of three main sections: input list, enrichment/statistics; and visualization. A connection is made to the UniProt-Gene Ontology Annotation (UniProt-GOA) and Kyoto Encyclopedia of Genes and Genomes (KEGG) databases, and then the content of the input list is analyzed. The enrichment/statistics section organizes and analyzes the data from the input list, retrieves information for all genes or proteins in the list, and the results are stored in files in .xlsx format. The National Center for Biotechnology Information (NCBI) logo indicates the source from which the reference proteomes are obtained by the KEGG database. The visualization section provides four types of graphical representation in .png format and high definition.

2.3. Input File

NeVOmics uses an input file in plain text containing a list of genes (KEGG gene ID) or proteins (UniProt Entry ID) obtained by any OMICs approach, such as comparative transcriptomics, proteomics, or proteome-specific methodologies (phosphoproteomics, acetylomics, etc.). The file can contain up to three columns (in Tabular format) depending on the results obtained in the "OMICs" experiment. The first column corresponds to the list of genes or proteins to be analyzed, for example a set of proteins showing abundance changes in a particular condition with respect to the control. The second column contains numerical values of expression/abundance or any other quantifiable value related to the study. The third column corresponds to a background list (e.g., all genes or proteins identified in the study) used as reference for statistics in the enrichment analysis; if the third column is absent in the input file the program automatically uses the entire proteome (UniProt-GOA) or KEGG pathways of the corresponding organism. In order to avoid ambiguities or inconsistencies in identifiers, NeVOmics allows only identifiers compatible with UniProtKB.

2.4. Annotations Sources

The Gene Ontology Annotation (GOA) database provides high quality electronic (mapping and automatic transfer of annotation to orthologous gene products) and manual (based on the literature) annotations to the UniProtKB (Swiss-Prot, TrEMBL, and PIR-PSD) using the standardized vocabulary of Gene Ontology. The GOA database contains information of nearly 60,000 species and more than 160,000 taxa, with more than 32 million annotations. NeVOmics uses the Gene Ontology of the Gene Ontology Consortium (GOC) [5], which is downloaded from: http://purl.obolibrary.org/obo/go.obo. It also uses the GOA association files, which are downloaded from File Transfer Protocol (FTP) in tab-delimited format (ftp://ftp.ebi.ac.uk/pub/databases/GO/goa/), and GO annotation in UniProtKB, which is provided as flat files that are downloaded from: http://www.ebi.ac.uk/uniprot/. In Swiss-Prot entries, only the manually annotated information is displayed. To view the complete GO annotation for a Swiss-Prot entry, the master copy of the data in the GOA association files is downloaded. GOA updates its annotation information weekly, while UniProtKB does it every four weeks. KEGG database covers information at different molecular levels. The KEGG Pathways database (https://www.genome.jp/kegg/pathway.html) is a collection of manually curated pathways, including information on molecular interactions, reactions, and network relationships. The KEGG database contains more than 24 million annotated genes and 530 pathways with more than six million pathway-linked genes.

2.5. Background

A background list of genes or proteins is essential for performing an adequate enrichment analysis and must be carefully chosen; a list with all the genes/proteins detected in any condition of the OMICs experiment is usually a valid background list. If the column of the background list (third column in the input file) is absent the program automatically uses the entire proteome in the case of UniProt-GOA or the entire KEGG database. For Gene Ontology analysis, NeVOmics builds background lists organized by category (Biological Process, Molecular Function, Cellular Component) of a specific organism for mapping the query protein list. NeVOmics downloads a GOA association file (ftp://ftp.ebi.ac.uk/pub/databases/GO/goa/proteomes/) from which it extracts all GO terms. For KEGG Pathways analysis NeVOmics builds a background file with all genes of a specific organism for mapping the query gene list.

2.6. Enrichment Analysis

NeVOmics analyzes the input list against the user's preferred background and retrieves over-represented terms in the three GO categories (molecular function, biological process, cellular component) or in the KEGG pathways, with the genes/proteins classified by term or pathway. The tool adopts GeneMerge statistical algorithm to obtain the over-represented functions or categories in the input list [13]. A hypergeometric distribution test is performed to calculate the discrete probability of x (term or pathway) in a sample (gene/protein input list) of size k drawn from a background of size N (entire proteome by default or a background list provided by the user in the third column of the input file), where m corresponds to the total number of genes/proteins associated with a term or pathway (Figure S2). An FDR or modified Bonferroni correction of p-value is applied to identify the statistically more represented function annotations.

2.7. Output Files

NeVOmics generates three output files, stored in specific directories, containing lists of particular functions or KEGG pathways over-represented with statistical significance (Figure 1): (I) An edges file with two columns as 'Source' and 'Target', which contain the values of individual nodes that are linked together. (II) A nodes file with information of nodes ID, expression values, p-value and additional information for building plots. (III) Finally, an .xlsx file with enrichment analysis results for plotting in other network-based tools like Cytoscape [14] and Gephi [15]. NeVOmics automatically provides four types of graphical representations in high definition to facilitate the analysis and interpretation of results, these are: circular and random network, chord diagram [16], and UpSet plot [17]. These graphics are built with R packages as tidyverse, tidygraph, ggraph, igraph, viridis, circlize, RcolorBrewer, cowplot, networkD3, gridBase, ComplexHeatmap, and UpSetR. The graphics are configured according to the amount and type of data that resulted over-represented in the enrichment analysis.

2.8. Comparison of NeVOmics with Other Functional Enrichment Bioinformatic Tools

Gene Ontology and annotation sources for the analyses with NeVOmics were indicated in Section 2.4. GO terms from UniProt-GOA correspond to the version released on 18 October 2018. GOrilla and WebGestalt use GOC (http://www.geneontology.org/) for GO terms and as annotation source (http://geneontology.org/page/go-consortium-contributors-list); the update of 20 October 2018 was used in our analyses. g:Profiler uses also GOC for GO terms, and the annotation sources are Ensembl v93 (http://jul2018.archive.ensembl.org/index.html) and Ensembl Genomes v40 (http://ensemblgenomes.org/).

It is recommended that NeVOmics users make reference to the date of the analysis when publicizing their results, so that other researchers have information on the version of the UniProt-GOA release used for the analysis.

3. Results

NeVOmics is an integral tool with two major features: it allows enrichment analysis from a given list with data from some 'OMICs' experiment, and it builds different graphical representations in network form from the enrichment results. To test the functionality of NeVOmics we performed two case studies, using datasets from two different experimental OMICs procedures recently published. The first dataset comes from a platelet proteome of patients with early-stage cancer [18], and the second comes from a transcriptome analysis of mutants in tail module subunits of Mediator in *Arabidopsis thaliana*, a model system for research in plant biology [19].

3.1. Case Study 1: Enrichment Analysis of Differentially Expressed Platelet Proteins in Early-Stage Cancer

Platelets play an important role in tumor angiogenesis, growth and metastasis [20]. A study from Sabrkhany et al. [18] identified 4384 unique proteins expressed in platelets, of which 85 showed a significant abundance change (criteria Fc > 1.5 and $p < 0.05$) in early-stage cancer as compared to the control. Samples from 12 cancer patients (eight with lung cancer and four with pancreatic cancer) and 11 controls were used in the study. We analyzed with NeVOmics an input list with 61/24 platelet proteins which were $\geq$1.5-fold more abundant/$\geq$1.5-fold less abundant, respectively, in platelets from individuals with early-stage cancer as compared to platelets from healthy individuals. As background we used a list with the total 4384 platelet-expressed proteins identified in this study. We carried out an enrichment analysis for Gene Ontology (with an FDR of 0.02) and KEGG Pathways (FDR 0.1). Nineteen biological processes were identified as enriched (Figure 2A), these fall mainly within the areas of inflammatory response, immune response, and cancer, according to UniProtKB annotations. In this category, 19 proteins were identified, of which nine (P08311, P06702, P05109, P04196, O75594, P03973, P06899, Q29960, and Q95365) are involved in more than two processes: P06702 has been shown to be differentially expressed in various types of cancer such as breast, colon, liver, gastric and non-small cell lung cancer, and is crucial for promoting cancer growth by recruitment of myeloid-derived suppressor cells. P06702 and P05109 usually form a heterodimer and are involved in inflammatory response and cancer development and progression, which explains their simultaneous presence in several enriched processes. In addition, five proteins were identified which are directly related to the angiogenesis process (GO:0001525), relevant in the early stages of cancer. For its part, in the category of molecular function, six functions resulted enriched (Figure 2A). In this category, 21 proteins were identified, of which two (P06702 and P05109) are involved in more than two functions. In the analysis with the KEGG database, 11 pathways were identified as enriched, mainly related to the immune response (Figure 2B). Only 12 proteins from the input list share pathways: Q95365/3106 and Q29960/3107 are involved in seven of the enriched pathways, whereas the other 10 are involved in one or two pathways.

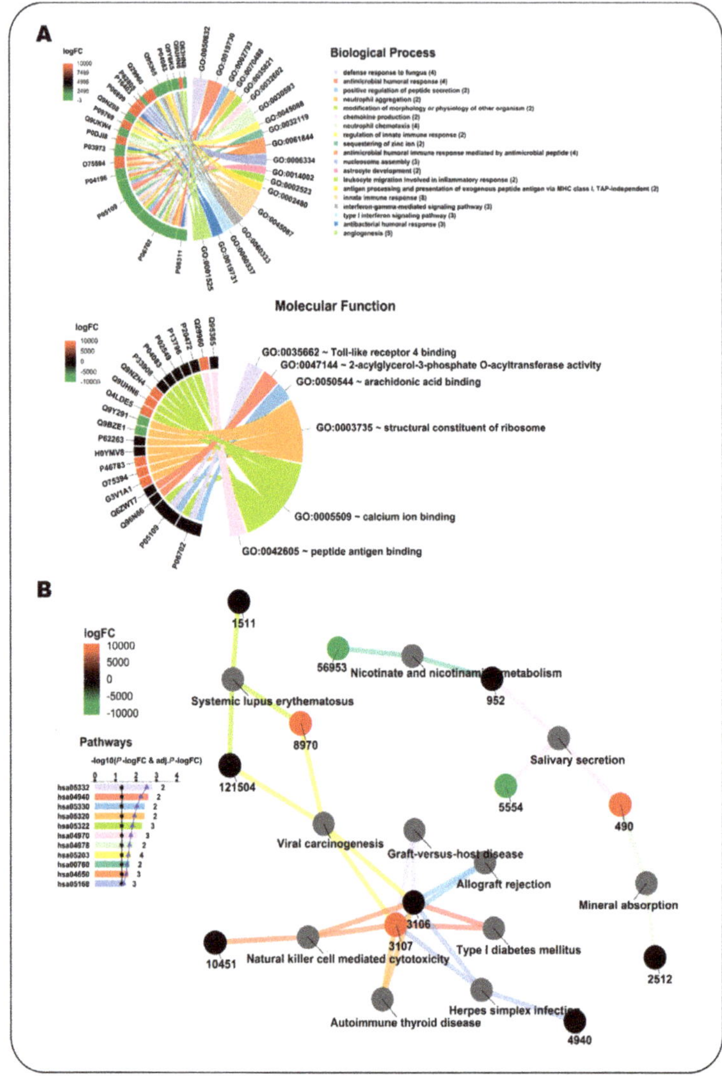

Figure 2. Enrichment analysis with NeVOmics of a dataset of platelet proteins showing significant abundance change (≥1.5-fold) in early-stage cancer vs. healthy condition (study conducted by Sabrkhany et al. [18]). (**A**) Chord diagram clustered by colors. Colors in the chords correspond to enriched biological processes (upper) or molecular functions (lower) (GO terms), linking each protein to the processes/functions to which it is related. Colors appearing in the sections of the outer circle beside each protein correspond to the abundance fold-change found in the study and are according to the heat map scale (logFC: log2 fold-change). (**B**) Network clustered by colors. Colors in the bar plot and the lines correspond to the identified enriched KEGG pathways, whereas colors in the protein identifier nodes correspond to the fold-change found in the study and are according to the heat map scale. The identifiers at the protein nodes are those provided by NCBI-GeneID. The bar plot indicates the −log10 of the *p*-value of each GO term, and the numbers at the end of the bars correspond to the total number of proteins detected for each term. The black line in the bar plot corresponds to the −log10 of 0.05, used as reference for *p*-value. The blue line on the bar plot corresponds to the −log10 of the adjusted *p*-value.

3.2. Case Study 2: Enrichment Analysis of Gene Datasets from a Transcriptomic Study of Arabidopsis thaliana *Mutants in Mediator Subunits*

The mediator complex is a central component of transcriptional regulation in *Eukaryotes*. Whitney and Clint [19] studied the role of four *Arabidopsis* Mediator tail module subunits (MED2, MED5a/b, MED16, and MED23) by analyzing the transcriptome of mutants in each of the subunit-encoding genes. We used NeVOmics to perform an enrichment analysis of GO terms and KEGG pathways in four gene datasets corresponding each to the up- and downregulated genes found in a tail subunit mutant: med2 (341 genes), med5ab (283), med16 (723), and med23 (35). The total number of expressed genes identified in the Whitney and Clint's study [19] was 18,842; we used this gene list to obtain compatible identifiers with the UniProtKB from The Arabidopsis Information Resource (TAIR) database, mapping 17,626 genes successfully. This list of 17,626 gene products was used as background in the enrichment analysis.

To compare our results with those obtained by Whitney and Clint [19], we used an FDR of 0.05 to consider a GO term enriched, the same they used when performing their enrichment test with DAVID v6.8 [21]. In total, 137 unique GO terms were enriched with NeVOmics in the four conditions, they are represented in a clustered heatmap that includes all up- or downregulated identified gene products (Figure S3 and Table S1). In Figure 3 is shown a clustered heatmap elaborated with 42 enriched GO terms that contained more than five proteins each (data also in Table S2). The raw data generated by NeVOmics are stored in several .xlsx format output files from which information can be extracted and used to make custom visualizations using other packages or programs. Whitney and Clint [19] identified only one enriched GO term (response to bacterium) in the condition of "downregulated genes in the med2 mutant", and none in the "upregulated genes", whereas with NeVOmics we were able to detect 54 GO terms in the "downregulated genes" condition and five in the "upregulated genes" one (Figure S3 and Table S1). Regarding the med5 mutant, Whitney and Clint [19] detected 16 GO terms in "downregulated genes", and five in "upregulated genes", whereas with NeVOmics we found 53 GO terms in "downregulated genes" and 12 GO terms in "upregulated genes". In all cases, all GO terms identified by Whitney and Clint [19] with DAVID v6.8 were also detected with NeVOmics. In the new GO terms identified by NeVOmics there are more processes associated with the regulation of transcription along with different response mechanisms, for example: regulation of transcription by RNA polymerase II, DNA-templated regulation of transcription, DNA-templated response to oxidative stress, response to water deprivation, response to salt stress, and response to abscisic acid (Figure S3 and Table S1). In the med16 mutant, NeVOmics detected 22 enriched GO terms in "downregulated genes" and 53 in "upregulated genes". In Figure 3, the cluster 1 contains several enriched GO terms detected mainly in the mutants med2, med5 ("downregulated genes") and med16 ("upregulated genes"). The GO terms included in cluster 1 represent processes involved in response and defense to several factors. Finally, in the med23 mutant, whose transcriptome revealed only 35 downregulated genes, NeVOmics identified several GO terms that are absent in the Whitney and Clint work (Figure S3 and Table S1). In summary, most of the terms detected by NeVOmics are related mainly to some response and defense to several factors, or related to small molecules such as auxins, gibberellins, abscisic acid, ethylene, brassinosteroids, jasmonic acid, and salicylic acid (Figure S3 and Table S1), which play a major role in seed maturation and germination, as well as in adaptation to abiotic environmental stresses [19].

Regarding the Pathway analysis, Whitney and Clint [19] identified enriched pathways in both "upregulated" and "downregulated genes" conditions from all med mutants, except in the med23 mutant. They found a total of eight enriched KEGG pathways: ath04075 (Plant hormone signal transduction) and ath00945 (Stilbenoid, diarylheptanoid and gingerol biosynthesis) were found in the condition of "downregulated genes in the med2 mutant"; ath01110 (Biosynthesis of secondary metabolites), ath00073 (Cutin, suberine and max biosynthesis) and ath00904 (Diterpenoid biosynthesis) were found in "downregulated genes in the med16 mutant"; ath00940 (Phenylpropanoid biosynthesis) was found in "upregulated genes in the med5 mutant"; and two other pathways were found additionally, ath04626 (Plant-pathogen interaction) and ath00592 (alpha-Linolenic acid metabolism).

In our analysis, NeVOmics identified all pathways described by Whitney and Clint [19] except for ath00945, and additionally it detected three more pathways: ath00270 (Cysteine and methionine metabolism), ath00130 (Ubiquinone and other terpenoid-quinone biosynthesis), and ath04016 (MAPK signaling pathway plant) (Figure 4).

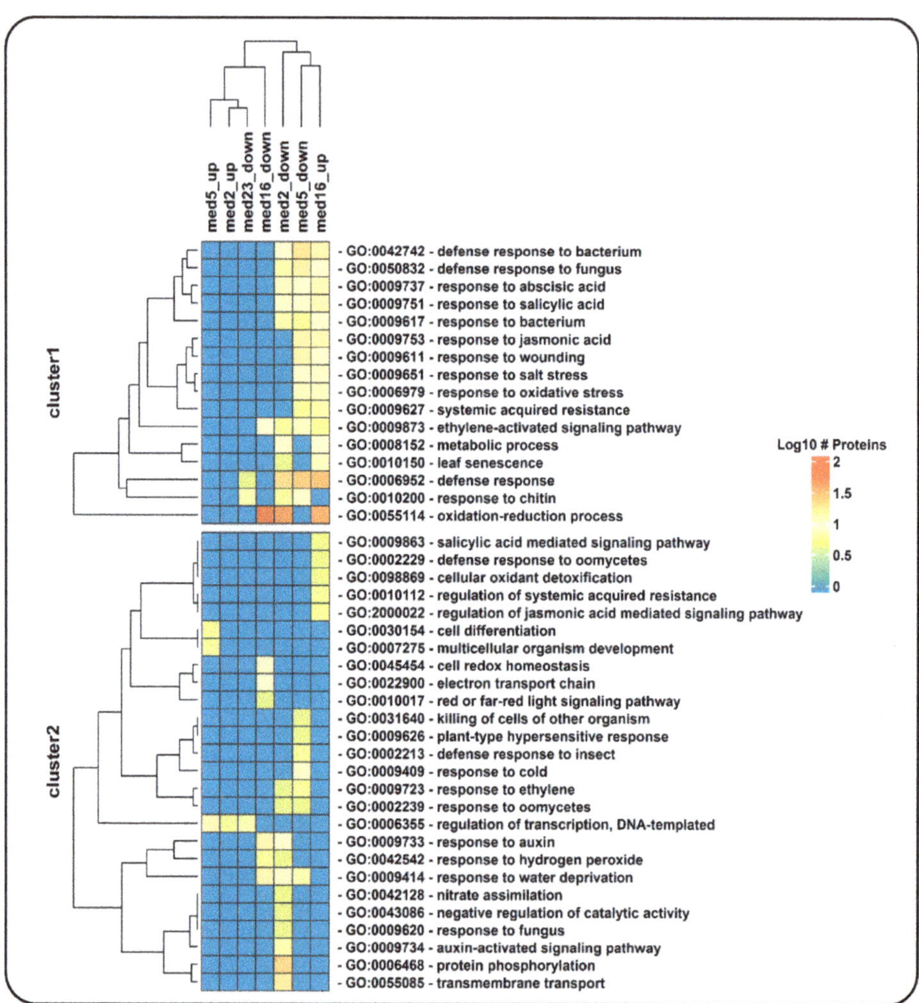

Figure 3. Hierarchical clustering of enriched GO terms in four gene datasets corresponding to up- and downregulated genes identified in a transcriptomic analysis of mutants of Mediator tail module subunits from *Arabidopsis* (study conducted by Whitney and Clint [19]). Only GO terms belonging to the category of 'biological process' are shown, and among them only those including more than five genes from the datasets. Each cell in the heatmap is colored according to the number of proteins (log10 of this number) detected by NeVOmics in the corresponding enriched GO term (rows) and in the specific condition (columns); "up" and "down" tags in column names correspond to "upregulated genes" and "downregulated genes", respectively.

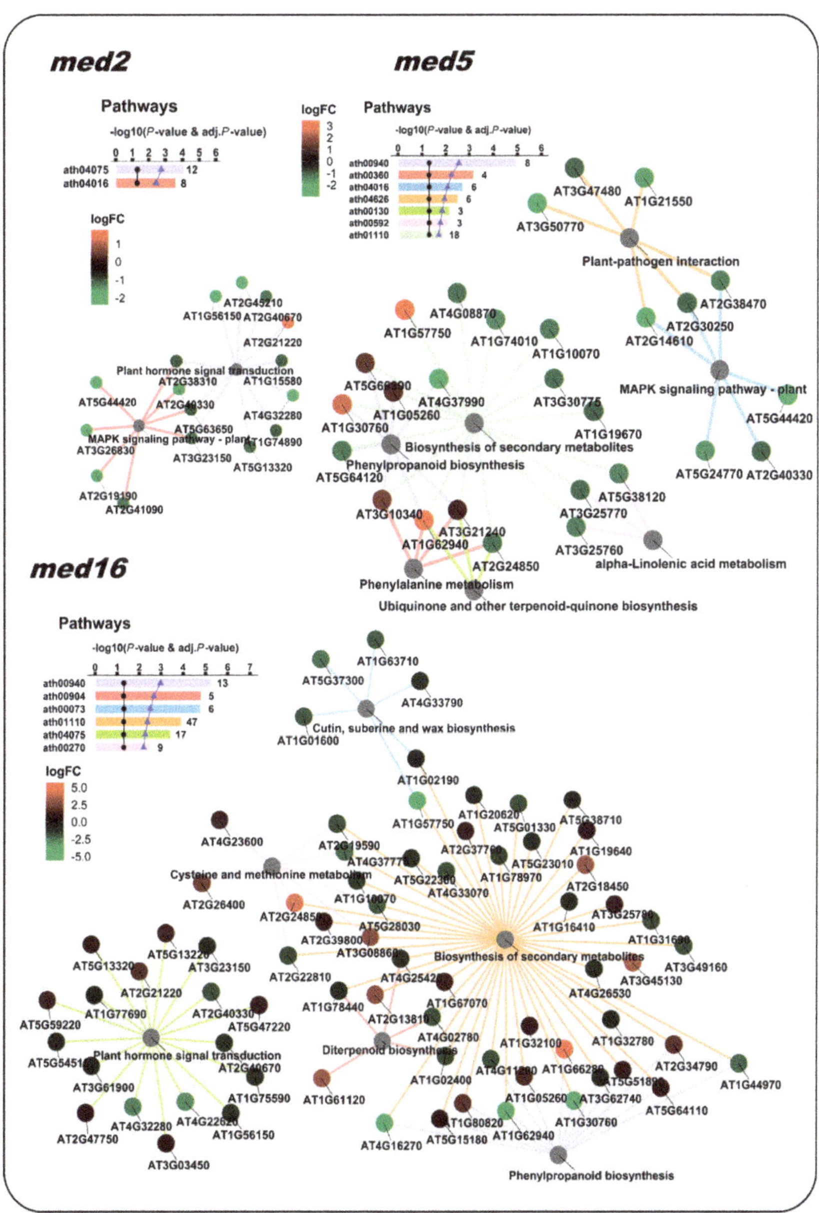

Figure 4. NeVOmics enrichment analysis of KEGG pathways using the same datasets as in Figure 3. med5 and med16 mutants shared several enriched pathways. These networks enable the effective representation of associated data, such as the number of elements in the aggregates and intersections, and the quick identification of proteins involved in more than one process or pathway. The ratio of each gene product abundance in the corresponding mutant over the control is indicated by the node color, according to the fold-change color scale (logFC: log2 fold-change). The bar plot indicates the −log10 of the *p*-value and the number of proteins found for each of the enriched pathways.

3.3. Comparison and Advantages of NeVOmics over Other Function Enrichment Analysis Tools

NeVOmics is designed to perform enrichment analysis using updated information of the two main annotation databases (GO and KEGG) automatically, making it unnecessary to have to update the information by other means. NeVOmics has the versatility to use UniProt-GOA and UniProtKB separately as databases, to obtain complete (electronic + manually-curated) and only manually-curated information, respectively, in order to get as much information as possible. There exist other available tools that can perform enrichment analysis on gene or protein datasets, however, these tools update their information on Gene Ontology less frequently and thus are less reliable for performing function enrichment analysis [22]. Some of these tools do not include more than two databases, or do not offer graphical representations to aid in the visualization of results (Table S3).

NeVOmics is also very versatile regarding the organisms that can be submitted to analysis, it supports all organisms deposited in UniProtKB (https://www.uniprot.org/proteomes/) and KEGG (https://www.kegg.jp/kegg/catalog/org_list.html) databases, and actually any organism can be used, even those lacking annotated genome information, due to NeVOmics functionality to assign pathways searching orthologous gene pathways deposited in the KEGG database (see Materials and Methods, Section 2.2). In contrast, other tools such as g:Profiler [23], GOrilla [24], GOEAST [25], and WebGestalt [26] have the limitation of excluding non-model organisms (Table S3).

To test whether NeVOmics is able to offer different and/or more complete/accurate results in enrichment analysis when compared to other tools, we performed a comparison test between NeVOmics and some publicly available enrichment tools: g:Profiler (version beta), Gorilla (version not available), GOEAST (version 1.20), and WebGestalt (version 2019 beta). Unlike NeVOmics and WebGestalt, Gorilla, and g:Profiler present flexibility limitations in the adjustment of some parameters during the analyses, for instance they control the p-value threshold not allowing to set up other custom-chosen FDR values. Despite these differences, we tried to make the analyses as comparable as possible. In the comparison test we searched for enriched GO terms in the category of Biological Processes using the data of Case Study 1, with an FDR of 0.05 for NeVOmics. Although WebGestalt allows to control several parameters, when selecting an FDR of 0.05 it does not detect enriched GO terms, therefore, in order to include this tool in the analyses we used its "TOP" option for 10 GO enriched terms. As shown in Figure 5, NeVOmics attained greater coverage of identified GO terms in comparison to other tools. With NeVOmics some GO terms were detected which went undetected to other tools, and the opposite also occurred in some cases. The terms detected by NeVOmics and absent in the analyses with other the tools were: GO:0002523 (leukocyte migration involved in inflammatory response), GO:0050727 (regulation of inflammatory response), GO:0030890 (positive regulation of B cell proliferation), GO:0030307 (positive regulation of cell growth), GO:0001525 (angiogenesis), GO:0070555 (response to interleukin-1), and GO:0060333 (interferon-gamma-mediated signaling pathway), all of them related to immune response processes and cancer. On the other hand, NeVOmics, GOrilla, and WebGestalt did not detect GO terms related to response to environmental stimulus (GO:0009607, GO:0043207, and GO:0009605), which appeared highly enriched in g:Profiler. Therefore, there are differences between tools in the number and identity of detected GO terms and also in the amount of proteins that each program identifies. Three possible reasons that could determine these differences are: the source of annotation and GO definitions, the regularity with which these databases are updated, and the statistical approach to obtain enriched terms. All the other tools detected fewer proteins than NeVOmics: WebGestalt covered 41% of the total number of proteins detected by NeVOmics, g:Profiler covered 52% and GOrilla 61%, which indicates that our tool can provide more complete functional information. In addition, NeVOmics automatically generates a set of graphs to facilitate the visualization of results. NeVOmics can generate four different graphics per analysis depending on the type and amount of data, whereas some of the analyzed tools generate only simple visualizations.

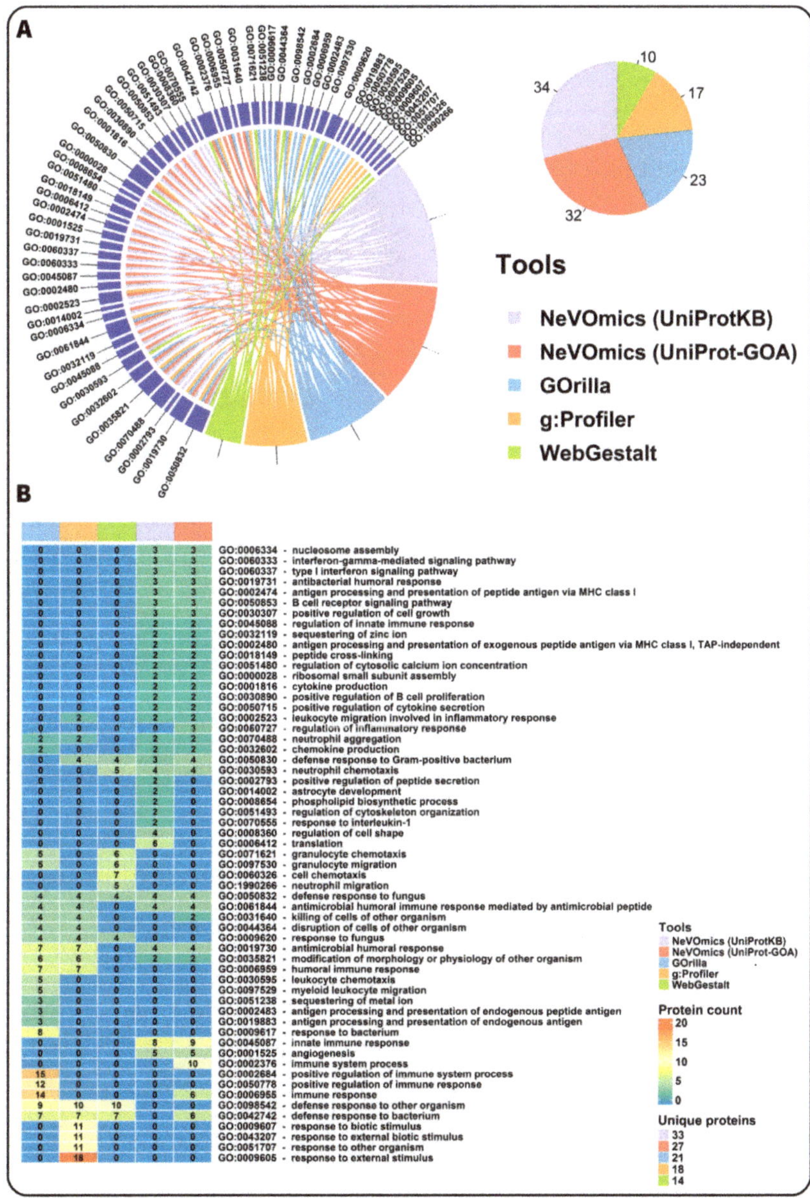

Figure 5. Comparison of NeVOmics with other enrichment tools. (**A**) Chord diagram showing a comparison of enriched GO terms in the category of biological process detected by NeVOmics and three other tools using the data of Case Study 1 previously described. GOEAST was also included in the analysis, however it did not recognize the identifiers and generated an error message. The pie chart shows the total number of GO terms identified by each tool. (**B**) Heatmap showing protein counts for each GO term (rows) obtained with the different tools (columns). Each cell is colored according to the number of proteins detected by a tool in the specific GO term. "Unique proteins" indicates the total number of unique proteins identified by each tool.

4. Discussion

NeVOmics is an enrichment analysis tool executable in the command line, able to analyze from five to thousands of identifiers in gene/protein datasets. It can perform three different types of analyses (described in Section 2.2), using automatically updated information from the main annotation resources for Gene Ontology and pathways (GOA and KEGG). There is no requirement to download additional databases or any other resources.

NeVOmics uses ORA methodology and has been designed to facilitate the analysis and interpretation of large amounts of data, such as those obtained by high-throughput OMICs techniques, from a very wide range of organisms. The tool also allows the inclusion of expression/abundance data or any other experimentally-obtained quantifiable value, such as phosphorylation or other protein modifications, which provides information about how genes/proteins are up/downregulated in different biological conditions, thus allowing a better understanding of their role in the processes and pathways that are enriched.

As proof of concept, we performed different tests with datasets from distantly related organisms such as humans and *Arabidopsis thaliana*, with similarly successful results. In addition, we compared our tool to other available enrichment tools by analyzing a protein dataset from platelets of early-stage cancer patients. NeVOmics proved to be more sensitive and was able to identify a higher number of cancer-related processes (GO terms) and proteins participating in these processes than any of the other tools. Nevertheless, we encourage researchers to use more than one tool when performing Gene Ontology and pathway enrichment analysis, in order to make comparisons and verification of results and thus be able to make more accurate conclusions.

In addition to its capabilities for functional annotation by enrichment analysis, NeVOmics builds and provides different types of network-based graphical visualizations to get a better comprehension of experimental results from OMICs technologies, and to illustrate and communicate the information in an integral form. In future versions, we aim to include the Reactome (https://reactome.org/) and NCG (http://ncg.kcl.ac.uk/) databases (only for *Homo sapiens*). We will also work to include new features such as enrichment of protein domains and building of networks based on protein-protein interactions.

Supplementary Materials: The following materials are available online at: http://www.mdpi.com/2073-4425/9/12/569/s1 and https://github.com/bioinfproject/bioinfo/. Figure S1: Description on how to run NeVOmics on the command line in Windows and Linux. Figure S2: Hypergeometric distribution. Figure S3: Heat map of all GO terms detected by NeVOmics generated with Table S1. Table S1: Data matrix in .csv format with 137 unique GO terms enriched by NeVOmics in Case Study 2. Table S2: Data matrix in .csv format with GO terms enriched by NeVOmics in Case Study 2, with more than five proteins each. Table S3: Comparison of NeVOmics features with other functional enrichment tools. The supplementary material also includes thirteen files (File_S1–File_S13) in .tsv format with protein lists from *Homo sapiens* and *Arabidopsis thaliana* used in Case Studies 1 and 2, respectively, to test NeVOmics.

Author Contributions: Conceptualization, E.Z.-L., U.C.-N., and F.F.; Formal analysis, E.Z.-L. and U.C.-N.; Methodology, E.Z.-L., U.C.-N., and F.F.; Resources, E.Z.-L.; Software, E.Z.-L.; Supervision, E.Z.-L., U.C.-N., and F.F.; Validation, E.Z.-L. and U.C.-N.; Visualization, E.Z.-L., U.C.-N., and F.F.; Writing—original draft, E.Z.-L., U.C.-N, and F.F.; Writing—review & editing, E.Z.-L., U.C.-N, and F.F.

Funding: This research was funded by the CONACyT through the research project CB-2013-01-222207. EZL received an individual grant from the CONACyT (scholarship no. 401684). UCN holds a postdoc grant associated to the CONACyT research project CB-2013-01-222207.

Conflicts of Interest: The authors declare no conflict of interest.

References

1. Karahalil, B. Overview of systems biology and omics technologies. *Curr. Med. Chem.* **2016**, *23*, 4221–4230. [CrossRef] [PubMed]

2. Villavicencio-Diaz, T.N.; Rodriguez-Ulloa, A.; Guirola-Cruz, O.; Perez-Riverol, Y. Bioinformatics tools for the functional interpretation of quantitative proteomics results. *Curr. Top. Med. Chem.* **2014**, *14*, 435–449. [CrossRef] [PubMed]

3. Tipney, H.; Hunter, L. An introduction to effective use of enrichment analysis software. *Hum. Genom.* **2010**, *4*, 202–206. [CrossRef]

4. Khatri, P.; Sirota, M.; Butte, A.J. Ten years of pathway analysis: Current approaches and outstanding challenges. *PLoS Comput. Biol.* **2012**, *8*, e1002375. [CrossRef] [PubMed]

5. Ashburner, M.; Ball, C.A.; Blake, J.A.; Botstein, D.; Butler, H.; Cherry, J.M.; Davis, A.P.; Dolinski, K.; Dwight, S.S.; Eppig, J.T.; et al. Gene Ontology: Tool for the unification of biology. *Nat. Genet.* **2000**, *25*, 25–29. [CrossRef] [PubMed]

6. Kanehisa, M.; Sato, Y.; Kawashima, M.; Furumichi, M.; Tanabe, M. KEGG as a reference resource for gene and protein annotation. *Nucleic Acids Res.* **2016**, *44*, D457–D462. [CrossRef] [PubMed]

7. Schmidt, A.; Forne, I.; Imhof, A. Bioinformatic analysis of proteomics data. *BMC Syst. Biol.* **2014**, *8*, S3. [CrossRef] [PubMed]

8. Thomas, P.D.; Campbell, M.J.; Kejariwal, A.; Mi, H.; Karlak, B.; Daverman, R.; Diemer, K.; Muruganujan, A.; Narechania, A. PANTHER: A library of protein families and subfamilies indexed by function. *Genome Res.* **2003**, *13*, 2129–2141. [CrossRef] [PubMed]

9. Ma, X.; Gao, L. Biological network analysis: Insights into structure and functions. *Brief. Funct. Genom.* **2012**, *11*, 434–442. [CrossRef] [PubMed]

10. Huntley, R.P.; Sawford, T.; Mutowo-Meullenet, P.; Shypitsyna, A.; Bonilla, C.; Martin, M.J.; O'Donovan, C. The GOA database: Gene Ontology annotation updates for 2015. *Nucleic Acids Res.* **2015**, *43*, D1057–D1063. [CrossRef] [PubMed]

11. Camon, E.; Magrane, M.; Barrell, D.; Lee, V.; Dimmer, E.; Maslen, J.; Binns, D.; Harte, N.; Lopez, R.; Apweiler, R. The Gene Ontology Annotation (GOA) Database: Sharing knowledge in Uniprot with Gene Ontology. *Nucleic Acids Res.* **2004**, *32*, 262D–266D. [CrossRef] [PubMed]

12. The UniProt Consortium. UniProt: The universal protein knowledgebase. *Nucleic Acids Res.* **2017**, *45*, D158–D169. [CrossRef] [PubMed]

13. Castillo-Davis, C.I.; Hartl, D.L. GeneMerge—Post-genomic analysis, data mining, and hypothesis testing. *Bioinformatics* **2003**, *19*, 891–892. [CrossRef] [PubMed]

14. Shannon, P.; Markiel, A.; Ozier, O.; Baliga, N.S.; Wang, J.T.; Ramage, D.; Amin, N.; Schwikowski, B.; Ideker, T. Cytoscape: A software environment for integrated models of biomolecular interaction networks. *Genome Res.* **2003**, *13*, 2498–2504. [CrossRef] [PubMed]

15. Bastian, M.; Heymann, S.; Jacomy, M. Gephi: An Open Source Software for Exploring and Manipulating Networks Visualization and Exploration of Large Graphs. In Proceedings of the Third International AAAI Conference on Weblogs and Social Media, San Jose, CA, USA, 17–20 May 2009; AAAI Publications.

16. Gu, Z.; Gu, L.; Eils, R.; Schlesner, M.; Brors, B. *circlize* Implements and enhances circular visualization in R. *Bioinformatics* **2014**, *30*, 2811–2812. [CrossRef] [PubMed]

17. Conway, J.R.; Lex, A.; Gehlenborg, N. UpSetR: An R package for the visualization of intersecting sets and their properties. *Bioinformatics* **2017**, *33*, 2938–2940. [CrossRef] [PubMed]

18. Sabrkhany, S.; Kuijpers, M.J.E.; Knol, J.C.; Olde Damink, S.W.M.; Dingemans, A.C.; Verheul, H.M.; Piersma, S.R.; Pham, T.V.; Griffioen, A.W.; Oude Egbrink, M.G.A.; et al. Exploration of the platelet proteome in patients with early-stage cancer. *J. Proteom.* **2018**, *177*, 65–74. [CrossRef] [PubMed]

19. Dolan, W.L.; Chapple, C. Transcriptome analysis of four *Arabidopsis thaliana* mediator tail mutants reveals overlapping and unique functions in gene regulation. *Genes Genomes Genet.* **2018**, *9*, 3093–3108.

20. Gay, L.J.; Felding-Habermann, B. Contribution of platelets to tumour metastasis. *Nat. Rev. Cancer* **2011**, *11*, 123–134. [CrossRef] [PubMed]

21. Huang, D.W.; Sherman, B.T.; Lempicki, R.A. Systematic and integrative analysis of large gene lists using DAVID bioinformatics resources. *Nat. Protoc.* **2009**, *4*, 44–57. [CrossRef] [PubMed]

22. Kellner, R. Proteomics. Concepts and perspectives. *Fresenius J. Anal. Chem.* **2000**, *366*, 517–524. [CrossRef] [PubMed]

23. Reimand, J.; Arak, T.; Adler, P.; Kolberg, L.; Reisberg, S.; Peterson, H.; Vilo, J. g:Profiler—A web server for functional interpretation of gene lists (2016 update). *Nucleic Acids Res.* **2016**, *44*, W83–W89. [CrossRef] [PubMed]

24. Eden, E.; Navon, R.; Steinfeld, I.; Lipson, D.; Yakhini, Z. GOrilla: A tool for discovery and visualization of enriched GO terms in ranked gene lists. *BMC Bioinformat.* **2009**, *10*, 48. [CrossRef] [PubMed]

25. Zheng, Q.; Wang, X.J. GOEAST: A web-based software toolkit for Gene Ontology enrichment analysis. *Nucleic Acids Res.* **2008**, *36*, W358–W363. [CrossRef] [PubMed]
26. Zhang, B.; Kirov, S.; Snoddy, J. WebGestalt: An integrated system for exploring gene sets in various biological contexts. *Nucleic Acids Res.* **2005**, *33*, W741–W748. [CrossRef] [PubMed]

 *genes*

Article

IPCT: Integrated Pharmacogenomic Platform of Human Cancer Cell Lines and Tissues

Muhammad Shoaib [1,2,†], **Adnan Ahmad Ansari** [1,2,†], **Farhan Haq** [3,*] **and Sung Min Ahn** [2,4,*]

1   Department of Biomedical Engineering, College of Medicine, University of Ulsan, Asan Medical Center, Seoul 100-011, Korea; muhemmed.shoaib@gmail.com (M.S.); adnanansa@gmail.com (A.A.A.)
2   Gachon Institute of Genome Medicine and Sciences, Incheon 400-011, Korea
3   Department of Biosciences, COMSATS University Islamabad, Islamabad 45710, Pakistan
4   Department of Genome Medicine and Science, College of Medicine, Gachon University, Seongnam 461-140, Korea
*   Correspondence: farhan.haq@comsats.edu.pk (F.H.); smahn@gachon.ac.kr (S.M.A.)
†   These authors equally contributed to this paper.

Received: 21 January 2019; Accepted: 18 February 2019; Published: 22 February 2019

Abstract: (1) *Motivation*: The exponential increase in multilayered data, including omics, pathways, chemicals, and experimental models, requires innovative strategies to identify new linkages between drug response information and omics features. Despite the availability of databases such as the Cancer Cell Line Encyclopedia (CCLE), the Cancer Therapeutics Response Portal (CTRP), and The Cancer Genome Atlas (TCGA), it is still challenging for biologists to explore the relationship between drug response and underlying genomic features due to the heterogeneity of the data. In light of this, the Integrated Pharmacogenomic Database of Cancer Cell Lines and Tissues (IPCT) has been developed as a user-friendly way to identify new linkages between drug responses and genomic features, as these findings can lead not only to new biological discoveries but also to new clinical trials. (2) *Results*: The IPCT allows biologists to compare the genomic features of sensitive cell lines or small molecules with the genomic features of tumor tissues by integrating the CTRP and CCLE databases with the REACTOME, cBioPortal, and Expression Atlas databases. The input consists of a list of small molecules, cell lines, or genes, and the output is a graph containing data entities connected with the queried input. Users can apply filters to the databases, pathways, and genes as well as select computed sensitivity values and mutation frequency scores to generate a relevant graph. Different objects are differentiated based on the background color of the nodes. Moreover, when multiple small molecules, cell lines, or genes are input, users can see their shared connections to explore the data entities common between them. Finally, users can view the resulting graphs in the online interface or download them in multiple image or graph formats. (3) *Availability and Implementation*: The IPCT is available as a web application with an integrated MySQL database. The web application was developed using Java and deployed on the Tomcat server. The user interface was developed using HTML5, JQuery v.3.1.0, and the Cytoscape Graph API v.1.0.4. The IPCT web and the source code are available in Sample Availability section.

Keywords: genomics; pharmacogenomics; cell lines; database; drug sensitivity

---

1. Introduction

Advancements in pharmacogenomics through comprehensive next-generation sequencing studies have paved the way for developing effective therapeutics against cancer. The omics data of cancer cell lines and cancer tissues are now readily used for categorizing genomic diversity and identifying anti-cancer drug responses [1]. However, in the era of big data, biologists face new challenges in dealing with the large amount of segregated data available in different cancer genomic repositories [2,3].

In the past decade, data scientists have developed numerous biological databases to help biologists analyze the underlying genetic mechanisms of cancer. NCI-60 [2], the first cancer cell line database, remained a unique resource of in vitro drug discovery for many years [4]. Recently, large pharmacogenomic databases such as the Cancer Cell Line Encyclopedia (CCLE) [5], Genomics of Drug Sensitivity in Cancer (GDSC), and the Cancer Therapeutics Response Portal (CTRP) have also emerged. The CCLE provides genomic and transcriptomic information on 947 human cancer cell lines and the drug response data for 24 compounds [5]. The CTRP provides drug response information for more than 860 cancer cell lines against 481 compounds [6]. Furthermore, in addition to cell line data, the omics data of thousands of cancer patients were also generated by The Cancer Genome Atlas (TCGA) and the European Molecular Biology Laboratory (EMBL) [7].

Unfortunately, the volume and heterogeneity of the data has prevented biologists from making effective use of these databases [8]. Therefore, an efficient and biologist-friendly integration of these omics and pharmacogenomics databases is needed. This integration would help biologists generate accurate and practical hypotheses for identifying anti-cancer drug responses. The prime objective of this study was to provide a uniquely user-friendly platform for cancer biologists that they can use to investigate interlinked pharmacogenomics and cancer genomics data.

In this study, we have developed the Integrated Pharmacogenomics Platform of Cancer Cell Lines and Tissues (IPCT), which integrates major drug response information from the CTRP with omics data from the CCLE, cBioPortal [8], REACTOME [9], and Expression Atlas [10] databases. The IPCT is a biologist-friendly platform with numerous novel features, highlighting:

(1) the genomic features sensitive to specific drugs;
(2) the percentage of affected cancer patients sensitive to a drug;
(3) the pathways associated with the drug response;
(4) cancer cell lines that are true representatives of cancer tissues;
(5) user-friendly single-click access to multiple datasets, which facilitates the generation of new and practical hypotheses.

2. Materials and Methods

The CTRP contains quantitatively measured sensitivity for 481 small molecules in 860 deeply characterized cancer cell lines. The IPCT (1) integrates the CTRP database with external biological databases and (2) allows biologists to query CTRP data in an integrated graphical fashion. Biologists can start querying by entering a list of cell lines or small molecules and use the context of the results of their search to generate new hypotheses.

The IPCT was developed in three different phases:

(1) construction of the database
(2) development of the database update pipeline
(3) design of the web application

The database is an essential component of the IPCT, storing all the data points and the connections between those data points from the CTRP, CCLE, cBioPortal, REACTOME, and Expression Atlas databases. The update pipeline is a script written in Python that is used to update the database in real time. The web application is a GUI-based application that will be used by the end users to explore the data points and their connections.

The IPCT is a biological database that integrates data about cancer cell lines, small molecules, human pathways, experimental results, and cancer somatic mutations. Figures 1 and 2 show the architecture of the IPCT database and demonstrate how multiple databases have been integrated in the IPCT database. We collected the cell line data from the CCLE dataset, the small molecule features from the CTRP dataset, the pathway data from REACTOME, the expression data from the Expression Atlas, the list of cancer genes from cancer genes census [11] and OncoKB [12], and the genomic features of cancer studies from cBioPortal. Our objective was to create an integrated database by connecting the data points in the above databases.

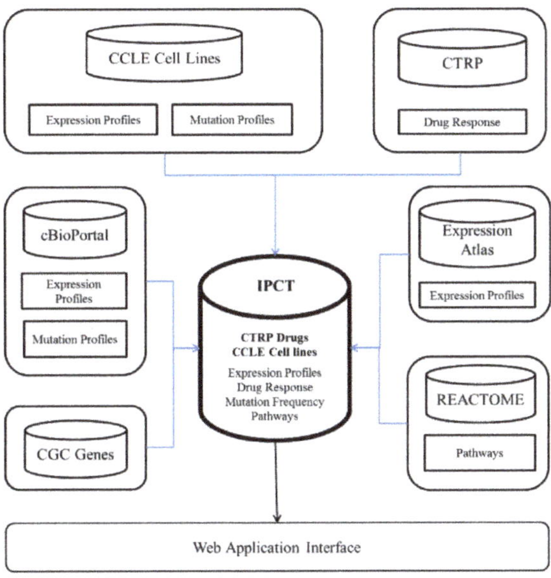

Figure 1. Entities in the IPCT database.

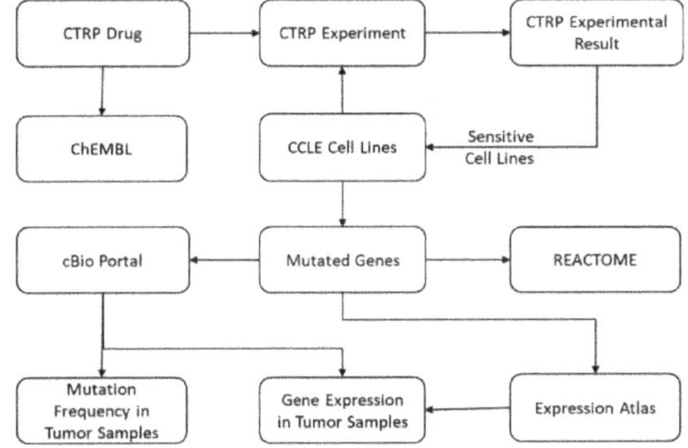

Figure 2. Connectivity map of CTRP drugs and CCLE cell lines with ChEMBL, REACTOME, the Expression Atlas, and cBioPortal.

The CTRP database contains experimental details and results reported as the area under the curve (AUC). Therefore, as a first step, we computed a sensitivity score for each small molecule against each cell line using these AUC values. The sensitivity scores were computed using the R package "extremevalues" in a similar manner to Speyer et al. [13]. After this step, we obtained a numeric score for each small molecule–cell line pair. A small molecule is sensitive to a cell line if its sensitivity score is below -1 and resistant if its sensitivity score is above 1 [13]. This sensitivity score was then used to construct a small molecule–cell line network, and the small molecules were connected to the ChEMBL database. A total of 158 small molecules were common between the CTRP and ChEMBL.

In the second step, we added genes to our network. To do this, a list of genes and their relationships to cell lines was required. We extracted genetic metadata from the NCBI website and connected the genes and cell lines based on genomic changes, which were present in the CCLE dataset in the form of mutations, copy number alterations, and gene expression. We extracted these for each cell line from the CCLE dataset and used this information to construct a small molecule–cell line–gene graph. A gene was included in the small molecule–cell line–gene network if it had mutations, copy number amplification, copy number deletion, high expression, or low expression in at least 10% of the cell lines sensitive to an input small molecule. The IPCT, by default, connects only genes with genomic aberration in 20% of the cell lines sensitive to an input small molecule. However, users can relax or tighten these criteria as needed. Having already constructed a small molecule–cell line network, in this step we only had to connect the cell lines with the genes. To do this, we connected cell lines with the genes that had mutations or copy number alterations in the given cell lines. This process was repeated for all cell lines in the CCLE, which resulted in a cell line–gene network with genes and cell lines as nodes and mutations or copy number alterations as edges. The cell line–gene graph was then merged with the small molecule–cell line graph, which resulted in a small molecule–cell line–gene network. After this step, we could identify genes with mutations or copy number alterations in the cell lines that are sensitive to a given set of small molecules.

Once we had identified the mutated genes, we added the pathways of the mutated genes to our small molecule–cell line–gene network, collecting pathway data from the REACTOME database. These pathways were connected using Entrez GeneIDs present in both databases.

The next step was to identify if the mutated genes had been reported as up-regulated or down-regulated in previous experiments. The Expression Atlas contains differential expression data from approximately 2500 experiments performed in different experimental conditions. However, the Expression Atlas uses Ensembl IDs instead of gene names or Entrez GeneIDs in its analyzed files. In the first step, we filtered only those experiments that were related to cancer, loaded them into the database, and removed insignificant records with p-value > 0.05 and log fold change >-1 and <1. Records with log fold change ≥ 1 or ≤ -1 and p-value < 0.05 were used for further processing. Next, we connected all Ensembl IDs with their Entrez GeneIDs using the R package "org.Hs.eg.db". We used this database to construct a gene–experiment network with genes and experiments as nodes and up-regulation or down-regulation as edges. This graph was then merged with the small molecule–cell line–gene network constructed in the previous step.

After construction of this pathway graph, our next task was to identify if the mutated genes had any potential relationship with any cancer types in published cancer studies. To do this, we extracted data from cBioPortal. For each gene, we computed what percentage of samples were mutated, altered, up-regulated, and down-regulated in each study, identifying mutation and copy number alteration frequencies for 30,000 genes in 151 cancer studies and 33 cancer types. The data from cBioPortal were not used in network construction but are available as a separate entity for further investigation.

The IPCT can be accessed via the web application, which allows users to explore the connections between the data points of five biological databases in an integrated graphical fashion. When a user enters a small molecule, cell line, or gene, a graph is displayed with the data points as nodes and the relationships between the data points as edges. Using this graph, the user can intuitively investigate the connectivity of the given small molecules, cell lines, and genes. When a user enters multiple cell lines, small molecules, or genes, the IPCT first independently constructs a graph for each element in the list. Next, it takes two random graphs from among those and merges them using the common data points. This step is repeated until all the graphs are merged into one graph, which is ultimately displayed to the user.

3. Results

The IPCT comprises two major components: (1) the IPCT database and (2) the IPCT web portal. The IPCT web portal provides an easy way to investigate the connections between the data points available in the CTRP, CCLE, Expression Atlas, REACTOME, and cBioPortal databases in an integrated fashion. The IPCT database currently contains 860 cell lines, 481 small molecules, ~2500 differential expression studies, 2000 human pathways, and 151 cancer studies. Moreover, the IPCT contains

8,214,573 unique connections between the different data points (Table S1). The overall database size is 20 GB. The distinctive functionality and features of the IPCT are as follows:

1. Users can input up to ten cell lines, small molecules, or genes to find potential connectivity with other data points.
2. Users can filter small molecules and cell lines sensitive to each other according to a minimum sensitivity score.
3. Users can apply a filter on genes if they want to view only cancer genes, exclude commonly mutated genes, or view all genes.
4. Users can apply filters if they want to see only mutated, copy number altered, or high- or low-expressed genes.
5. Users can check the mutation frequencies and differential expression frequencies in different cancer studies.
6. Users can highlight genes of their interest by applying a gene filter to the network.
7. Users can select if they want to show all connections or only shared connections when multiple cell lines, small molecules, or genes are entered.
8. Users can view the output in the web browser as a graph or table. Alternatively, users can download the graph and view it with Cytoscape version 1.0.4 or graph viewing tools that show JSON and CSV files.
9. Users can save the graphs in JSON, PNG, or PDF formats and table in CSV format.

3.1. Data Exploration

Users can start exploring the IPCT by entering small molecules, cell lines, or genes. If users enter a list of cell lines, the IPCT outputs graphs with small molecules that are sensitive to the queried cell lines and genes that are mutated or altered in the given cell lines. If users enter a small molecule, the IPCT outputs a graph containing the cell lines sensitive to the given small molecule and genes mutated in the sensitive cell lines. If users enter genes, the IPCT outputs a graph of cell lines with mutations or copy number alterations in the given genes and the small molecules sensitive to those cell lines. Users can then expand their search by expanding the graph to include data points from the Expression Atlas or REACTOME. User can apply filters as explained in Table 1 and reduce number of entities in graph. Figure 3 illustrates the output generated by the IPCT for lapatinib with the shared pathway filter. By default, the IPCT shows the pathways associated with more than 20% of genes connected with the input drug, but users can modify this option to show all pathways if they want to see the pathways of connected genes. Supplementary Figure S1 shows the result of same query with all pathways. The IPCT also allows users to apply different filters to define the context of their search.

Table 1. Database filters that can be applied to searches in the IPCT.

Database Filter	Applicable Object	Function
Compound sensitivity	Small molecules	Allow users to set thresholds for small molecule sensitivity
Mutation frequency	Genes	Allow users to set mutation frequency
Gene filter	Genes	Allow users to show only cancer genes, exclude commonly mutated genes, or see all genes
Pathway filter	REACTOME	Allows users to select metabolic and signaling pathways
Genomic aberration	Genes	Allow users to filter gene relationships based on mutations, copy number alterations, and gene expression

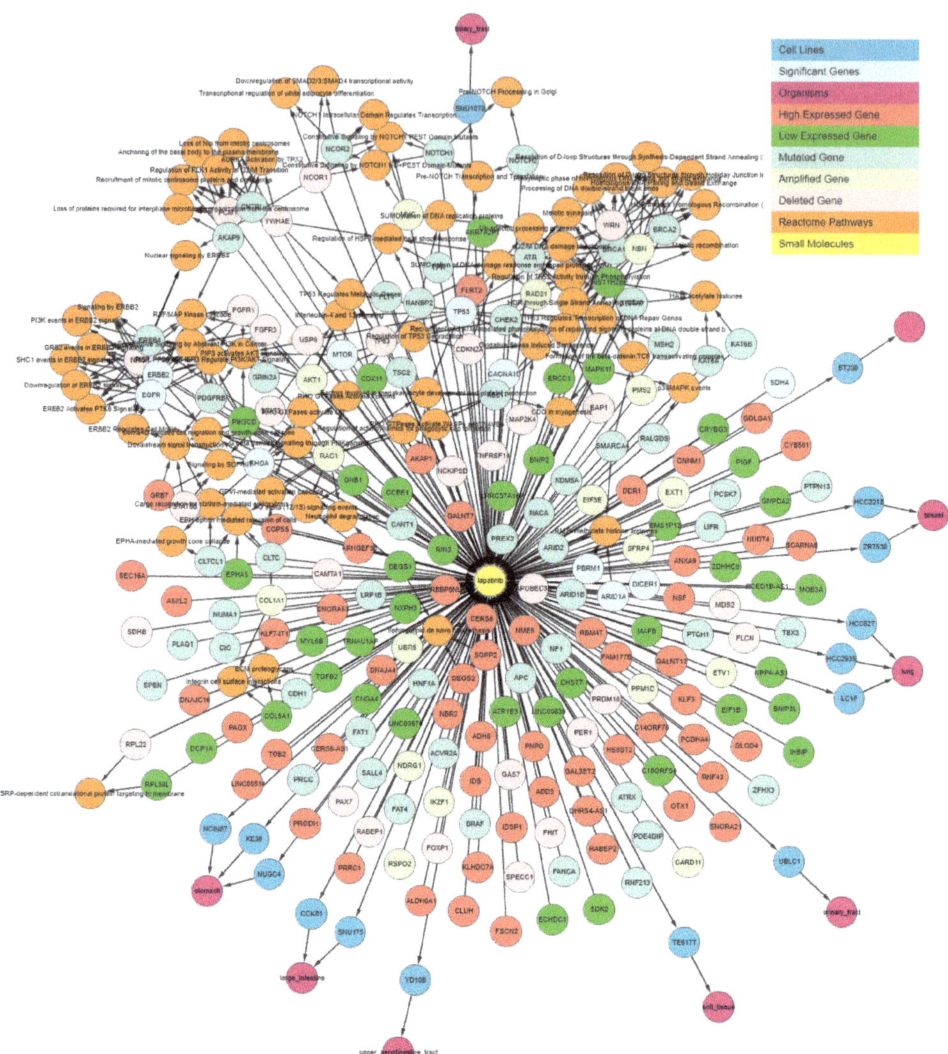

Figure 3. IPCT output for small molecule user query lapatinib. The graph shows all data points connected with lapatinib. Yellow nodes represent small molecules; blue nodes show cell lines sensitive to lapatinib; sky-blue nodes represent significant genes (those with multiple genomic aberrations); green and red nodes represent genes that are up-regulated and down-regulated in the sensitive cell lines, respectively; light green and light red represent the amplified and deleted genes in the sensitive cell lines, respectively; white nodes represent mutated genes; and orange nodes represent the REACTOME pathways of mutated genes.

3.2. Comparison Between Cell Lines and Real Tissues

As not all cancer cell lines have equal values to the tumor models, comparison between the genetic profiles of cell lines and real tumors is of importance. For example, when a mutation is found in a cell line, the first question might be if the specific mutation has also been reported in any cancer studies, followed by whether the given gene has any reported differential expression. The IPCT, by integrating data from cBioPortal and the Expression Atlas, provides answers to both questions. When a user clicks on a mutated gene's node, he can explore the cancer studies in which the given gene is up-regulated or down-regulated and observe its mutation or copy number alteration percentages in all cBioPortal cancer studies. For example, in the previously illustrated query (Figure 3), by investigating lapatinib, sorafenib, gefitinib, and sunitinib together, we identified that *ERBB4* is mutated in 38% of cell lines sensitive to sorafenib, 33% of cell lines sensitive to gefitinib, 31% of cell lines sensitive to sunitinib, and 27% of cell lines sensitive to lapatinib. Users could then further investigate its frequency in real tumors. Figure 4 illustrates the results for the mutations and the differential expression frequency of *ERBB4* in different cancer studies.

A

ERBB4 EXPRESSION IN CBIOPORTAL					
Gene Symbol	Cancer Type	Cancer Study	Measurement Type	Up Regulated	Down Regulated
ERBB4	Cervical Squamous Cell Carcinoma	cesc_tcga	rna seq v2 mrna median Zscores	8%	0%
ERBB4	Glioblastoma Multiforme	gbm_tcga_pub2013	rna seq v2 mrna median Zscores	14%	0%
ERBB4	Uterine Carcinosarcoma/Uterine Malignant Mixed Mullerian Tumor	ucs_tcga	rna seq v2 mrna median Zscores	12%	0%
ERBB4	Invasive Breast Carcinoma	brca_tcga	rna seq v2 mrna median Zscores	9%	0%
ERBB4	Bladder Urothelial Carcinoma	blca_tcga	rna seq v2 mrna median Zscores	10%	0%
ERBB4	Cutaneous Melanoma	skcm_tcga	rna seq v2 mrna median Zscores	4%	0%
ERBB4	Thymoma	thym_tcga	rna seq v2 mrna median Zscores	14%	0%
ERBB4	Esophageal Adenocarcinoma	esca_tcga	rna seq v2 mrna median Zscores	2%	0%
ERBB4	Bladder Urothelial Carcinoma	blca_tcga_pub	rna seq v2 mrna median Zscores	12%	0%
ERBB4	Lung Squamous Cell Carcinoma	lusc_tcga_pub	rna seq mrna median Zscores	8%	0%

B

ERBB4 MUTATIONS IN CBIOPORTAL				
Gene Symbol	Cancer Type	Cancer Study	Mutations	Alterations
ERBB4	Bladder Urothelial Carcinoma	blca_dfarber_mskcc_2014_sequenced	34%	34%
ERBB4	Cutaneous Squamous Cell Carcinoma	cscc_dfarber_2015_sequenced	31%	31%
ERBB4	Desmoplastic Melanoma	desm_broad_2015_sequenced	30%	30%
ERBB4	Small Cell Lung Cancer	sclc_clcgp_sequenced	21%	21%
ERBB4	Cutaneous Melanoma	skcm_broad_dfarber_sequenced	20%	20%
ERBB4	Cutaneous Melanoma	skcm_broad_sequenced	16%	16%
ERBB4	Mixed Cancer Types	cellline_nci60_cnaseq	15%	15%
ERBB4	Stomach Adenocarcinoma	stad_tcga_cnaseq	12%	14%
ERBB4	Esophagogastric Adenocarcinoma	stes_tcga_pub_cnaseq	11%	13%
ERBB4	Colorectal Adenocarcinoma	coadread_genentech_sequenced	11%	11%
ERBB4	Cutaneous Melanoma	skcm_yale_sequenced	11%	11%
ERBB4	Esophageal Adenocarcinoma	esca_tcga_cnaseq	9%	13%

Figure 4. ERBB4's genetic profile in real tumors extracted from cBioPortal. (**A**) ERBB4's differential expression in different cancer studies. (**B**) ERBB4's mutation and copy number alteration frequency in different cancer studies.

3.3. Filtering Genes

Cell lines harbor mutations in many genes; however, not all mutated genes are of interest to biologists, who mostly value mutations in oncogenes or tumor suppressor genes because of their well-defined role in cancer. In light of this, the IPCT gene filter has the following three options:

1. Cancer genes: Construct graphs in the context of only oncogenes or tumor suppressor genes.
2. Exclude commonly mutated genes: Construct graphs in the context of all genes, but exclude genes that have mutations in more than 90% of cell lines.
3. All genes: Disable the filter and construct graphs in the context of all genes.

This filter enables users to focus on mutations in cancer genes and further explore only cancer genes as well as to focus on rarely mutated genes by allowing them to exclude genes that are mutated in more than 90% of cell lines. Figure 4 illustrates the effect of applying a gene filter. Figure 5A shows only cancer genes that are mutated in sensitive cell lines, and Figure 5B shows the network with all genes excluding frequently mutated genes, i.e., genes that are mutated in more than 90% of overall cell lines.

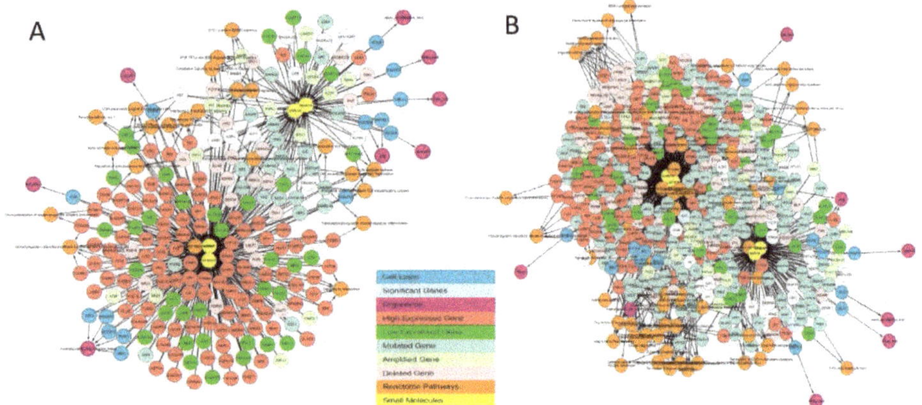

Figure 5. IPCT output for small molecule user query lapatinib, sorafenib, gefitinib, and sunitinib after disabling the REACTOME and Expression Atlas databases and enabling cell lines and mutated genes only. (**A**) Gene filter = only cancer genes. (**B**) Gene filter = exclude common mutations.

3.4. Finding Shared Connections

Another important feature of the IPCT for biologists is identifying shared connections between data entities, such as hidden direct or indirect relationships between two cell lines and small molecule sensitivity or between small molecule sensitivity and gene mutations.

The shared connection filter in the IPCT allows users to investigate unknown or hidden relationships between the data entities of the five connected databases. For example, when a user inputs more than two small molecules, the IPCT constructs a graph with cell lines sensitive to the input small molecules, their mutated genes, and the data entities connected with the mutated genes. By enabling the shared connection filter, users can restrict the results to cell lines sensitive to both small molecules. Similarly, users can restrict the graph to genes mutated in more than one cell line or to pathways that are common between mutated genes.

3.5. Case Study

Lapatinib and afatinib are two tyrosine kinase inhibitors that are effective in breast cancer. These drugs are usually effective in *HER2* (*ERBB2*) mutation-positive patients [14,15]. In this section, we demonstrate how the IPCT can be used to identify the mechanism of action of these two kinase inhibitors. For this purpose, in the first stage, we query lapatinib and afatinib in the IPCT. Figure 6 shows the graph containing the sensitive cell lines, associated genes, and their pathways generated by the IPCT as result of the query, without applying any filter. Genes associated with these drugs are colored and shaped based on their relationship; each color and shape represent a unique relationship between the genes and the cell lines sensitive to the input drug. As such, genes with certain colors and shapes can be classified as more important than other genes.

Next, we apply a filter to shortlist our gene set. We first apply the shared connection filter to see if any genes are associated with both drugs. Genes can have different associations with each drug, and the more important genes will be those that have the same association with both drugs. Figure 7 shows the resulting graph. The sky-blue genes are the most important ones, whereas those with a white background are the least important. The circled genes can be classified as the most relevant gene set due to the pathway clusters. Figure 7 shows that *EGFR* is amplified, *NRG1* and *FGFR1* are deleted, and *AKAP9* and *TP53* have mutations in cell lines sensitive to both drugs. These genes have been found to be relevant to lapatinib and afatinib in the literature [15–17]. *ERBB2* is amplified and highly expressed in 95% of afatinib-sensitive cell lines and 100% of lapatinib-sensitive cell lines. Finally, we apply relationship filters to identify the most relevant results. These filters are designed to filter genes that have multiple genomic aberrations with the queried drugs. Finally, Figure 8 demonstrates the relationship of *ERBB2* with lapatinib and afatinib.

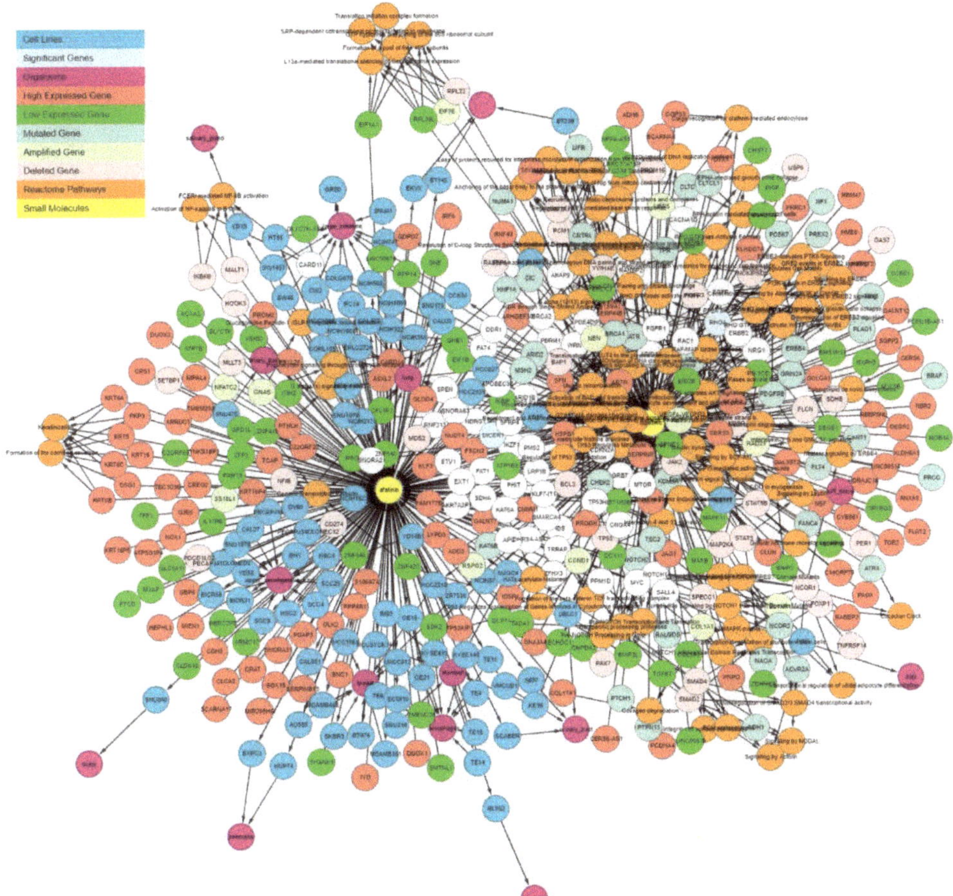

Figure 6. IPCT output for small molecule user query lapatinib and afatinib. The graph shows all data points connected with lapatinib and afatinib. Yellow nodes represent small molecules; blue nodes show cell lines sensitive to lapatinib and afatinib; sky-blue nodes represent significant genes (those with multiple genomic aberrations); green and red nodes represent genes that are up-regulated and down-regulated in the sensitive cell lines, respectively; light green and light red represent the amplified and deleted genes in the sensitive cell lines, respectively; white nodes represent mutated genes; and orange nodes represent the REACTOME pathways of mutated genes.

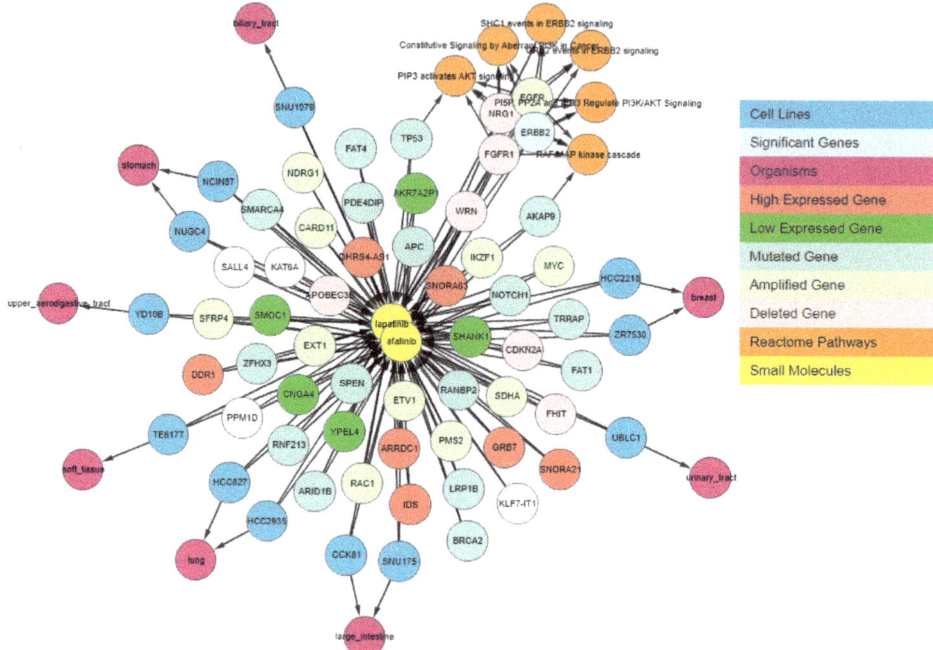

Figure 7. IPCT output for small molecule user query lapatinib and afatinib with the shared connection filter enabled. The graph shows all data points connected with lapatinib and afatinib. Yellow nodes represent small molecules; blue nodes show cell lines sensitive to lapatinib and afatinib; sky-blue nodes represent significant genes (those with multiple genomic aberrations); green and red nodes represent genes that are up-regulated and down-regulated in the sensitive cell lines, respectively; light green and light red represent the amplified and deleted genes in the sensitive cell lines, respectively; white nodes represent mutated genes; and orange nodes represent the REACTOME pathways of mutated genes.

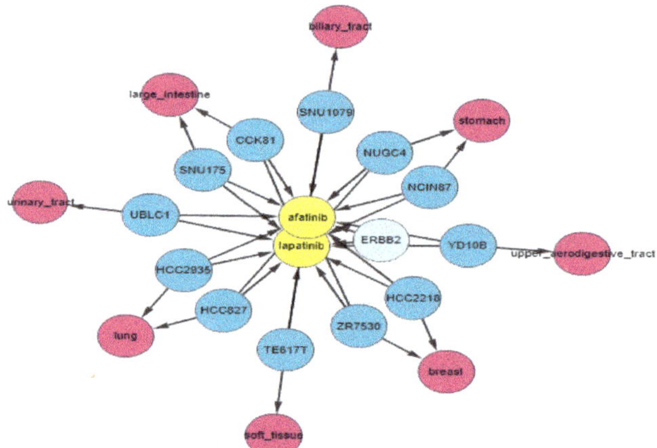

Figure 8. IPCT output for small molecule user query lapatinib and afatinib with the shared connection filter and the relationship filter enabled. The graph shows all data points connected with lapatinib and afatinib. Yellow nodes represent small molecules; blue nodes show cell lines sensitive to lapatinib and afatinib; and sky-blue nodes represent significant genes (those with multiple genomic aberrations).

3.6. Download Graph

The web interface of the IPCT allows users to navigate the data entities connected with CTRP small molecules or CCLE cell lines. In addition to this, users can also download the output connectivity maps for future reference as image or PDF files. Furthermore, users can also download connectivity maps in the GraphML version 1.0, JSON, PNG, PDF or CSV formats for further analysis or generate high-quality graphs using external graph-making tools such as Cytoscape. In these formats, the nodes represent data entities and the edges represent the connections between them. Each edge is identified with a unique identifier and contains information in the form of a label, database identifier, and URL. The labels store the display information, and the database identifiers store the color coding. Nodes representing small molecules contain additional information about their sensitivity to the connected cell lines. Similarly, nodes representing genes contain additional information about their mutation frequency in the sensitive cell lines.

4. Discussion and Conclusions

Recent advancements in pharmacogenomics through high-throughput sequencing have necessitated that big-data scientists should develop innovative strategies to deal with the rapidly increasing amount of available biological data. The two major limitations to making effective use of all this information are the extensive heterogeneity of the data and a lack of integration. To overcome these limitations, data scientists have developed integrated, biologist-friendly databases. For instance, the European Bioinformatics Institute's RDF platform is a state-of-the-art example that has enabled the integration of six different biological databases, including UniProt, the Expression Atlas, REACTOME, ChEMBL, BioModels, and BioSamples. However, to the best of our knowledge, no large-scale efforts have been made to integrate the pharmacogenomic features of cancer cell lines with the cancer-related genomic features of real cancer patients.

One of the key questions for any biologist is whether the genomic features of cancer cell lines that are sensitive to drugs are also relevant in real cancer tissues. To address this, biologists previously had to search through multiple heterogeneous databases, which is a challenging job even for researchers with advanced computer skills. Recently, Elena Piñeiro-Yáñez et al. [18] developed PanDrugs to prioritize anticancer drug treatments depending on patients' genomic profiles. PanDrugs mostly focuses on the clinical aspects of cancer genomics, whereas the IPCT is designed to help researchers in the generation and in silico testing of hypotheses on the pharmacogenomics data of human cell lines and the genomic data of human tumor samples. The IPCT enables data integration and interoperability between the CTRP, CCLE, Expression Atlas, REACTOME, and cBioPortal databases, allowing users to investigate the connectivity maps of cell lines, small molecules, and genes of interest in a user-friendly fashion.

In summary, the IPCT enables biologists to investigate the connectivity of small molecules and genomics features in relationship with cancer cell lines and real cancer tissues. It also highlights the genomic features sensitive to a specific drug and the percentage of cancer patients affected by that drug. Notably, IPCT can also identify cancer cell lines that are truly representative of real cancer tissues. In conclusion, the integration of these five major databases in a biologist-friendly manner will help researchers generate new and tangible hypotheses, leading to further clinical trials in the quest for better cancer treatment.

Supplementary Materials: The following are available online at http://www.mdpi.com/2073-4425/10/2/171/s1, Figure S1: IPCT output for small molecule user query lapatinib with all pathways levels, Table S1: User Input query for Figure 3, Table S2: User Input query for Figure 4, Table S3: User Input query for Figure 5, Table S4: No of Connections in IPCT, Table S5: Comparison with other Tools.

Author Contributions: S.M.A. and F.H. conceived and supervised the study; M.S. and A.A.A. conducted bioinformatics analyses and database development; S.M.A., M.S., and A.A.A. drafted the article; all authors were involved in manuscript preparation and review.

Funding: This research was supported by a grant from the Korea Health Technology R&D Project though the Korea Health Industry Development Institute funded by the Ministry of Health and Welfare (HI16C1985).

Conflicts of Interest: The authors declare no conflict of interest

References

1. Gillet, J.-P.; Varma, S.; Gottesman, M.M. The clinical relevance of cancer cell lines. *JNCI J. Natl. Cancer Inst.* **2013**, *105*, 452–458. [CrossRef] [PubMed]
2. Shoemaker, R.H. The NCI60 human tumour cell line anticancer drug screen. *Nat. Rev. Cancer* **2006**, *6*, 813–823. [CrossRef] [PubMed]
3. Stransky, N.; Ghandi, M.; Kryukov, G.V.; Garraway, L.A.; Lehár, J.; Liu, M.; Sonkin, D.; Kauffmann, A.; Venkatesan, K.; Edelman, E.J.; et al. Pharmacogenomic agreement between two cancer cell line data sets. *Nature* **2015**, *528*, 84. [CrossRef] [PubMed]
4. Weinstein, J.N. Spotlight on molecular profiling: "Integromic" analysis of the NCI-60 cancer cell lines. *Mol. Cancer Ther.* **2006**, *5*, 2601–2605. [CrossRef] [PubMed]
5. Barretina, J.; Caponigro, G.; Stransky, N.; Venkatesan, K.; Margolin, A.A.; Kim, S.; Wilson, C.J.; Lehár, J.; Kryukov, G.V.; Sonkin, D.; et al. The Cancer Cell Line Encyclopedia enables predictive modelling of anticancer drug sensitivity. *Nature* **2012**, *483*, 603–607. [CrossRef] [PubMed]
6. Basu, A.; Bodycombe, N.E.; Cheah, J.H.; Price, E.V.; Liu, K.; Schaefer, G.I.; Ebright, R.Y.; Stewart, M.L.; Ito, D.; Wang, S.; et al. An interactive resource to identify cancer genetic and lineage dependencies targeted by small molecules. *Cell* **2013**, *154*, 1151–1161. [CrossRef] [PubMed]
7. Tomczak, K.; Czerwińska, P.; Wiznerowicz, M. The Cancer Genome Atlas (TCGA): An immeasurable source of knowledge. *Contemp. Oncol.* **2015**, *19*, A68–A77. [CrossRef] [PubMed]
8. Gao, J.; Aksoy, B.A.; Dogrusoz, U.; Dresdner, G.; Gross, B.; Sumer, S.O.; Sun, Y.; Jacobsen, A.; Sinha, R.; Larsson, E.; et al. Integrative analysis of complex cancer genomics and clinical profiles using the cBioPortal. *Sci. Signal.* **2013**, *6*, pl1. [CrossRef] [PubMed]
9. Croft, D.; Mundo, A.F.; Haw, R.; Milacic, M.; Weiser, J.; Wu, G.; Caudy, M.; Garapati, P.; Gillespie, M.; Kamdar, M.R.; et al. The Reactome pathway knowledgebase. *Nucleic Acids Res.* **2014**, *42*, D472–D477. [CrossRef] [PubMed]
10. Petryszak, R.; Burdett, T.; Fiorelli, B.; Fonseca, N.A.; Gonzalez-Porta, M.; Hastings, E.; Huber, W.; Jupp, S.; Keays, M.; Kryvych, N.; et al. Expression Atlas update—A database of gene and transcript expression from microarray- and sequencing-based functional genomics experiments. *Nucleic Acids Res.* **2014**, *42*, D926–D932. [CrossRef] [PubMed]
11. Futreal, P.A.; Coin, L.; Marshall, M.; Down, T.; Hubbard, T.; Wooster, R.; Rahman, N.; Stratton, M.R. A census of human cancer genes. *Nat. Rev. Cancer* **2004**, *4*, 177–183. [CrossRef] [PubMed]
12. Chakravarty, D.; Gao, J.; Phillips, S.; Kundra, R.; Zhang, H.; Wang, J.; Rudolph, J.E.; Yaeger, R.; Soumerai, T.; Nissan, M.H.; et al. OncoKB: A precision oncology knowledge base. *JCO Precis. Oncol.* **2017**, *1*, 1–16. [CrossRef] [PubMed]
13. Speyer, G.; Mahendra, D.; Tran, H.J.; Kiefer, J.; Schreiber, S.L.; Clemons, P.A.; Dhruv, H.; Berens, M.; Kim, S. Differential pathway dependency discovery associated with drug response across cancer cell lines. *Pa. Symp. Biocomput.* **2016**, *22*, 497–508.
14. Rimawi, M.F.; Aleixo, S.B.; Rozas, A.A.; Nunes de Matos Neto, J.; Caleffi, M.; Figueira, A.C.; Souza, S.C.; Reiriz, A.B.; Gutierrez, C.; Arantes, H.; et al. A neoadjuvant, randomized, open-label phase II trial of afatinib versus trastuzumab versus lapatinib in patients with locally advanced HER2-positive breast cancer. *Clin. Breast Cancer* **2015**, *15*, 101–109. [CrossRef] [PubMed]
15. Li, D.; Ambrogio, L.; Shimamura, T.; Kubo, S.; Takahashi, M.; Chirieac, L.R.; Padera, R.F.; Shapiro, G.I.; Baum, A.; Himmelsbach, F.; et al. BIBW2992, an irreversible EGFR/HER2 inhibitor highly effective in preclinical lung cancer models. *Oncogene* **2008**, *27*, 4702–4711. [CrossRef] [PubMed]
16. Forster, J.A.; Paul, A.B.; Harnden, P.; Knowles, M.A. Expression of NRG1 and its receptors in human bladder cancer. *Br. J. Cancer* **2011**, *104*, 1135–1143. [CrossRef] [PubMed]

17. Leech, A.O.; Vellanki, S.H.; Rutherford, E.J.; Keogh, A.; Jahns, H.; Hudson, L.; O'Donovan, N.; Sabri, S.; Abdulkarim, B.; Sheehan, K.M.; et al. Cleavage of the extracellular domain of junctional adhesion molecule-A is associated with resistance to anti-HER2 therapies in breast cancer settings. *Breast Cancer Res.* **2018**, *20*, 140. [CrossRef] [PubMed]

18. Piñeiro-Yáñez, E.; Reboiro-Jato, M.; Gómez-López, G.; Perales-Patón, J.; Troulé, K.; Rodríguez, J.M.; Tejero, H.; Shimamura, T.; López-Casas, P.P.; Carretero, J.; et al. PanDrugs: A novel method to prioritize anticancer drug treatments according to individual genomic data. *Genome Med.* **2018**, *10*, 41. [CrossRef] [PubMed]

Sample Availability: The IPCT is available as a web application with an integrated MySQL database. The web application was developed using Java and deployed on the Tomcat server. We developed the user interface using HTML5, JQuery, and the Cytoscape Graph API. The IPCT can be accessed at http://ipct.ewostech.net. The source code is available at https://github.com/muhammadshoaib/ipct.

 *genes*

Article

miRmapper: A Tool for Interpretation of miRNA–mRNA Interaction Networks

Willian A. da Silveira [1], Ludivine Renaud [2,3], Jonathan Simpson [1], William B. Glen Jr. [1], Edward. S. Hazard [1,4], Dongjun Chung [5] and Gary Hardiman [1,2,3,5,6,*]

[1] Center for Genomic Medicine, Bioinformatics, Medical University of South Carolina (MUSC), Charleston, SC 29425, USA; silveira@musc.edu (W.A.d.S.); jondsimp@gmail.com (J.S.); glen@musc.edu (W.B.G.); hazards@musc.edu (E.S.H.)
[2] Division of Nephrology, Department of Medicine, Medical University of South Carolina (MUSC), Charleston, SC 29425, USA; renaudl@musc.edu
[3] Laboratory for Marine Systems Biology, Hollings Marine Laboratory, Charleston, SC 29412, USA
[4] Academic Affairs Faculty, Medical University of South Carolina (MUSC), Charleston, SC 29425, USA
[5] Department of Public Health Sciences, Medical University of South Carolina (MUSC), Charleston, SC 29425, USA; chungd@musc.edu
[6] Institute for Global Food Security, Queens University Belfast, Stranmillis Road, Belfast BT9 5AG, UK
* Correspondence: G.Hardiman@qub.ac.uk

Received: 28 August 2018; Accepted: 7 September 2018; Published: 14 September 2018

Abstract: It is estimated that 30% of all genes in the mammalian cells are regulated by microRNA (miRNAs). The most relevant miRNAs in a cellular context are not necessarily those with the greatest change in expression levels between healthy and diseased tissue. Differentially expressed (DE) miRNAs that modulate a large number of messenger RNA (mRNA) transcripts ultimately have a greater influence in determining phenotypic outcomes and are more important in a global biological context than miRNAs that modulate just a few mRNA transcripts. Here, we describe the development of a tool, "miRmapper", which identifies the most dominant miRNAs in a miRNA–mRNA network and recognizes similarities between miRNAs based on commonly regulated mRNAs. Using a list of miRNA–target gene interactions and a list of DE transcripts, miRmapper provides several outputs: (1) an adjacency matrix that is used to calculate miRNA similarity utilizing the Jaccard distance; (2) a dendrogram and (3) an identity heatmap displaying miRNA clusters based on their effect on mRNA expression; (4) a miRNA impact table and (5) a barplot that provides a visual illustration of this impact. We tested this tool using nonmetastatic and metastatic bladder cancer cell lines and demonstrated that the most relevant miRNAs in a cellular context are not necessarily those with the greatest fold change. Additionally, by exploiting the Jaccard distance, we unraveled novel cooperative interactions between miRNAs from independent families in regulating common target mRNAs; i.e., five of the top 10 miRNAs act in synergy.

Keywords: bioinformatics pipelines; algorithm development for network integration; miRNA–gene expression networks; multiomics integration; network topology analysis

---

1. Introduction

Mature microRNAs (miRNAs) are ~22-nucleotide-long single-stranded noncoding RNAs which function as translational repressors in all known animal and plant genomes [1,2]. It is estimated that 30% of all genes in the mammalian cells are regulated by miRNAs [3]. Each miRNA can regulate the expression of hundreds of messenger RNAs (mRNAs), and each mRNA can be targeted by various miRNAs, with multiple miRNA-binding sites being required for the efficient repression of a target mRNA [2,3].

The traditional paradigm regarding the mode of silencing of miRNAs is that (1) most animal miRNAs bind their target mRNAs with mismatches, promoting repression of mRNA translation with little or no influence on mRNA abundance; and that (2) most plant miRNAs bind their targets with near-perfect complementarity, allowing Ago-catalyzed cleavage and degradation of the mRNA strand [4]. The scientific dogma was that perfect complementarity excluded translational repression because it enabled cleavage, and this contributed to the notion that plant and animal miRNAs behaved in fundamentally different ways. However, several reports have demonstrated that animal miRNAs induce significant degradation of target mRNAs [5–8] and that translational repression also occurs in plants [9].

These studies initiated a debate regarding the mode of action of miRNAs, a discourse that remains active in the scientific community [10–12], highlighting the complexity of miRNA-induced *translational repression and degradation*. This sparked yet more unanswered questions such as "is degradation an independent mechanism by which silencing is accomplished?", or "is it a consequence of a primary effect on translation?" In 2008, Brodersen et al. suggested that translational repression is the default mechanism by which miRNAs repress gene expression in both animals and plants [9], a study that was followed by a contradicting study by Guo et al. [13] stating that mammalian micRNAs predominantly act to decrease target mRNA levels. This work was based on the fact that only a small fraction of repression observed by ribosome profiling (11–16%) is attributable to reduced translational efficiency, whereas at least 84% of the repression is attributable instead to decreased mRNA levels [13].

Several studies have made important advances in elucidating the relative contributions of translational repression and mRNA degradation by animal microRNAs and have further characterized how translational repression is accomplished: inhibition of translation initiation; inhibition of translation elongation; cotranslational protein degradation; and premature termination of translation [14]. Regarding miRNA-induced mRNA degradation, it appears that the extent of degradation is specified by the mRNA target, and not by the miRNA itself, because the same miRNA can either repress translation or induce mRNA decay in a target-specific manner [6]. It remains unclear why some targets are degraded and others are not. It has been suggested that the number, type, and position of mismatches in the miRNA/mRNA duplex plays an important role in triggering degradation or translational arrest [15]. Although defining how miRNAs mediate their repressive effects has been a controversial subject over the past two decades, current evidence suggests that target mRNA degradation contributes largely to the miRNA-induced silencing effects. Given, however, that many of these studies were conducted *in vitro* with cultured mammalian cells rapidly dividing, it is necessary to confirm this shift in paradigm using other cell types and in *in vivo* studies.

In studying networks, including miRNA–mRNA interaction networks, one of the most relevant metrics is *"centrality"*. Simply described, centrality is a measure of the degree, i.e., the number of edges connected to a vertex (Figure 1a) [16]; the assumption is that vertices with the highest degrees (with the most connections) play important roles in the functioning of the system, making the degree of centrality a useful guide for focusing attention on the system's most crucial elements. In directional networks, vertices have two different degrees, an "in-degree" and an "out-degree", corresponding to the number of edges pointing inward to and outward from these vertices [16]. In the context of social networks, individuals who have connections to many others may be perceived as having greater influence, more access to information, or higher prestige than those who have fewer connections [17,18]. The same can be applied to the evaluation of scientific publications: the count of how many times a paper has been cited, equivalent to the "in-degree" in the citation network (Figure 1b), provides a measure of whether the paper has been influential or not. This is widely used as a metric for judging the impact of scientific research [19,20]. Centrality, when applied to miRNA–mRNA interaction networks, can highlight which miRNAs are more important than others in a specific context such as disease or biological processes by defining how many in-degrees and out-degrees each miRNA possesses [21,22]; as an example, the number of transcription factors (TF) regulating an miRNA characterizes the "in-degree", and the number of mRNA targets of this miRNA for silencing is the "out-degree" (Figure 1c), and both metrics can greatly contribute to the determination of the importance of a specific miRNA in a given system [23].

In a network, vertices with an unusually high degree of centrality become *"hubs"*. Even if few hubs exist within a network, they can be very informative and play a central role in the functioning of the system. For example, social networks often contain a few central individuals with many acquaintances. Few websites exist, for instance, with an extraordinarily large number of links. In a cellular context, there are few metabolites that take part in almost all metabolic processes.

Another important metric in network analysis is that of *"structural equivalence"* between vertices, i.e., a measure of *similarity* [16]; two vertices in a network are structurally equivalent if they share many of the same network neighbors (Figure 1d). Online dating sites compute similarity measures to match users to one another by using descriptions of people's interests, background, likes, and dislikes [24,25]. In the context of miRNA–mRNA interaction networks, measuring structural equivalence could help in identifying groups of collaborative miRNAs based on the number of similar mRNA targets they share [26,27].

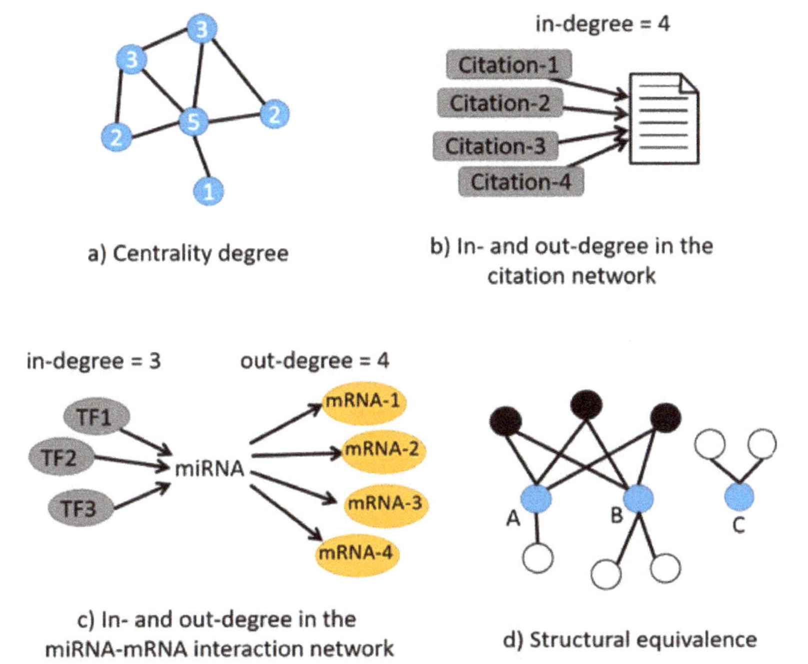

Figure 1. (**a**) The degree of centrality defines the number of edges (black lines) connected to a vertex (blue dots). The number inside the dots represents the centrality degree of each vertex; (**b**) The in-degree of a scientific publication is the number of other papers citing it (citations in grey boxes); (**c**) In a microRNA (miRNA)–messenger RNA (mRNA) interaction network, the number of transcription factors (TF) regulating an miRNA characterizes the in-degree, and the number of mRNA targets of this miRNA for silencing is the out-degree; (**d**) Structural equivalence between 3 vertices, A, B, and C: A and B share, in this case, 3 of the same neighbors (black dots), although both also have other neighbors that are not shared (white dots). Vertex C is not similar to A and B because it does not share any neighbors with them.

Comparison with Available Tools

According to the increasing experimental evidence supporting target mRNA degradation rather than translational repression as the main silencing mechanism used by miRNAs, the integration of target predictions with miRNA and gene expression profiles based on high-throughput sequencing

(HTS) analyses from the same sample would greatly improve the characterization of functional miRNA–mRNA relationships. Several online tools that aim to identify miRNA–mRNA interactions exist: (1) MicroRNA and mRNA integrated analysis (MMIA) [28] is a versatile web server that permits query of miRNA–mRNA interactions. It applies systems level analysis to identify pathways and diseases in which the miRNAs of interest may be involved. However, MMIA ignores the network of collaborative miRNAs that work together to silence genes; (2) **miRror-Suite** [29] uses a list of miRNAs in a contextual manner to predict the most likely set of regulated genes in a cell line or tissue, or from a list of genes. However, the input is either a miRNA list or a gene list, but cannot be both. Additionally, it relies only on public datasets, does not let users provide their own paired miRNA–gene expression datasets, and fails to provide a metric in which miRNA is the most important variable; (3) DIANA-mirExTra [30] uses repository information to build a network with miRNA–gene targets from miRNA and gene expression datasets. However, it does not classify the importance of the miRNA based on interaction (it only considers fold change) and the networks do not provide a metric of miRNA similarity; (4) miRGator [31] is a mining data and hypothesis generating tool that uses big data from public datasets combined with data from miRNA–target repositories and a negative correlation algorithm to define miRNA regulatory networks. It allows enquiries regarding where the expression of the miRNAs is more relevant and the most commonly affected biological functions. However, it does not let users input their own data and lacks biological contextual information for tissue-specific miRNAs; (5) In 2010, the web tool MAGIA (miRNA and genes integrated analysis) was designed, allowing integration of target predictions with gene expression profiles using different relatedness measures for matched and unmatched expression profiles, using miRNA–mRNA bipartite network reconstruction, gene functional enrichment, and pathway annotations for browsing results [32]. In 2012, it was updated to MAGIA^2, which now focuses on mixed regulatory circuits involving miRNAs, transcription factors (TFs, in-degree measure), and mRNA targets (out-degree measure) [23]. Nevertheless, MAGIA and MAGIA^2 do not calculate network metrics as degrees of centrality and structural similarity. (6) NetworkAnalyzer [33] is a software plug-in that assesses several network topological parameters, including centrality, and represents them graphically. However, this tool was designed with protein–protein interactions in mind, not miRNA–mRNA networks. Additionally, the shared neighbor measure is not suited to define similarity in miRNA–mRNA networks, and the graphs generated focus on the network parameters and not on the importance, or identification, of each node—i.e., miRNA—in the network. (7) A more recent tool, SpidermiR, allows evaluation of the degree of centrality of miRNA–gene target networks and capabilities that resemble the structural equivalence analysis [34]. Although the graphical output of SpidermiR helps in simplifying network interpretation, the tool is focused on the analysis of public available data, principally from The Cancer Genome Atlas (TCGA) repository, and does not allow the researcher to input locally generated data [34]. None of these tools mentioned above offer a measure for centrality and structural equivalence combined with out-degrees that represent useful metrics to determine the system's most crucial miRNAs and collaborations between miRNAs. Finally none provide output graphs that help the user focus on these important conclusions (Table 1).

Here, we present miRmapper, an open-source application that researchers can use to identify the most important miRNA and mRNAs, identified in their own experimental design or produced by publicly available data, in a miRNA–mRNA interaction network by leveraging the centrality and similarity metrics. Based on the assumption that miRNAs with the highest number of target genes are probably the most important ones, and that the genes being targeted by numerous miRNAs are probably the most crucial ones, miRmapper users can easily visualize collaborative miRNAs in relation to their mRNA targets as a result of graphical outputs such as dendrograms and heatmaps. This ultimately allows the user to focus their attention on the system's most crucial elements. Note that miRmapper is designed for miRNA–mRNA interactions in the context of mRNA degradation and measures the centrality, similarity, and out-degree of each miRNA based on the topology of the miRNA–mRNA interaction network created from same-sample miRNA-seq and mRNA-seq datasets.

This novel tool will provide information that can drive further research by uncovering potential biomarkers and drug targets.

Table 1. Tool comparison. Each column represents a feature and each row represents a software tool.

Tools	Input Your Own Data	Output Contextualized with Your Experimental Design	Calculate the Centrality of miRNAs in the Network	Calculate Centrality of Genes in the Network	Calculate the Structural Equivalence of miRNA Interactions	Graphical Depiction of miRNAs Organized by Centrality	Graphical Depiction of miRNA Clusters by Structural Equivalence
miRmapper	X	X	X	X	X	X	X
MMIA	-	-	-	-	-	-	-
miRror-Suite	X	X	-	-	-	-	-
DIANA-mirExTra	X	X	-	-	-	-	-
miRGator	-	-	-	-	-	-	-
MAGIA	X	X	-	-	-	X	-
MAGIA^2	X	X	-	-	-	X	-
NetworkAnalyzer	X	X	X	X	-	-	-
SpidermiR	-	-	X	X	-	X	-

MMIA: MicroRNA and mRNA integrated analysis

2. Materials and Methods

The method presented here is based on the following assumptions: (1) miRNAs tend to act via the downregulation of their gene targets, in an inverse correlation relationship (i.e., miRNA canonical function) [35]; (2) the regulatory effect of miRNAs is dependent on the cellular context [35,36]; (3) miRNAs regulating the greatest number of targets have a greater impact on the phenotype (network centrality) [16,37]; (4) in a given context, the list of common targets of two miRNAs can be used to infer how similar their effects are, independent of their nucleotide sequence similarity (network similarity by structural equivalence) [16,38]; and (5) a gene being regulated by the greatest number of miRNAs is probably a key gene in the system studied [16,39].

This package was conceived to be used downstream of paired miRNA and differential gene expression analyses, and it also requires a list of interactions of the DE miRNAs and target genes. For sequencing experiments, DE analysis can be performed using DESeq2, EdgeR, and Limma programs for mRNA sequencing [40,41]. DE results for miRNA sequencing can be obtained from the CAP-miRSeq pipeline, mirPRo, and miARma-Seq [42–44]. Any form of DE analysis that permits the acquisition of a list of DE mRNAs and DE miRNAs—such as high-throughput sequencing, microarray technology, quantitative PCR (qPCR) arrays, etc.—can be used as input. Predicted targets of miRNA can be collected from databases such as microRNA.org, TargetScan, and the multiMiR R Package [45–47]. The user needs to be aware that repositories provide the entire list of predicted genes and that only those that are in the DE gene list are of interest. Packages such as multiMiR have functionalities to select only the appropriate interactions [47]. Consequently, users using an interaction list directly from other repositories will have to use the intersection of genes between the interaction list and the list of DE genes in their experiment. Taking into account our first assumption, we considered the analysis to be more insightful if only downregulated mRNAs are selected as possible targets for upregulated miRNAs and vice versa.

miRmapper provides simple and effective metrics to analyze the predicted influence of miRNAs on gene expression; a workflow of the method is shown in Figure 2. Starting with the postulate that DE miRNAs that impact a larger number of DE genes are of greater importance for gene regulation in the context of the experiment [16], the percentage of predicted target genes over the total targets is calculated for each miRNA to indicate its level of centrality (Equation (1)). Similarly, we calculate the proportion of predicted targets for each miRNA relative to all differentially expressed genes (Equation (2)); this second calculation not only provides us with the information about miRNA centrality, but adds the overall impact of the miRNA expression in the regulation of a given gene's expression. The package also provides as output the degrees of centrality for each gene target.

These calculations are provided in both a tabular form and a bar plot of publication quality. The proportions are given by the following formulas, where t is the number of predicted target genes for miRNA m, T is the number of total gene targets, and G is the number of total DE genes:

$$Influence_{DE}^{m} = \frac{t}{T} \tag{1}$$

$$Influence_{Total}^{m} = \frac{t}{G} \tag{2}$$

We represent the predicted interactions in the form of an adjacency matrix. The adjacency matrix is a convenient data structure for detecting miRNAs that target the same genes. We then apply the Jaccard distance formula to measure dissimilarity between miRNAs (Equation (3)) [48,49]. With this metric, we calculate and visualize miRNA clustering with an identity plot and dendrogram for a hierarchical representation, i.e., network similarity. The Jaccard distance is given by the following formula, where D_{ij}, also known as the Jaccard distance, is the proportion of gene targets that are not shared between miRNAs i and j relative to the total number of genes targeted by these two miRNAs:

$$D_{ij} = 1 - \frac{|t_i \cap t_j|}{|t_i \cup t_j|}, \tag{3}$$

where t_i and t_j are the genes targeted by miRNAs i and j, $|t_i \cap t_j|$ is the shared gene targets of t_i and t_j, and $|t_i \cup t_j|$ is the total gene targets of t_i and t_j.

The Jaccard index has the advantage in that it only counts the mutual presence of gene targets in its calculations [49]. In the context of multiple DE miRNAs with large and non-overlapping lists of interactions, a method that takes into consideration only the presence of the interactions in a list will be the one with the greater biological meaning. Methods such as the simple matching coefficient and the chi-square statistic will consider two miRNA as being highly similar if they have no common gene target, but have a large list of genes that both do not target [50].

The software is implemented as an R package, "miRmapper". As input, the package requires a table with miRNAs and their targets and an optional list of the total differentially expressed genes. The tool then produces an adjacency matrix describing all miRNA–target interactions and the additional information of the number of miRNAs regulating each gene, i.e., degree of centrality for the genes. From this matrix is calculated the impact that each miRNA has on the list of genes, i.e., degree of centrality for the miRNAs, and the results are depicted as a boxplot ordered by miRNAs with the greatest centrality. Also from the matrix, the Jaccard distance is calculated between the miRNAs based on their targets, i.e., similarity, and a dendrogram and an identity plot are generated to identify how closely related the miRNAs in the study are. More details about the package installation and dependencies can be found in the package vignette.

To illustrate the usefulness of our method to interpret miRNA–target interactions in a biological application, we used transcriptomic (i.e., mRNA and miRNA) data from the human bladder cancer cell lines T24 (poorly metastatic) and FL4 (its metastatic derivative) [47]. Both datasets are available at the ArrayExpress repository and can be found under the accession numbers E-MTAB-2610 and E-MTAB-2611, for mRNA and miRNA respectively. The processed data and probe-to-gene annotation were downloaded from the ArrayExpress repository, probe IDs were annotated to gene symbols as designated by the Human Genome Organization (HUGO) Gene Nomenclature Committee, and where multiple probes were present for a given gene the highest expression value was selected; finally, differential expression (DE) analysis was performed using Limma [51] Bioconductor R Package version 3.32.10, and a p value of 0.05 and a linear fold change of two were used as the threshold for statistical significance. The correlation of miRNA–gene targets for the upregulated DE miRNAs and downregulated DE genes were acquired using the multiMiR [47] Bioconductor R Package, considering only the top 35% of predicted interactions.

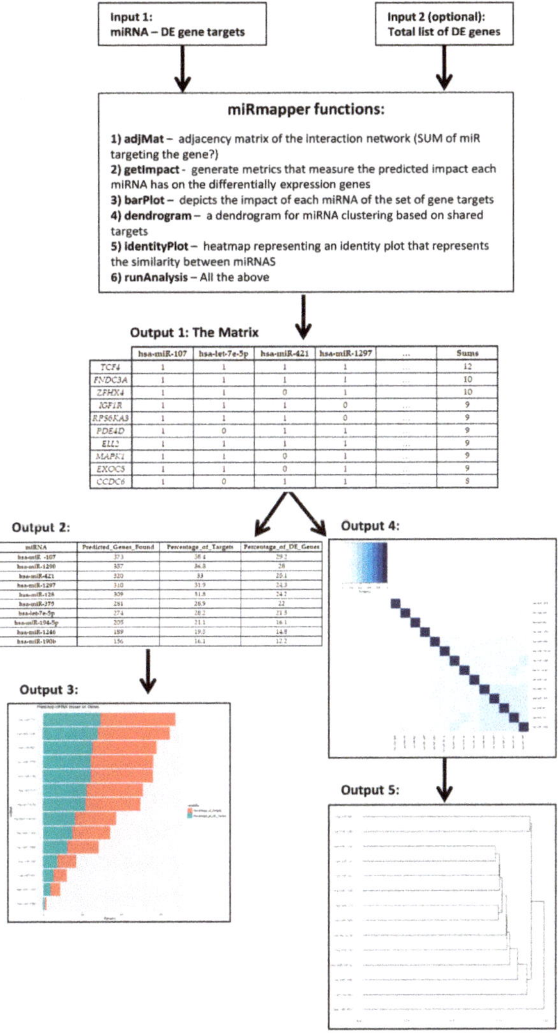

Figure 2. The miRmapper workflow. An miRNA-gene interaction data frame is the required input for the tool (Input 1), additionally a list of total differentially expressed (DE) genes can be used in conjunction (Input 2). The use of the miRmapper functions will provide an adjacency matrix of the miRNA-genes interactions with gene centrality (Output 1), from this a table is generated with the miRNA impact on gene expression (Output 2) and the graphical representation of that impact (Output 3). Also from Output 1, the structural similarity of miRNAs networks is calculated and graphically represented as an identity plot (Output 4) and as a dendogram (Output 5).

3. Results

In this section, we demonstrate the usage of miRmapper when applied to biological data and discuss the functionalities of the software and its outputs. We analyzed differential transcriptomic data—miRNA and gene expression—from cell lines T24 and FL4 and built the table with the miRNA–target interaction and DE genes (Tables 2 and 3 and Supplementary Tables S1 and S2).

Table 2. miRmapper input. miRNA-gene interaction data frame, no headers.

hsa-miR-107	N4BP1
hsa-let-7e-5p	FNDC3A
hsa-let-7e-5p	HAND1
hsa-let-7e-5p	IGF1R
hsa-let-7e-5p	OSBPL3
hsa-let-7e-5p	RRM2
hsa-let-7e-5p	STX3
hsa-miR-107	ASH1L
hsa-miR-107	CAPZA2
hsa-miR-107	YWHAH
hsa-miR-421	AFF4
. . .	. . .

Table 3. miRmapper inputs. List of total differentially expressed genes; this is an optional input.

IFI16
COL5A2
GJA1
ALCAM
TXNIP
PLS3
CXCL8
SPARC
FBN1
CDH2
TMEM158
. . .

Two data frames containing these data are available within the package. The template data are loaded into an R environment as follows:

```
R > data ("interaction.matrix.miR.up")
R > interact <- (interaction.matrix.miR.up)
R > data ("DE.gene.dn")
R > DEgene <- ("DE.gene.dn")
```

The input tables, as they contain all the information of the miRNA–gene target network, have a size that do not allow the researcher to interpret it. It is necessary first to organize it in a way that enable it to be read. We first generate a mirMapper object, as described below:

```
R > miRm <- miRmapper (interactions = interact, DEgenes = DEgene)
```

The next step is to generate an adjacent matrix (Table 4 and Supplementary Table S3) using Supplementary Table S1 as input, as described below:

Table 4. Adjacency matrix of top 10 regulated genes.

	hsa-miR-107	hsa-let-7e-5p	hsa-miR-421	hsa-miR-1297	. . .	Sums
TCF4	1	1	1	1	. . .	12
FNDC3A	1	1	1	1	. . .	10
ZFHX4	1	1	0	1	. . .	10
IGF1R	1	1	1	0	. . .	9
RPS6KA3	1	1	1	0	. . .	9
PDE4D	1	0	1	1	. . .	9
ELL2	1	1	1	1	. . .	9
MAPK1	1	1	0	1	. . .	9
EXOC5	1	1	0	1	. . .	9
CCDC6	1	0	1	1	. . .	8

The interaction between a miRNA and gene is depicted as binary: "1" means the gene is a target for the miRNA; "0" means it is not.

R > adjMat(miRm)

The adjacency matrix provides two results: first, the data organization allows the user to perform downstream analysis; second, it defines the gene targets with the greatest degree of centrality; in this case, the gene Transcription Factor 4 (*TCF4*). *TCF4* (log 2-fold change = -1.22, $p = 0.001$, Supplementary Table S2) has a greater degree of centrality than the serglycin gene, *SRGN* (log 2-fold change = -6.0, $p = 4.68 \times 10^{-5}$, Supplementary Table S2), the most downregulated transcript that has a degree of centrality = 2 (Supplementary Table S3).

This matrix is used as input to also define the centrality of the miRNA itself, depicting it as a table (Table 5 and Supplementary Table S4) and graphically in a bar plot (Figure 3a).

Table 5. miRNA impact on the gene expression; upregulated miRNA affecting downregulated genes.

miRNA	Predicted_Genes_Found	Percentage_of_Targets	Percentage_of_DE_Genes
hsa-miR-107	373	38.4	29.2
hsa-miR-1290	357	36.8	28
hsa-miR-421	320	33	25.1
hsa-miR-1297	310	31.9	24.3
hsa-miR-128	309	31.8	24.2
hsa-miR-375	281	28.9	22
hsa-let-7e-5p	274	28.2	21.5
hsa-miR-194-5p	205	21.1	16.1
hsa-miR-1246	189	19.5	14.8
hsa-miR-190b	156	16.1	12.2

DE: Differentially expressed

The miRNA hsa-miR-146a (log 2-fold change = 4.7, DE $p = 6.39 \times 10^{-5}$, Supplementary Table S5), with the greatest fold change, had no impact on its targets, whereas hsa-miR-107 (log 2-fold change = 1.4, $p = 0.007$), with a linear expression 8 times smaller than miR-146a, has an impact on 29.21% of all downregulated genes and regulates 38.41% of all the genes being targeted by a miRNA in the dataset (Table 5).

The miRNA impact matrix (Table 5 and Supplementary Table S4) can be made using the command:

R > getImpact(miRm);
and the barplot (Figure 3) with the command:
R > barPlot(miRm).

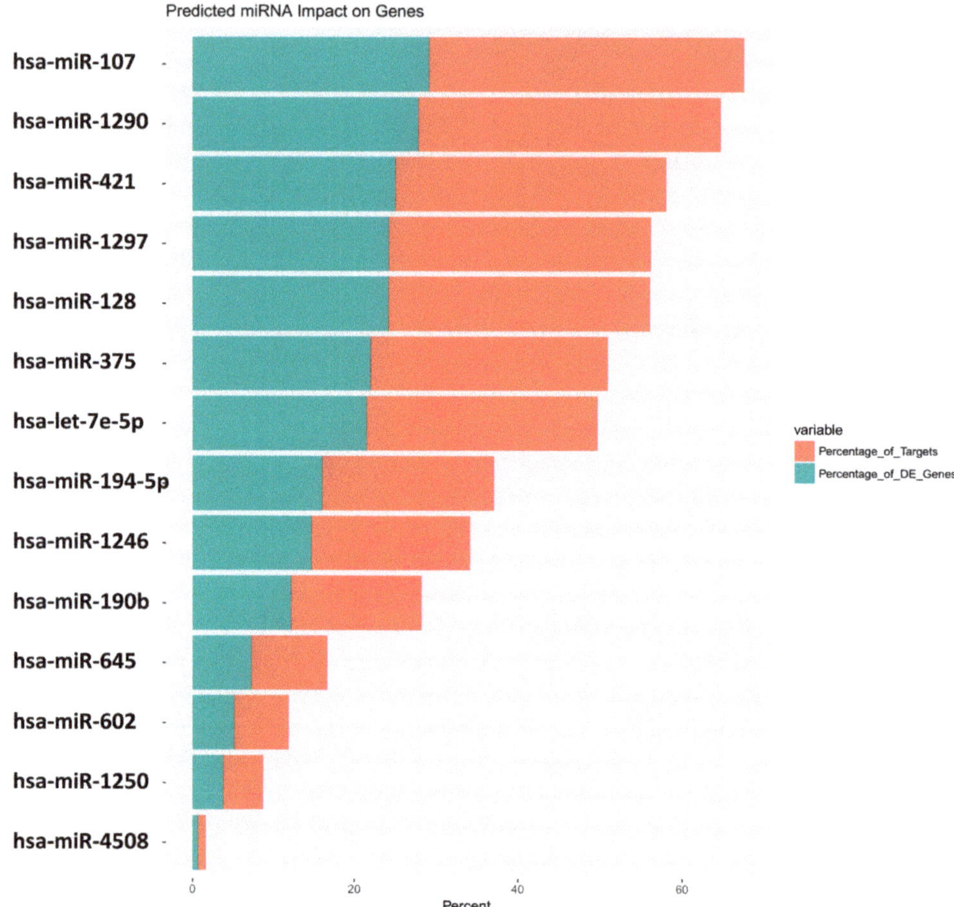

Figure 3. miRmapper output: miRNA boxplot. Data are presented in the order of the greatest number of impacted genes to the lowest, with the percentage of total targets affected by the miRNA in red and the percentage of total DE genes affected by the miRNA in blue.

The miRmapper approach allows the user to rapidly identify those miRNAs which are working synergistically (Figures 4 and 5), as it is normally necessary for more than one miRNA to act on a target to cause a significant impact in the transcript levels [52]. In our case, we found that hsa-miR-107, hsa-miR-1290, hsa-miR-421, hsa-miR-1297, and hsa-miR-375 were clustered as having similarly modulated mRNA targets, which allows us to infer that they are working cooperatively. These five miRNAs belong to five distinct miRNA families [53], and we would not be able to infer that they are working together with their sequence analysis only.

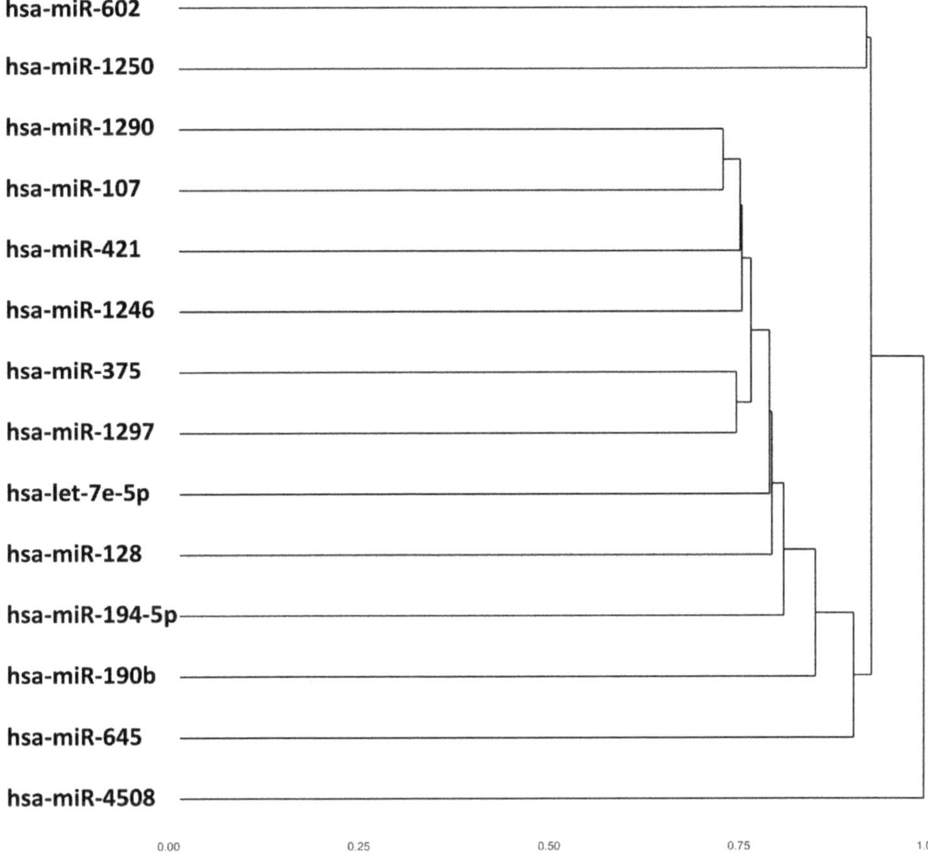

Figure 4. miRmapper output: dendrogram. The cluster is based on the similarity of the miRNAs' Jaccard index values to each other.

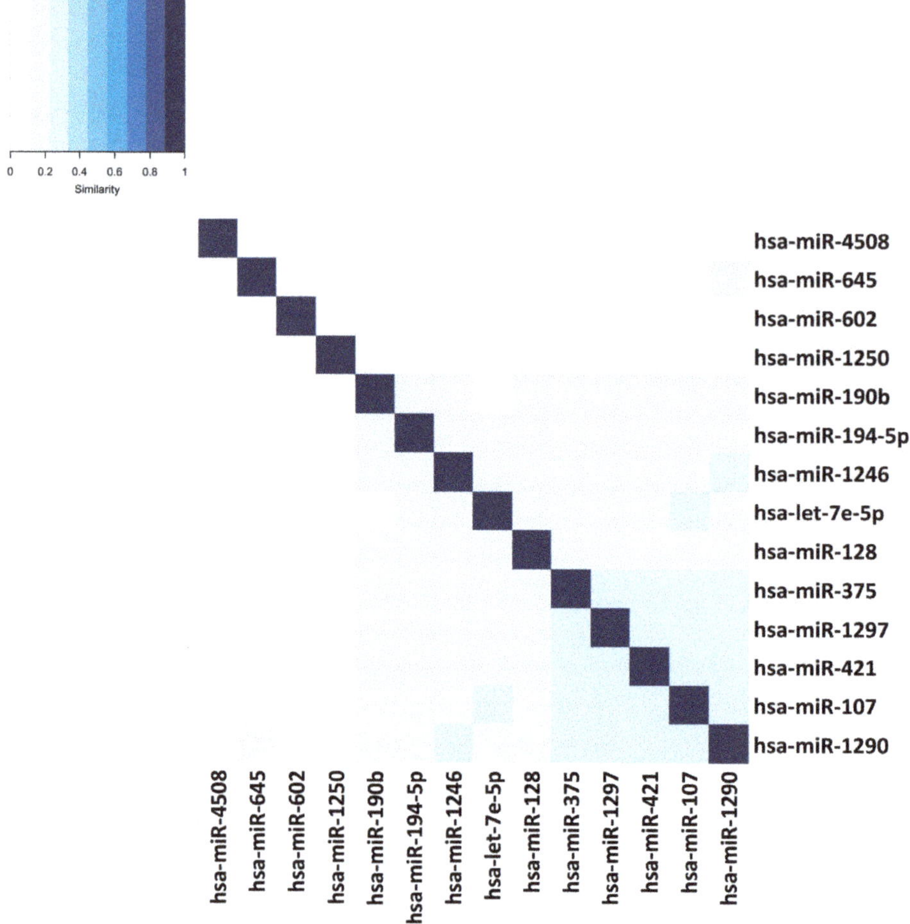

Figure 5. miRmapper output: identity plot with the miRNAs clustered by mRNA target similarity. The distances were based on the similarity of the miRNAs' Jaccard index values to each other.

The dendrogram (Figure 4) can be made using the command:

R > dendrogram(miRm);
and the identity plot (Figure 5) with the command:
R > identityPlot(miRm).

The package has the capability of running all of the above functions at the same time and saves the outputs in the working directory using the function below:

R > runAnalysis(miRm).

4. Discussion

When changes occur in a transcriptional network, it is not only important to know which genes are changed the most with regard to the level of their gene expression, but also which are the most relevant changes in the context of the network [39]. A miRNA that exhibits greatly upregulated expression between two biological conditions, but with none of its transcribed targets being downregulated, can be seen as a potential biomarker on its own, but this isolationist approach does not demonstrate far-reaching biological relevance. On the other hand, a miRNA that is mildly upregulated, yet causes a deep impact in the downregulation of its targets, would be of global relevance for the cell. In our miRNA DE analysis and miRmapper analyses, we found 31 upregulated miRNAs (Supplementary Table S5), but only 14 of them showed an impact in the downregulation of their targets (Supplementary Table S4).

Our analysis recognized miR-107 as the most important DE miRNA, based on the number of affected targets. miR-107 was shown to promote migration and invasion in osteosarcoma, hepatocellular carcinoma, and pancreatic ductal adenocarcinoma [54–56]. There is no report in the literature about the role of miR-107 in bladder cancer; the original analysis of the dataset elected mir-146a as a possible metastasis inducer, and although data of mir-146a in bladder cancer is also scarce, mir-146a overexpression has been reported to inhibit migration, invasion, and metastasis in bladder cancer [57]. The lack of downregulated mir-146a targets in our analysis agrees with the report and provides more support for the hypothesis that miR-146 can have different roles in different tissue types [58].

Similarly, our analysis recognized *TCF4* as the most regulated gene. Although TCF4 was reported to promote cancer cell stemness and metastasis in breast cancer patients [59,60] and in clear cell renal cell carcinoma [61], its role in invasive bladder cancer was described to be beneficial, participating in the inhibition of tumor growth [62]. This can be an indication that, as with miR-146a, the roles of TCF4 are tissue-specific.

We also identified hsa-miR-107, hsa-miR-375, hsa-miR-421, hsa-miR-1290, and hsa-miR-1297 as working synergistically. Although hsa-miR-375, hsa-miR-1290, and hsa-miR-1297 were identified to target the same transcription factor cluster [63], the literature has no report of these five miRNAs influencing gene expression together and of their roles in bladder cancer. miR-375 was shown to play a role in epithelial-to-mesenchymal transition and in the recurrence of breast cancer [64,65]. miR-421 was found to induce cell migration and metastasis in neuroblastoma, osteosarcoma, and gastric cancer [66–68]. In the context of breast cancer, both were described as capable of inducing and inhibiting metastasis [69,70], emphasizing again the context-dependent role of regulatory elements in gene expression. hsa-miR-1290 was demonstrated to have a role in cancer stem cell formation and metastasis in non-small cell lung cancer [71], to promote metastasis in esophageal squamous cell carcinoma [72], and as a prognostic marker for a poor outcome in colorectal cancer [73]. As with miR-421, miR-1297 was reported to induce migration and invasion of colorectal cancer cells [74], but to inhibit invasion in prostate and hepatocellular carcinoma [75,76].

Both miR-107 and *TCF4* are part of the Wnt signaling pathway [62,77]. The Wnt pathway plays a key role in regulating development and stemness and pathway members are typically altered in aggressive cancers, including bladder cancer [78,79]. Considering also the description of the synergistic properties of hsa-miR-107, hsa-miR-1290, hsa-miR-421, hsa-miR-1297, and hsa-miR-375, we, for the first time, identified a possible pivotal axis in the development of bladder cancer metastasis that can be tested in the laboratory. This discovery was only made possible through the use of our tool, miRmapper, which allowed evaluation of network topographic properties of miRNA–mRNA target networks in a simple and visual way.

5. Conclusions

The miRmapper tool identifies the most dominant miRNAs in a miRNA–mRNA network and recognizes functional similarities between miRNAs based on their commonly regulated mRNAs.

The miRmapper software uncovers novel cooperative interactions between miRNAs from independent families in regulating common target mRNAs. We showed here that miRmapper identified miRNAs and regulated mRNAs involved in a known pathway for cancer metastasis, i.e., the Wnt signaling pathway. This highlights the utility of miRmapper to interpret miRNA–gene networks and to identify key elements and possible biomarkers and drug targets. Future improvements of the methodology will address noncanonical miRNA functions. The source code of the package and the tutorials are available on GitHub at http://github.com/ MUSC-CGM/miRmapper. Installation documentation and a detailed vignette are provided.

Supplementary Materials: The following are available online at http://www.mdpi.com/2073-4425/9/9/458/s1: Supplementary Table S1: input list of miRNA–DE gene target, Supplementary Table S2: input complete list of DE genes, Supplementary Table S3: complete output of adjacency matrix, Supplementary Table S4: complete miRNA impact on the gene expression, Supplementary Table S5: complete upregulated DE miRNA.

Author Contributions: Conceptualization, W.A.d.S., L.R., J.S., E.S.H., D.C., and G.H.; Methodology, W.A.d.S., D.C., and G.H.; Software, J.S. and W.B.G.Jr.; Formal Analysis, W.A.d.S. and L.R.; Resources, E.S.H., D.C., and G.H.; Writing—Original Draft Preparation, W.A.d.S., L.R., and G.H.; Writing—Review & Editing, L.R., E.S.H., and G.H.; Supervision, G.H.; Project Administration, G.H.; Funding Acquisition, D.C. and G.H.

Funding: This work was supported by the funding from SC EPSCoR (Gary Hardiman, 2017); the National Institute of General Medical Sciences (NIGMS) under grant R01 GM122078: Statistical Methods for Genetic Studies, Using Network and Integrative Analysis (Dongjun Chung, 2016–2021); the National Cancer Institute (NCI) under grant R21 CA209848: Algorithms for Literature-Guided Multi-Platform Identification of Cancer Subtypes (Dongjun Chung, 2016–2018); and the start-up grants from the Department of Public Health Sciences (Dongjun Chung) and Department of Medicine (Gary Hardiman).

Conflicts of Interest: The authors declare no conflict of interest.

References

1. Chen, K.; Rajewsky, N. The evolution of gene regulation by transcription factors and microRNAs. *Nat. Rev. Genet.* **2007**, *8*, 93–103. [CrossRef] [PubMed]

2. Voskarides, K. Plasticity vs mutation. The role of microRNAs in human adaptation. *Mech. Ageing Dev.* **2017**, *163*, 36–39. [CrossRef] [PubMed]

3. Osada, H.; Takahashi, T. MicroRNAs in biological processes and carcinogenesis. *Carcinogenesis* **2007**, *28*, 2–12. [CrossRef] [PubMed]

4. Carthew, R.W.; Sontheimer, E.J. Origins and mechanisms of miRNAs and siRNAs. *Cell* **2009**, *136*, 642–655. [CrossRef] [PubMed]

5. Bagga, S.; Bracht, J.; Hunter, S.; Massirer, K.; Holtz, J.; Eachus, R.; Pasquinelli, A.E. Regulation by let-7 and lin-4 miRNAs results in target mRNA degradation. *Cell* **2005**, *122*, 553–563. [CrossRef] [PubMed]

6. Eulalio, A.; Rehwinkel, J.; Stricker, M.; Huntzinger, E.; Yang, S.-F.; Doerks, T.; Dorner, S.; Bork, P.; Boutros, M.; Izaurralde, E. Target-specific requirements for enhancers of decapping in miRNA-mediated gene silencing. *Genes Dev.* **2007**, *21*, 2558–2570. [CrossRef] [PubMed]

7. Giraldez, A.J.; Mishima, Y.; Rihel, J.; Grocock, R.J.; Van Dongen, S.; Inoue, K.; Enright, A.J.; Schier, A.F. Zebrafish MiR-430 promotes deadenylation and clearance of maternal mRNAs. *Science* **2006**, *312*, 75–79. [CrossRef] [PubMed]

8. Wu, L.; Belasco, J.G. Micro-RNA regulation of the mammalian lin-28 gene during neuronal differentiation of embryonal carcinoma cells. *Mol. Cell. Biol.* **2005**, *25*, 9198–9208. [CrossRef] [PubMed]

9. Brodersen, P.; Sakvarelidze-Achard, L.; Bruun-Rasmussen, M.; Dunoyer, P.; Yamamoto, Y.Y.; Sieburth, L.; Voinnet, O. Widespread translational inhibition by plant miRNAs and siRNAs. *Science* **2008**, *320*, 1185–1190. [CrossRef] [PubMed]

10. Eulalio, A.; Huntzinger, E.; Izaurralde, E. Getting to the root of miRNA-mediated gene silencing. *Cell* **2008**, *132*, 9–14. [CrossRef] [PubMed]

11. Filipowicz, W.; Bhattacharyya, S.N.; Sonenberg, N. Mechanisms of post-transcriptional regulation by microRNAs: Are the answers in sight? *Nat. Rev. Genet.* **2008**, *9*, 102. [CrossRef] [PubMed]

12. Wu, L.; Belasco, J.G. Let me count the ways: Mechanisms of gene regulation by miRNAs and siRNAs. *Mol. Cell* **2008**, *29*, 1–7. [CrossRef] [PubMed]

13. Guo, H.; Ingolia, N.T.; Weissman, J.S.; Bartel, D.P. Mammalian microRNAs predominantly act to decrease target mRNA levels. *Nature* **2010**, *466*, 835. [CrossRef] [PubMed]

14. Huntzinger, E.; Izaurralde, E. Gene silencing by microRNAs: Contributions of translational repression and mRNA decay. *Nat. Rev. Genet.* **2011**, *12*, 99. [CrossRef] [PubMed]

15. Aleman, L.M.; Doench, J.; Sharp, P.A. Comparison of siRNA-induced off-target RNA and protein effects. *RNA* **2007**, *13*, 385–395. [CrossRef] [PubMed]

16. Newman, M.E.J. *Networks: An introduction*; Oxford University Press: Oxford, UK; New York, NY, USA, 2010; p. 772.
17. Borgatti, S.P.; Mehra, A.; Brass, D.J.; Labianca, G. Network analysis in the social sciences. *Science* **2009**, *323*, 892–895. [CrossRef] [PubMed]
18. Brass, D.J. Being in the right place: A structural analysis of individual influence in an organization. *Adm. Sci. Q.* **1984**, *29*, 518–539. [CrossRef]
19. González-Pereira, B.; Guerrero-Bote, V.P.; Moya-Anegón, F. A new approach to the metric of journals' scientific prestige: The SJR indicator. *J. Inf.* **2010**, *4*, 379–391. [CrossRef]
20. Yan, E.; Ding, Y. Applying centrality measures to impact analysis: A coauthorship network analysis. *J. Am. Soc. Inf. Sci. Technol.* **2009**, *60*, 2107–2118. [CrossRef]
21. Hsu, C.W.; Juan, H.F.; Huang, H.C. Characterization of microRNA-regulated protein-protein interaction network. *Proteomics* **2008**, *8*, 1975–1979. [CrossRef] [PubMed]
22. Ragusa, M.; Statello, L.; Maugeri, M.; Majorana, A.; Barbagallo, D.; Salito, L.; Sammito, M.; Santonocito, M.; Angelica, R.; Cavallaro, A.; et al. Specific alterations of the microRNA transcriptome and global network structure in colorectal cancer after treatment with MAPK/ERK inhibitors. *J. Mol. Med.* **2012**, *90*, 1421–1438. [CrossRef] [PubMed]
23. Bisognin, A.; Sales, G.; Coppe, A.; Bortoluzzi, S.; Romualdi, C. Magia^2: From miRNA and genes expression data integrative analysis to microRNA–transcription factor mixed regulatory circuits (2012 update). *Nucleic Acids Res.* **2012**, *40*, W13–W21. [CrossRef] [PubMed]
24. Chen, L. A Social Matching System: Using Implicit and Explicit Information for Personalized Recommendation in Online Dating Service. Ph.D. Thesis, Queensland University of Technology, Brisbane, Australia, 2013.
25. Lorrain, F.; White, H.C. Structural equivalence of individuals in social networks. *J. Math. Sociol.* **1971**, *1*, 49–80. [CrossRef]
26. Beckman, J.D.; Chen, C.; Nguyen, J.; Thayanithy, V.; Subramanian, S.; Steer, C.J.; Vercellotti, G.M. Regulation of heme oxygenase-1 protein expression by miR-377 in combination with miR-217. *J. Biol. Chem.* **2011**, *286*, 3194–3202. [CrossRef] [PubMed]
27. Wu, S.; Huang, S.; Ding, J.; Zhao, Y.; Liang, L.; Liu, T.; Zhan, R.; He, X. Multiple microRNAs modulate p21Cip1/Waf1 expression by directly targeting its 3' untranslated region. *Oncogene* **2010**, *29*, 2302. [CrossRef] [PubMed]
28. Nam, S.; Li, M.; Choi, K.; Balch, C.; Kim, S.; Nephew, K.P. MicroRNA and mRNA integrated analysis (MMIA): A web tool for examining biological functions of microRNA expression. *Nucleic Acids Res.* **2009**, *37*, W356–W362. [CrossRef] [PubMed]
29. Friedman, Y.; Naamati, G.; Linial, M. MiRror: A combinatorial analysis web tool for ensembles of microRNAs and their targets. *Bioinformatics* **2010**, *26*, 1920–1921. [CrossRef] [PubMed]
30. Alexiou, P.; Maragkakis, M.; Papadopoulos, G.L.; Simmosis, V.A.; Zhang, L.; Hatzigeorgiou, A.G. The DIANA-mirExTra web server: From gene expression data to microRNA function. *PLoS ONE* **2010**, *5*, e9171. [CrossRef] [PubMed]
31. Nam, S.; Kim, B.; Shin, S.; Lee, S. MiRGator: An integrated system for functional annotation of microRNAs. *Nucleic Acids Res.* **2008**, *36*, D159–D164. [CrossRef] [PubMed]
32. Sales, G.; Coppe, A.; Bisognin, A.; Biasiolo, M.; Bortoluzzi, S.; Romualdi, C. MAGIA, a web-based tool for miRNA and genes integrated analysis. *Nucleic Acids Res.* **2010**, *38*, W352–W359. [CrossRef] [PubMed]
33. Doncheva, N.T.; Assenov, Y.; Domingues, F.S.; Albrecht, M. Topological analysis and interactive visualization of biological networks and protein structures. *Nat. Protoc.* **2012**, *7*, 670–685. [CrossRef] [PubMed]
34. Cava, C.; Colaprico, A.; Bertoli, G.; Graudenzi, A.; Silva, T.C.; Olsen, C.; Noushmehr, H.; Bontempi, G.; Mauri, G.; Castiglioni, I. SpidermiR: An R/bioconductor package for integrative analysis with miRNA data. *Int. J. Mol. Sci.* **2017**, *18*, 274. [CrossRef] [PubMed]
35. Maute, R.L.; Dalla-Favera, R.; Basso, K. RNAs with multiple personalities. *Wiley Interdiscip. Rev. RNA* **2014**, *5*, 1–13. [CrossRef] [PubMed]
36. Erhard, F.; Haas, J.; Lieber, D.; Malterer, G.; Jaskiewicz, L.; Zavolan, M.; Dolken, L.; Zimmer, R. Widespread context dependency of microRNA-mediated regulation. *Genome Res.* **2014**, *24*, 906–919. [CrossRef] [PubMed]
37. Winterbach, W.; Van Mieghem, P.; Reinders, M.; Wang, H.; de Ridder, D. Topology of molecular interaction networks. *BMC Syst. Biol.* **2013**, *7*, 90. [CrossRef] [PubMed]

38. Bracken, C.P.; Scott, H.S.; Goodall, G.J. A network-biology perspective of microRNA function and dysfunction in cancer. *Nat. Rev. Genet.* **2016**, *17*, 719–732. [CrossRef] [PubMed]

39. Wang, E. *Cancer Systems Biology*; CRC Press: Boca Raton, FL, USA, 2010.

40. Davis-Turak, J.; Courtney, S.M.; Hazard, E.S.; Glen, W.B., Jr.; da Silveira, W.A.; Wesselman, T.; Harbin, L.P.; Wolf, B.J.; Chung, D.; Hardiman, G. Genomics pipelines and data integration: Challenges and opportunities in the research setting. *Expert Rev. Mol. Diagn.* **2017**, *17*, 225–237. [CrossRef] [PubMed]

41. Robinson, M.D.; McCarthy, D.J.; Smyth, G.K. edgeR: A bioconductor package for differential expression analysis of digital gene expression data. *Bioinformatics* **2010**, *26*, 139–140. [CrossRef] [PubMed]

42. Sun, Z.; Evans, J.; Bhagwate, A.; Middha, S.; Bockol, M.; Yan, H.; Kocher, J.P. CAP-miRSeq: A comprehensive analysis pipeline for microRNA sequencing data. *BMC Genomics* **2014**, *15*, 423. [CrossRef] [PubMed]

43. Shi, J.; Dong, M.; Li, L.; Liu, L.; Luz-Madrigal, A.; Tsonis, P.A.; Del Rio-Tsonis, K.; Liang, C. MirPRo-a novel standalone program for differential expression and variation analysis of miRNAs. *Sci. Rep.* **2015**, *5*, 14617. [CrossRef] [PubMed]

44. Andres-Leon, E.; Nunez-Torres, R.; Rojas, A.M. MiARma-Seq: A comprehensive tool for miRNA, mRNA and circRNA analysis. *Sci. Rep.* **2016**, *6*, 25749. [CrossRef] [PubMed]

45. Betel, D.; Wilson, M.; Gabow, A.; Marks, D.S.; Sander, C. The microRNA. Org resource: Targets and expression. *Nucleic Acids Res.* **2008**, *36*, D149–D153. [CrossRef] [PubMed]

46. Agarwal, V.; Bell, G.W.; Nam, J.W.; Bartel, D.P. Predicting effective microRNA target sites in mammalian mRNAs. *Elife* **2015**, *4*, e05005. [CrossRef] [PubMed]

47. Ru, Y.; Kechris, K.J.; Tabakoff, B.; Hoffman, P.; Radcliffe, R.A.; Bowler, R.; Mahaffey, S.; Rossi, S.; Calin, G.A.; Bemis, L.; et al. The multimiR R package and database: Integration of microRNA-target interactions along with their disease and drug associations. *Nucleic Acids Res.* **2014**, *42*, e133. [CrossRef] [PubMed]

48. Leskovec, J.; Rajaraman, A.; Ullman, J.D. *Mining of Massive Datasets*; Cambridge University Press: Cambridge, UK, 2014.

49. Fuxman Bass, J.I.; Diallo, A.; Nelson, J.; Soto, J.M.; Myers, C.L.; Walhout, A.J. Using networks to measure similarity between genes: Association index selection. *Nat. Methods* **2013**, *10*, 1169–1176. [CrossRef] [PubMed]

50. Podani, J. *Introduction to the Exploration of Multivariate Biological Data*; Backhuys Publishers: Kerkwerv, NL, USA, 2000.

51. Ritchie, M.E.; Phipson, B.; Wu, D.; Hu, Y.; Law, C.W.; Shi, W.; Smyth, G.K. Limma powers differential expression analyses for RNA-sequencing and microarray studies. *Nucleic Acids Res.* **2015**, *43*, e47. [CrossRef] [PubMed]

52. Peter, M.E. Targeting of mRNAs by multiple miRNAs: The next step. *Oncogene* **2010**, *29*, 2161–2164. [CrossRef] [PubMed]

53. Kozomara, A.; Griffiths-Jones, S. Mirbase: Annotating high confidence microRNAs using deep sequencing data. *Nucleic Acids Res.* **2014**, *42*, D68–D73. [CrossRef] [PubMed]

54. Jiang, R.; Zhang, C.; Liu, G.; Gu, R.; Wu, H. MicroRNA-107 promotes proliferation, migration, and invasion of osteosarcoma cells by targeting tropomyosin 1. *Oncol. Res.* **2017**, *25*, 1409–1419. [CrossRef] [PubMed]

55. Su, S.G.; Yang, M.; Zhang, M.F.; Peng, Q.Z.; Li, M.Y.; Liu, L.P.; Bao, S.Y. MiR-107-mediated decrease of *HMGCS2* indicates poor outcomes and promotes cell migration in hepatocellular carcinoma. *Int. J. Biochem. Cell Biol.* **2017**, *91*, 53–59. [CrossRef] [PubMed]

56. Xiong, J.; Wang, D.; Wei, A.; Lu, H.; Tan, C.; Li, A.; Tang, J.; Wang, Y.; He, S.; Liu, X.; et al. Deregulated expression of miR-107 inhibits metastasis of PDAC through inhibition PI3K/Akt signaling via caveolin-1 and PTEN. *Exp. Cell Res.* **2017**, *361*, 316–323. [CrossRef] [PubMed]

57. Xiang, W.; Wu, X.; Huang, C.; Wang, M.; Zhao, X.; Luo, G.; Li, Y.; Jiang, G.; Xiao, X.; Zeng, F. *PTTG1* regulated by miR-146a-3p promotes bladder cancer migration, invasion, metastasis and growth. *Oncotarget* **2017**, *8*, 664–678. [CrossRef] [PubMed]

58. Ferracin, M.; Veronese, A.; Negrini, M. Micromarkers: MiRNAs in cancer diagnosis and prognosis. *Expert Rev. Mol. Diagn.* **2010**, *10*, 297–308. [CrossRef] [PubMed]

59. Chen, C.; Cao, F.; Bai, L.; Liu, Y.; Xie, J.; Wang, W.; Si, Q.; Yang, J.; Chang, A.; Liu, D.; et al. IKKβ enforces a LIN28B/*TCF7L2* positive feedback loop that promotes cancer cell stemness and metastasis. *Cancer Res.* **2015**, *75*, 1725–1735. [CrossRef] [PubMed]

60. Vijaya Kumar, A.; Salem Gassar, E.; Spillmann, D.; Stock, C.; Sen, Y.P.; Zhang, T.; Van Kuppevelt, T.H.; Hulsewig, C.; Koszlowski, E.O.; Pavao, M.S.; et al. *HS3ST2* modulates breast cancer cell invasiveness via

MAP kinase- and tcf4 (Tcf7l2)-dependent regulation of protease and cadherin expression. *Int. J. Cancer* **2014**, *135*, 2579–2592. [CrossRef] [PubMed]

61. Kojima, T.; Shimazui, T.; Horie, R.; Hinotsu, S.; Oikawa, T.; Kawai, K.; Suzuki, H.; Meno, K.; Akaza, H.; Uchida, K. *FOXO1* and *TCF7L2* genes involved in metastasis and poor prognosis in clear cell renal cell carcinoma. *Genes Chromosomes Cancer* **2010**, *49*, 379–389. [PubMed]

62. Tang, Y.; Simoneau, A.R.; Liao, W.X.; Yi, G.; Hope, C.; Liu, F.; Li, S.; Xie, J.; Holcombe, R.F.; Jurnak, F.A.; et al. *WIF1*, a Wnt pathway inhibitor, regulates *SKP2* and c-myc expression leading to G1 arrest and growth inhibition of human invasive urinary bladder cancer cells. *Mol. Cancer Ther.* **2009**, *8*, 458–468. [CrossRef] [PubMed]

63. Sengupta, D.; Bandyopadhyay, S. Participation of microRNAs in human interactome: Extraction of microRNA-microRNA regulations. *Mol. Biosyst.* **2011**, *7*, 1966–1973. [CrossRef] [PubMed]

64. Giricz, O.; Reynolds, P.A.; Ramnauth, A.; Liu, C.; Wang, T.; Stead, L.; Childs, G.; Rohan, T.; Shapiro, N.; Fineberg, S.; et al. Hsa-miR-375 is differentially expressed during breast lobular neoplasia and promotes loss of mammary acinar polarity. *J. Pathol.* **2012**, *226*, 108–119. [CrossRef] [PubMed]

65. Zehentmayr, F.; Hauser-Kronberger, C.; Zellinger, B.; Hlubek, F.; Schuster, C.; Bodenhofer, U.; Fastner, G.; Deutschmann, H.; Steininger, P.; Reitsamer, R.; et al. Hsa-miR-375 is a predictor of local control in early stage breast cancer. *Clin. Epigenetics* **2016**, *8*, 28. [CrossRef] [PubMed]

66. Li, Y.; Li, W.; Zhang, J.G.; Li, H.Y.; Li, Y.M. Downregulation of tumor suppressor menin by miR-421 promotes proliferation and migration of neuroblastoma. *Tumor Biol.* **2014**, *35*, 10011–10017. [CrossRef] [PubMed]

67. Zhou, S.; Wang, B.; Hu, J.; Zhou, Y.; Jiang, M.; Wu, M.; Qin, L.; Yang, X. MiR-421 is a diagnostic and prognostic marker in patients with osteosarcoma. *Tumor Biol.* **2016**, *37*, 9001–9007. [CrossRef] [PubMed]

68. Yang, P.; Zhang, M.; Liu, X.; Pu, H. MicroRNA-421 promotes the proliferation and metastasis of gastric cancer cells by targeting claudin-11. *Exp. Ther. Med.* **2017**, *14*, 2625–2632. [CrossRef] [PubMed]

69. Pan, Y.; Jiao, G.; Wang, C.; Yang, J.; Yang, W. MicroRNA-421 inhibits breast cancer metastasis by targeting metastasis associated 1. *Biomed. Pharmacother.* **2016**, *83*, 1398–1406. [CrossRef] [PubMed]

70. Zhang, W.; Shi, S.; Jiang, J.; Li, X.; Lu, H.; Ren, F. LncRNA MEG3 inhibits cell epithelial-mesenchymal transition by sponging miR-421 targeting E-cadherin in breast cancer. *Biomed. Pharmacother.* **2017**, *91*, 312–319. [CrossRef] [PubMed]

71. Kim, G.; An, H.J.; Lee, M.J.; Song, J.Y.; Jeong, J.Y.; Lee, J.H.; Jeong, H.C. Hsa-miR-1246 and hsa-miR-1290 are associated with stemness and invasiveness of non-small cell lung cancer. *Lung Cancer* **2016**, *91*, 15–22. [CrossRef] [PubMed]

72. Li, M.; He, X.Y.; Zhang, Z.M.; Li, S.; Ren, L.H.; Cao, R.S.; Feng, Y.D.; Ji, Y.L.; Zhao, Y.; Shi, R.H. MicroRNA-1290 promotes esophageal squamous cell carcinoma cell proliferation and metastasis. *World J. Gastroenterol.* **2015**, *21*, 3245–3255. [CrossRef] [PubMed]

73. Imaoka, H.; Toiyama, Y.; Fujikawa, H.; Hiro, J.; Saigusa, S.; Tanaka, K.; Inoue, Y.; Mohri, Y.; Mori, T.; Kato, T.; et al. Circulating microRNA-1290 as a novel diagnostic and prognostic biomarker in human colorectal cancer. *Ann. Oncol.* **2016**, *27*, 1879–1886. [CrossRef] [PubMed]

74. Chen, P.; Wang, B.L.; Pan, B.S.; Guo, W. Mir-1297 regulates the growth, migration and invasion of colorectal cancer cells by targeting cyclo-oxygenase-2. *Asian Pac. J. Cancer Prev.* **2014**, *15*, 9185–9190. [CrossRef] [PubMed]

75. Liang, X.; Li, H.; Fu, D.; Chong, T.; Wang, Z.; Li, Z. MicroRNA-1297 inhibits prostate cancer cell proliferation and invasion by targeting the AEG-1/Wnt signaling pathway. *Biochem. Biophys. Res. Commun.* **2016**, *480*, 208–214. [CrossRef] [PubMed]

76. Liu, Y.; Liang, H.; Jiang, X. MiR-1297 promotes apoptosis and inhibits the proliferation and invasion of hepatocellular carcinoma cells by targeting HMGA2. *Int. J. Mol. Med.* **2015**, *36*, 1345–1352. [CrossRef] [PubMed]

77. Zhang, Z.; Wu, S.; Muhammad, S.; Ren, Q.; Sun, C. MiR-103/107 promote ER stress-mediated apoptosis via targeting the Wnt3a/βcatenin/ATF6 pathway in preadipocytes. *J. Lipid Res.* **2018**, *59*, 843–853. [CrossRef] [PubMed]

78. Zhan, T.; Rindtorff, N.; Boutros, M. Wnt signaling in cancer. *Oncogene* **2017**, *36*, 1461–1473. [CrossRef] [PubMed]
79. Ahmad, I.; Sansom, O.J.; Leung, H.Y. Exploring molecular genetics of bladder cancer: Lessons learned from mouse models. *Dis. Model Mech.* **2012**, *5*, 323–332. [CrossRef] [PubMed]

MDPI

St. Alban-Anlage 66

4052 Basel

Switzerland

Tel. +41 61 683 77 34

Fax +41 61 302 89 18

www.mdpi.com

Genes Editorial Office

E-mail: genes@mdpi.com

www.mdpi.com/journal/genes